Tourism in Bangladesh: Investment and Development Perspectives

Azizul Hassan
Editors

Tourism in Bangladesh: Investment and Development Perspectives

Springer

Editor
Azizul Hassan
Tourism Consultants Network
The Tourism Society
London, UK

ISBN 978-981-16-1857-4 ISBN 978-981-16-1858-1 (eBook)
https://doi.org/10.1007/978-981-16-1858-1

This Springer imprint is published by the registered company Springer Nature Singapore Pte Ltd.
The registered company address is: 152 Beach Road, #21-01/04 Gateway East, Singapore 189721, Singapore

Contents

Contributors

Tanvir Abir is an author, industry expert, researcher. Dr. Abir currently works at the College of Business Administration, International University of Business Agriculture and Technology (IUBAT) as an Associate Professor. Earlier, he worked at the School of Science and Health, Western Sydney University, Australia. Also, he worked as a Sessional Lecturer in Academies Australasia Polytechnic under Federation University, Australia. Dr. Abir researches Business analytics, Business Tourism, Market Research, Healthcare Marketing, Applied Statistics, Applied Mathematics, and Applied Health Science.

Ayesha Afrin is a PhD student in the Department of Chemistry, University of Chittagong, Bangladesh. Afrin is an Assistant Professor in the Department of Applied Chemistry and Chemical Engineering in the same university. She received her Bachelor degree, Master degree, and an MPhil degree. Her main research area is analytical chemistry. She has published articles in national and international journals.

Jamee Ahmad is a Senior Lecturer at the Business School of International University of Business Agriculture & Technology (IUBAT) Dhaka, Bangladesh. He is now pursuing his PhD in China. Earlier, he completes MBA from Australia. His main research areas are Human resource policy, employee satisfaction, corporate leadership, performance appraisal, and training. Ahmad has authored several research articles on employee satisfaction and work conditions.

Shahida Akhter is currently working as an Associate Officer at the administration department in BRAC Head Office. During her working time, she has gathered over-priced quality knowledge of sustainable development from various programs of the organization. Her educational background is Bachelor in Social Science (Sociology). Also, she has two Master degrees. The first Master degree is in Sociology from the National University, Bangladesh and the other Master degree is in Development Studies from the BRAC Institute of Governance and Development of BRAC University. She has attended many trainings in areas like climate change and the environment.

Nazia Akthar is a Lecturer at the School of Business in Presidency University. She has a research interest in human resource management, organizational psychology, entrepreneurship, organizational behavior, and tourism management. She has worked as an HR business partner in a conglomerate. She also works as a training research consultant with corporates. She has one article proceeding in a peer-reviewed journal.

Md Rabiul Alam is a PhD candidate at the Faculty of Education, University of Malaya, Malaysia. He did his MEd in English as a Second Language with distinction from the said faculty. Before continuing his higher study at the University of Malaya, he served as a faculty in the Department of English, the Asian University of Bangladesh, for nearly eight years. Besides his study at the University of Malaya, he has been working as a Graduate Research Assistant (GRA) since 2018. His research interest includes TESOL, teacher education, teacher identity, and professional development, financial education, digital financial inclusion, and so forth. Alam has published a few of his works in reputed peer-reviewed journals. Apart from this, he has presented papers in various national and international conferences.

Md. Alauddin has been working as an Assistant Professor (Study Leave) of Marketing in International Islamic University Chittagong, (IIUC) Bangladesh. He has completed his BBA, MBA & MPhil degree in Marketing from the University of Chittagong, Bangladesh. He is now doing PhD in marketing from the City University of Hong Kong, Hong Kong. His research contribution and interest are in the field of Service Quality, Tourism Marketing, Strategic Marketing, Branding, and Customer Engagement.

Rashed Al Karim is currently working as an Associate Professor at the School of Business, East Delta University, Chattogram Bangladesh. He has completed his PhD from Universiti Utara Malaysia and MBA from the University of Wales, UK. His research interests are in tourism and hospitality, service marketing, consumer behavior, sustainable development. He has a number of publications in reputed peer-reviewed journals. He has also presented his research works at international conferences. He has reviewed several articles from reputed journals such as International Journal of Contemporary Hospitality Management, Journal of Hospitality and Tourism Insights, GeoJournal of Tourism and Geosites, Journal of Financial Services Marketing, and Current Journal of Applied Science and Technology.

Sadia Afrin Ananya is an Assistant Professor at the Department of Tourism and Hospitality Management, BSMRSTU (Bangabandhu Sheikh Mujibur Rahman Science and Technology University), Bangladesh. Ananya earned her Bachelor and Master degrees from the University of Dhaka. As the recognition of her academic feat, she earned Achieved Dean's Honor and Dean's Merit Award 2017 from FBS, DU, and Scholarship from the University of Dhaka. She has research interests in tourism, travel, hospitality, business, marketing, and statistical analysis. She has contributed to a good number of publications that are on the way to be published.

Ashraful Alam Anas has more than four years of working experience as a Researcher in different National & International NGO and Consultancy firms. He has studied Anthropology at Jahangirnagar University and has a clear understanding of gender, social exclusion, migration, international development, and politics. As a research officer in GRM, Anas has impressive experience in conducting qualitative method and collecting data from the field through Focus Group Discussion (FGD), In-depth Interview (IDI), Key Informant Interview (KII), Participatory Rural Appraisal (PRA), Observation, Home/workplace visit, and Case study. He is experienced in making Qualitative & Quantitative questionnaires. He has experience in working with youth and children, especially all around in the country through "Risk and Vulnerability Assessment and Baseline and Trend Analysis of sexual Exploitation of Children online and in Travel and Tourism" implemented by Ain o Salish Kendra (ASK) and supported by the Terre des hommes as a Researcher.

C.M. Atiqur Rahman is working as an Assistant Professor of Management at Southern University Bangladesh. He has completed his BBA and MBA degree in Management and International Management from the University of Chittagong, Bangladesh. His research contribution and interest are in Service Quality, Tourism Marketing, Human Resource Management, and Islamic Management.

Hasanul Banna is working at the Ungku Aziz Centre for Development Studies, Faculty of Economics and Administration, University of Malaya, Malaysia. Dr. Banna earned a Bachelor degree in Economics (major in Finance) from International Islamic University, Malaysia, with a Full-Scholarship and a PhD in Finance and Banking from the University of Malaya, Malaysia, with the most competitive Bright-Spark Scholarship. His research spans finance & banking, financial inclusion, digital finance, development economics, financial economics, managerial ability, and micro-finance. He has conducted a number of workshops on Research Methodology in various universities in the United Kingdom, Turkey, Scotland, Malaysia, Bangladesh, and Indonesia. His research has been published in various peer-reviewed journals, including Journal of International Development, Singapore Economic Review, Poverty and Public Policy, Journal of Islamic Monetary Economics and Finance, Economics, Total Quality Management, and Business Excellence, Journal of Asia Pacific Economy, Journal of Asian Finance, Economics and Business, Cities; also been featured in leading Newspapers including The Business Standard and Dhaka Tribune. He has presented his research works and delivered keynote speeches on finance, banking, and financial economics in several conferences and seminars worldwide, including United Kingdom, Scotland, Japan, Thailand, Malaysia, Vietnam, India, and Indonesia.

Kamrul Hasan Bhuiyan is a Lecturer of Tourism and Hospitality Management at the National University, Gazipur, Bangladesh. He holds MBA and BBA in Tourism and Hospitality Management from the University of Dhaka. His research interests include marketing, technology, and innovation in hospitality and Tourism, destination management, and development. He has published four papers in peer-reviewed journals and presented research works in international conferences. He was an Associate of the Pacific Asia Travel Association (PATA), Bangkok, Thailand.

Md. Asad Iqbal Chowdhury has been working as a Lecturer of Finance at International Islamic University Chittagong, Bangladesh. He has completed his BBA in finance and banking, MBA in finance from the University of Chittagong, and MDS with major in development economics from the University of Dhaka. His research contribution and interest are in business, finance, banking, and development studies.

Maria Fekete-Farkas is a Professor at the Institute of Economic Sciences, the Hungarian University of Agriculture and Life Sciences, Gödöllő, Budapest, Hungary. She is also a Professor of the University of Debrecen. Her research areas are sustainable development, behavior economics, new market structures, and pricing, the economic and social effect of industry 4.0. She is a member of the organizing committees of several international conferences and serves some international journals as a member of the editorial board, reviewer, and author.

Azizul Hassan is a member of the Tourism Consultants Network of the UK Tourism Society. Dr. Hassan has been working for the tourism industry as a consultant, academic, and researcher for over 20 years. His research interest areas are technology-supported marketing for tourism and hospitality, immersive technology applications in the tourism and hospitality industry, and technology-influenced marketing suggestions for sustainable tourism and hospitality industry in developing countries. Dr. Hassan has authored over 150 articles and book chapters in leading tourism outlets. He is also part of the editorial team of 25 book projects from Routledge, Springer, CAB International, and Emerald Group Publishing Limited. He is a regular reviewer of a number of international journals.

Md. Wasiul Islam holds a PhD in Tourism from the University of Queensland, Australia. Professor Islam is a forester working at Forestry and Wood Technology Discipline of Khulna University, Bangladesh. Previously he used to work in Bangladesh Forest Department to conduct research activities on tourism and wildlife of the Sundarbans mangrove forest. His research interests are focused on nature-based tourism, shared governance, participatory management of protected areas, co-management approach, and sustainability. He has published his research findings in several international peer-reviewed journals like Journal of Sustainable Tourism, Tourism Recreation Research, Journal of Hospitality and Tourism Management, Journal of Ecotourism, Tourism Review International, South Asian Journal of Tourism and Heritage, among others.

Shahriar Tanjimul Islam received his Bachelor degree and MSc degree in finance and business from the University of West England, Bristol, and the University of Ulster, both from the United Kingdom. In 2016, he joined as Lecturer at the College of Business Administration of International University of Business Agriculture and Technology (IUBAT). His research mainly centered on finance, accounting and business-related fields. Islam teaches several courses on accounting, business, and finance. Some names of the courses he specifically teaches are cost accounting, international finance, taxation, business communication and a few more at IUBAT. He has supervised many students during their final year practicum defence. He has attended a number of national and international conferences and presented several papers. He has been invited for talks and panel discussions by different and universities to talk about finance.

Yeasmin Islam is an Assistant Professor in the School of Business and Economics, United International University. She has research interests in management, human resource, entrepreneurship, international business, and tourism management. Islam has presented a few papers in reputed international business conferences at home and abroad. She has also published several articles in peer-reviewed journals, book reviews, and book chapters.

Md. Aktar Kamal has been working as an Assistant Professor of Management in Bangladesh University of Professionals. He has completed his BBA, MBA, and MPhil degree in Management Studies from the University of Chittagong, Bangladesh. He is now doing MSc in International Tourism and Convention Management from The Hong Kong Polytechnic University, Hong Kong. His research contribution and interest are in the field of Service Quality, Hospitality and Tourism Management, Human Capital Engagement, Social Capital.

Md Yusuf Hossein Khan is currently working as an Assistant Professor at the College of Tourism and Hospitality Management (CTHM) of the International University of Business Agriculture and Technology (IUBAT), Dhaka, Bangladesh. He is also a PhD Researcher at the faculty of Economics at the University of Algarve, Portugal. He has a good number of experience working within the tourism and hospitality industry in several countries. Khan has completed his MSc in International Tourism Management from the Cardiff Metropolitan University, UK, and MBA International from the Anglia Ruskin University, UK. Besides being a member of CinTurs – Research Centre for Tourism, Sustainability and Well-being of Ualg, Faro, Portugal, he is also serving several International Journals as an Editorial Board Member and reviewer. Khan is an expert in curriculum design and currently is an active member of CTHM curriculum task force team of IUBAT. His current research interests are Safety and Risk in Tourism, Travelers loyalty, Tourist motivation and behavior, and Tourism destination image and development. He has a good number of publications in these areas.

Bapon Chandra Kuri is an Assistant Professor, Department of Tourism and Hospitality Management, BSMRSTU (Bangabandhu Sheikh Mujibur Rahman Science and Technology University), Bangladesh. He received BBA and MBA degrees from the Department of Tourism and Hospitality Management, University of Dhaka. He has a good number of articles published in both national and international journals. His areas of research interest are CBT, Rebranding, and identifying employee satisfaction levels. Before joining the university, he was working in the Service Department of Radisson Blue Water Garden Hotel and in the Housekeeping Department of Hotel the Cox Today, Cox's Bazar, Bangladesh.

Md Aslam Mia is currently working as a Senior Lecturer at the School of Management, Universiti Sains Malaysia (USM). He researches productivity, market structure, poverty, and development economics with a special focus on microfinance. Dr. Mia has actively researched and published numerous articles in WoS and Scopus indexed journals related to microfinance institutions' productivity and efficiency, market structure, international trade, and urban economics. Among some of the journals he has published are, The European Journal of Development Research, Business History, Journal of the Asia Pacific Economy, Journal of Cleaner Production, Social Indicators Research, International Journal of Social Economics, Cities, Singapore Economic Review, Economic Analysis and Policy, Strategic Change, Journal of Financial Regulation and Compliance, Journal of Public Affairs, Journal of Economics and Administrative Sciences and other internationally reputed journals. Apart from journal articles, he has also published several book chapters published by Springer and Taylor & Francis. Nonetheless, his research findings are presented in around 15 international conferences. He also actively serves as a reviewer for ISI and Scopus indexed journals.

Md. Mohiuddin is currently working as an Assistant Professor at the Department of Geography and Environment, Jagannath University, Dhaka. His research interest is in the field of Disaster and Vulnerability Management, Migration, Tourism, Physical Geography, and Environmental Science. He has more than 12 research papers in his specialized arenas. He is working as a consultant with some national and international organizations. He has participated in many national and international seminars and conferences as a Research Fellow.

Nikola (Nick) Naumov is a Senior Lecturer in Hospitality & Tourism Management at the Department of Events, Tourism & Hospitality at the University of Northampton, United Kingdom. He has undergraduate and postgraduate degrees in tourism and hospitality management and a PhD in human geography completed at King's College London. His research interests include cultural and historical geographies of Eastern Europe, critical heritage studies, heritage tourism, gastronomic tourism, service quality in hospitality, and innovation in tourism and hospitality management. Nick is a visiting professor at several academic institutions in France, Germany, India, and Japan, an experienced tourism consultant with experience in Europe, Asia, and the Middle East, and a regular ad-hoc reviewer for a great number of geography, tourism, and hospitality academic journals.

Md. Nekmahmud is a PhD Research Fellow of Management and Business Administration Sciences at the Doctoral School of Economic and Regional Sciences, the Hungarian University of Agriculture and Life Sciences, Gödöllő, Hungary. He was awarded the CAS-Belt and Road Master Research Fellowship Programme Scholarship in Corporate Management at the University of Chinese Academy of Sciences, funded by the world's top research institution Chinese Academy of Sciences, China, and Sino-Danish Centre, Denmark. Moreover, he has completed his MBA in Marketing from Bangladesh. Nekmahmud published some research works in prominent scientific journals under Elsevier, Springer, Emerald, MDPI, and attended several international conferences as a paper presenter. He participated at the Ninth World Science Forum 2019 conference in Europe session and Prague European Summit-2020, Prague. He published book chapters in Springer Nature, Routledge, Taylor & Francis, Emerald, and Springer. He involved himself as a responsible reviewer in the Journal of Cleaner Production, Journal of Foodservice Business Research, SAGE Open, International Journal of Organization Theory & Behaviour, and Future Business Journal. He is also the article editor of SAGE Open Journal. His research interest areas include consumer behavior, environmental marketing, sustainable tourism, sustainable consumption behavior, sustainable business, ecotourism, and renewable energy.

Samshad Nowreen is currently an Associate Professor at the Department of Tourism and Hospitality Management, University of Dhaka. She holds a Master degree specialized in Environmental Science and Sustainable Enterprises from the USA. Her core subjects are Sustainable Tourism, Entrepreneurship, Tourism Planning and Development, Heritage Tourism, Climate Change, Environmental Science, and Disaster Management. She is a member of four international bodies, attended several international conferences, and presented five papers. She has been invited to many talks/panel discussions with various national and international organizations.

Mohammad Fazle Rabbi is a PhD Research Fellow in Károly Ihrig Doctoral School of Management and Business, Faculty of Economics and Business, University of Debrecen, Debrecen, Hungary. He was awarded by prestigious Hungarian Government scholarship, "Stipendium Hungaricum scholarship". Rabbi completed MBA major in Human Resource, BBA major in Marketing from Bangladesh, and Diploma in Business Administration (DBA) from the UK. He has eight years of humanitarian work experience in a managerial position that helped him understand how poverty, social injustice, and gender inequality conspire to make life miserable as a leader of the distressed community and an NGO activist. He has a lot of goals and plans. These goals are improving the distressed community, alleviating poverty, ensuring food security and achieving environmental sustainability. He is particularly interested in exploring research in food, energy and environmental securities, and sustainability.

Muhammad Khalilur Rahman is working as a faculty member in the Faculty of Entrepreneurship and Business at University Kelantan Malaysia. He obtained his PhD in operation management from the University of Malaya (UM). He has been actively involved in research activities and published more than 50 journal articles in internationally peer-reviewed journals with Web of Science and SCOPUS indexed. He has a wide interest in tourism and service management research, including medical tourism, Eco-tourism, Muslim-friendly tourism, service quality, brand equity, supply chain management, and green and sustainable development.

Md. Sohel Rana is currently pursuing a PhD in the Faculty of Business and Accountancy, University of Malaya, Malaysia. He holds a Bachelor's degree in Business Administration and an MBA in Management Studies from the University of Rajshahi, Bangladesh. His area of expertise in research includes Microfinance &

Financial Inclusion, Islamic Microfinance, Poverty and Development Economics, Urban Economics, Finance & Banking, Productivity & Efficiency, Halal Tourism, Halal Food, Insurance, HRM, Strategic Management, Environmental Economics etc.

Mallika Roy is a Research Student and Teaching Assistant at the Department of Economics and Finance of the City University of Hong Kong. She got Hong Kong UGC funded scholarship for her research study. She achieved the Prestigious Prime Minister Gold Medal Award Bangladesh for her scholastic academic result. She works as an assistant professor (on study leave) at the Department of Economics, University of Chittagong, Bangladesh. Her research interests include Macroeconomics, Economic Growth, and Development. She has a teaching experience of 9 years. She worked in Islamia University College and BGC Trust University Bangladesh as a lecturer. After joining the University of Chittagong, she worked at Bangladesh University of Professional (BUP) Bangladesh, Premier University, and BGC Trust University Bangladesh as an adjunct faculty. She published many of her research papers in national and international journals. She also published several newspaper articles. Recently, she attended many webinars as an invited speaker and plenary speaker She participated in an international webinar as a moderator and plenary speaker recently. She also attended a good number of national and international workshops. Recently, she attended an international workshop organized by CLIMB project, Winrock international.

Sanjoy Kumar Roy is currently working as Lecturer in the Tourism and Hospitality Management department of Noakhali Science and Technology University (NSTU), Bangladesh. He completed his BBA and MBA degree major in Tourism and Hospitality Management for the University of Dhaka in the year 2016 and 2017, respectively. The research areas which are interesting to the author involve travel motivations, service evaluation of travel businesses. The author has experience of working at the Bangladesh Parjatan Corporation and also at the National Hotel and Tourism Training Institute as his part of the MBA program. He was also involved in working with a domestic travel agency named "Vision Tours & Travels" as an intern marketing executive after completing his BBA degree.

Polin Kumar Saha completed BSc (Hons.) and Double MS in Soil, Water and Environment and Strategic Leadership towards Sustainability respectively from the University of Dhaka, and Blekinge Institute of Technology, Sweden. Also completed Post Graduate Diploma in Research Methods and Skills from Maastricht School of Management, The Netherlands; certificate courses on Essential Skills for Young Scientist and Research and Technology Policy from Lund University, Sweden. Saha has substantial experience of more than fourteen years working in Research, Publications, and NGO activities with different organizations. His keen interest is to conduct project implementation (research planning, designing, training module development, research training facilitation, data analysis, writing and evaluation of research papers and proposal), Monitoring and Evaluation of project (social, environmental, and economic sustainability), Develop multi-criteria decision-making concepts towards sustainability (SDGs), Conduct sustainability assessment/

management for product development (sustainable product innovation) Develop consultancy report for EIA (Environmental Impact Assessment), IEE (Initial Environmental Examination) and EMP (Environmental Management Plan).

Abul Hasnat Muhammed Salimullah is an Associate Professor in the Department of Economics at the University of Chittagong, Chattogram, Bangladesh. Dr. Salimullah graduated with a BSS (Honors) and MSS degree in Economics from the University of Chittagong in 2003, did a second MA degree in Financial Economics from Middle Tennessee State University (MTSU) in 2014, and completed his Ph.D. in Econometrics and Quantitative Economics from the same university, Middle Tennessee State University in 2018. His present areas of interest are Labor Economics, Health Economics, and Financial Economics. He published ten scholarly articles in national and international journals. Besides, he presented several papers at different national and international conferences. The extraordinary meetings that he attended and presented his research were the Academy of Economics and Finance (AEF) Conference in 2018 in Houston, Texas, and Dr. Gregory P. Domin Graduate Conference in 2017 at Columbus State University, Columbus, Georgia. He worked as a reviewer for a journal named *Scientia et Humanitas*, sponsored by the University Honors College at MTSU. He is a proud winner of the Deans' Distinguished Essay Award for his scholarly publication in the Honors College journal of MTSU. Besides, he got two first prize during scholars' week university-wide poster presentation at MTSU. He has been chosen as a proud member of the MTSU Chapter of Beta Gamma Sigma due to his scholarly achievement.

Israt Jahan Shathi has completed MS and BSc (Hons.) in Criminology and Police Science from Mawlana Bhashani Science and Technology University. She has more than two years of working experience as a Researcher in different National & International NGO and Consultancy firms. As a Research Assistant in TIB, Sathi has impressive expertise in conducting qualitative research method and collecting data from the field through Focus Group Discussion (FGD), In-depth Interview (IDI), Key Informant Interview (KII), Participatory Rural Appraisal (PRA), Observation, Home/workplace visit, and a Case study in her assigned MP Block allocation and fire safety of old Dhaka study in TIB. She is experienced in constructing a Qualitative & Quantitative questionnaire. She has experience as a research assistant in working with youth and children, especially worldwide, through "Importance of Volunteer police cadet in schools" implemented by the Police Headquarter.

Muhammad Shoeb-Ur-Rahman is an Associate Professor of Tourism and Hospitality Management at the University of Dhaka, Bangladesh. He holds a PhD in Tourism Management from Lincoln University, New Zealand, and MBA in Tourism and Hospitality Management from the University of Dhaka. His research interests include sustainable tourism, tourism management, tourism planning and policies, tourism governance, crisis and resilience in tourism systems, and destination development. Dr. Rahman has presented his papers in reputed international tourism conferences and published 14 refereed papers and nine book chapters. He also co-edited a book titled Tourism Policy and Planning in Bangladesh (Springer, 2020).

Mohammad Rahim Uddin has been working as an Associate Professor in International Islamic University Chittagong, Bangladesh. He has completed his Doctoral research on performance appraisal and leadership development from Cukurova University, Turkey. His research contribution and interest are in the field of management, human resource management, and Islamic management.

Chapter 1
Introduction

Azizul Hassan

Tourism is an industry that covers many social and economic fields, creating it a great growth vector. For the overall growth and development of the tourism industry, both domestic and foreign direct investment (FDI) play critical roles. On the ground of general understanding of investment Hayes (2021) explains that, an investment can be featured as a financial asset purchased with the hope to increase value or to earn money in the future. An investment typically comprises the existing spending of some assets (time, effort, money, etc.) with the prospect of a better return in the future than what was initially invested. A business, stocks, bonds, real estate holdings, for example, are some of the investment types. The key objective to invest is for generating income and increasing the value of an asset over time. Thus, any such mechanism to earn future income can be referred to as an investment. This can involve, for example, the acquisition of real estate property, stocks, bonds etc. Buying a business or property that can be used for manufacturing things or for offering services for commercial purposes can also be judged as investments. In addition, an activity that is made with the hopes to increase future revenue can be viewed as an investment, in general. When it comes for pursuing extra education, for example, the aim is frequently for broadening one's knowledge and improving one's capabilities (with the hope to ultimately produce more income). Because investing is based on the expectation of future income or growth, there is almost always some risks involved. It is possible that an investment can fail for paying off or losing money over time. An investor may, for example, put money into a business or company that goes bankrupt or a project that never gets off the ground. Saving is the practice of accumulating money for future use with no risk, whereas investing is the act of leveraging money for a potential future profit with risk. However, and not unexpectedly, surveys show, that tourism is relatively one of the higher priority

A. Hassan (✉)
Tourism Consultants Network, The Tourism Society, London, UK
e-mail: azizulhassan00@gmail.com

A. Hassan (ed.), *Tourism in Bangladesh: Investment and Development Perspectives*, https://doi.org/10.1007/978-981-16-1858-1_1

industries for attracting investment promotion agencies (IPAs) all over the world (Endo, 2006). Successful promotion is crucial for attracting investors in the industry, given the increasing international competitiveness between the destinations in tourism and the investment projects' greater contestability. Tourist destinations seek to draw financial capital from abroad and market and grow local attractions through human capital, as entrepreneurs, international franchises, and hospitality and travel managers. Such activities are often quite successfully undertaken, while raising foreign capital investment for tourism projects very often becomes a challenge. In the tourism industry, capital investment is negatively affected by the COVID-19 pandemic.

The COVID-19 pandemic has had a substantial effect on tourism investments. The tourism sector has experienced one of the steepest declines, especially when compared to global FDI, which fell by 42% in 2020, according to United Nations Conference on Trade and Development (UNCTAD) (2021). As a result, the sector has lost roughly USD 910 billion to USD 1.2 trillion in export revenues in 2020 compared to 2019, implying that roughly 100–120 million direct tourist employment are at risk in 2021 if global FDI stays sluggish, declining by another 5–10%. In such context, the World Tourism Organization (UNWTO) (2021) has identified investment as one of the most important sectors to focus on as the tourism industry is progressing toward economic recovery and believes that sustainable investments and their cross-sectoral impacts have the potentials to strengthen not only strategically important large regional projects, but also investments that encourage innovation and entrepreneurship among micro, small, and medium-sized enterprises (MSMEs). The focus is thus on MSMEs that along with tech startups can appear as an important driver of job creation, and innovation to help minimizing the ill effects of capital shortages.

Capital shortages are a significant obstacle to the growth of tourism. Many countries are increasingly looking to foreign investors, particularly in the developing world, to provide capital to help their tourism industries grow. The tourism industry's transnational corporations (TNCs) have also positively affected host destinations. In addition to investing money, global TNCs can support the host economies by diversifying tourism product supply and by improving the quality of local service, among other things. However, it is often difficult to attract FDI in the tourism sector, and there is an increasing demand among IPAs for help in this region. The value of tourism is now generally recognized as a factor in economic growth. The rise of international tourist commerce is widely acknowledged as a source of foreign exchange revenues, as well as a source of employment and profit. It is not surprising that the World Bank's economic reviews and evaluations have regularly taken the tourism industry into account for some time now. As one of the most exciting growth drivers for the world economy, tourism can play an essential role in pushing the transition to a Green economy and contributing to more sustainable and inclusive growth. With strong ties to various sectors at a destination and international level, there will be significant impacts on even minor changes towards greater tourism sustainability. Opportunities are wide-ranging, including public and private investment in low-carbon transport options and the building of resource-efficient tourism infrastructure, as well as innovation-friendly policies, facilitating the implementation of responsible business practices and supporting the incorporation of tourism companies into low-carbon and sustainable supply chains for tourism (United Nations Conference on Trade and Development, 2010). Several countries including France and China have considered taking necessary steps to capitalize such opportunities.

Cró and Martins (2020) apply a panel gravity model for bilateral inward FDI stock between investor countries and France and in the restaurant industry and hotels. Results find that France is specifically successful to attract FDI from the French-speaking country' hospitality industry having cultural proximity to France and a common border. Bangladesh has some successful neighbors in the region. South and Southeast Asia are already featured as popular tourist destinations for a substantial time. Bangladesh visibly can capitalize a negligible portion of these regions' huge market potentials. The most common tourist destinations in these regions are India, Malaysia, Indonesia, Thailand, Singapore and Vietnam, Nepal, Sri Lanka, the Philippines, Cambodia, etc. Bangladesh can follow these Asian countries' success and invest more to develop the tourism industry. The government's supportive role can benefit the tourism industry, and it is essential to focus on this research agenda in the context of Bangladesh. The example of China is relevant. According to Liu et al. (2020), China's central government is found for playing a steering role to guide rural tourism towards anticipated directions. They also believe the local government can help by monitoring tourism practices and collaborating with companies and people to provide services and resolve problems. Thus, the central and local governments' synergistic interaction in China stimulates rural tourism's rapid development. Like China, the Government of Bangladesh (GoB) can be hardly believed to have a focus on developing the environment of capital investment in the country.

In FY2019-20, the present Government of Bangladesh (GoB) allocated BDTk. 34 billion for Civil Aviation (Biman). At the same time, the allocation for the Ministry of Civil Aviation and Tourism (MoCAT) was doubled than the earlier fiscal year (National Board of Revenue, 2020). This was the glimpse of the GoB's willingness to invest and develop the tourism industry. The present status of the tourism industry in Bangladesh indicates an industry that rises and comprises a total GDP's 4.4% in 2018 (The World Bank, 2020). However, the efforts and arrangements are inadequate and insufficient considering her neighbors. Like almost all countries globally, the effects of the COVID-19 pandemic in the Bangladesh tourism industry are severe. Considering the present highly competitive and the COVID-19 pandemic hit the global tourism market, the Bangladeshi tourism industry's investment and development context need thorough and sufficient research.

A research gap thus exists in investment and development perspectives in Bangladesh's tourism industry. This book is the very early contribution in this identified research area. This book aims to show the past, present, and future of the investment and development contexts in Bangladesh's tourism industry. The tourism industry in this book is represented with relevant ancillary facilities (i.e. transportation, policy, products and services, hospitality, sustainability etc.). Chapters in this book are thus diverse to cover this research area comprehensively. Contents of this edited book are classified into seven parts (i.e. theoretical and contextual synopsis, the present context, institutional involvement, product and services, facility offers, sustainability practices, challenges, and the future). These parts of the book are accommodated by both primary and secondary data and information based on 25 chapters including introduction. Brief descriptions of each of these chapters are followed:

Chapter 1 is the introduction of the book. This chapter briefly highlights the investment and development perspectives of Bangladesh with the summary of abstract of all chapters.

In Chap. 2, Rahman, Bhuiyan and Hassan explain the conceptual grounds of investment decision-making and FDI in tourism and hospitality in Bangladesh. The theoretical perspective of FDI and tourism growth and their link-up is discussed in this chapter. With empirical examples from literature, it also addresses various methods and hypotheses about investment decision-making and the policy of the host country. The chapter discusses the contexts of investment decision-making, FDI and its role in advancing the tourism and hospitality industry. The theoretical aspects of investment decision-making and FDI are explained in this chapter and apply to Bangladesh's specific context.

Chapter 3 is contributed by Salimullah. This chapter contends the opportunities, nature, culture, and research & development of tourism in Bangladesh. The chapter outlines the scope of investment and further advancements of the tourism industry of Bangladesh. The author understands that at present, Bangladesh shows technology-relied approaches for attracting more investments for the expansion of the tourism industry.

In Chap. 4, Afrin and Hassan discuss the existing scenario, prospects, development, and investment in the tourism industry of Bangladesh. The authors inform that tourism in any country majorly relies on its history, historical settings, archaeology, ancient heritage, natural beauty, etc. The present landmass that constitutes Bangladesh is enriched in its historical background, with many tourist attractions spread across the country. The authors believe that the overall and sound administration in tourism can be achieved by establishing effective institutions that can address both prosperity and stability. The economy of Bangladesh can be integrated with tourism. This research thus outlines the development and investment of tourism.

In Chap. 5, Khan, Islam, and Hassan determine the factors that influence capital investment in the Bangladesh tourism industry. The authors argue that tourism is one of the most promising and potential industries of any country. This industry plays an important role in the national economy by helping to earn foreign exchanges. Peoples' making and living standards in Bangladesh have improved in the last decades, indicating the demand for the developed tourism industry. However, in Bangladesh, this is hard to understand the tourism industry's position that needs huge investments. As a result, this research concentrates on whether the tourism industry in Bangladesh can get sufficient investments from local and foreign sources. This literature review based research also examines the reasons for lower investments in the country and identifies the likely risks.

In Chap. 6, Rahman and Hassan analyze the investment situation for tourism product and service development. Tourism is viewed as an element of economic growth in Bangladesh. In this chapter, the authors define tourism products and services at the beginning of their discussions. This conceptual chapter then debates the investment context of tourism products and service development in Bangladesh and views them as the elements for gross economic growth.

In Chap. 7, Kuri, Ananya, Roy, and Hassan inform that one of the major Sustainable Development Goals' (SDG) is to promote economic growth that is inclusive, sustained, and sustainable followed by productive, decent, and full employment for all. This goal prioritizes sustainable tourism investment and development's path. Bangladesh can reach this important economic goal by ensuring a sustainable tourism environment with Green investment and development in the tourism industry. Major public tourism agencies in Bangladesh like Bangladesh Parjatan Corporation (BPC), Bangladesh Tourism Board (BTB), Biman Bangladesh, and MoCAT continuously update their constitutional plans, roles, and activities for promoting this industry's growth and GDP. Thus, this chapter confers public agencies' roles for investment and development in Bangladesh's tourism and hospitality industry.

In Chap. 8, Banna, Mia, and Rana explain banking and financial institutions' role in the Bangladeshi tourism and hospitality industry. The authors argue that the tourism industry can significantly contribute to Bangladesh's national economy. In recent years, this industry, natural tourism, beach tourism, historical tourism, apparel tourism, agricultural tourism, medical tourism, education tourism, entertainment tourism, and sports tourism, etc. emerge as potential sources of income for the scenic and natural beauties of the country. Still, such emerging tourism niches need exceptional attention and investment from private and public stakeholders. For promoting the tourism industry across the world, this is vital for improving pertinent infrastructure and logistics, including banking and financial services. Therefore, this chapter, by analyzing secondary sources of data, reviews and explores the role of banking and financial intermediaries for facilitating and investing in the tourism and hospitality industry to flourish rapidly and contribute significantly to the socio-economic development process of Bangladesh.

In Chap. 9, Rahman, Rana and Hassan write about developing and investing in core niche tourism products and services in Bangladesh. The chapter identifies strategies, policies, and investment challenges in Bangladesh's tourism industry. There are several niche tourism products and services in the country like archaeological sites, historic places, hills, beaches, islands, and forests that can attract tourists. The country also has tourism service facilities as tour operators, travel agents, tourist guides, visas, immigration services, hotels, catering, and restaurants to meet tourists' demands and satisfaction. The chapter demands distinct strategies for the investment and development of the country's niche tourism products and services. This research explores the investment and development perspectives of core niche tourism products and services in Bangladesh on the ground of different data and information.

In Chap. 10, Islam informs the investment and development in nature-based tourism (NBT) in Bangladesh that can be an element of sustainable development. This country has diverse natural resources and attractions of NBT. Still, a lack of research on investment and development in NBT in Bangladesh exists acutely. Thus, this chapter aims to study the overall investment and development status of NBT in the country with the support of content analysis based on reviewing various secondary information and data.

In Chap. 11, Naumov and Hassan write about the opportunities and challenges of gastronomic tourism in Bangladesh. According to the authors, food and food traditions are often considered useful markers of culture, identity, and local heritage. Local culinary and gastronomy include the places and spaces and contribute towards memorable and meaningful interaction between the "guests" and "hosts". As an inseparable element of the community and intangible heritage, food is also important for branding and destination image. The authors define that food and beverage attributes can play unique roles in the Bangladeshi heritage and culture.

In Chap. 12, Mia and Banna contend that even with the rapid tourism services growth in recent years, medical tourism remains an attractive business for many developing countries globally. Specifically, the developed world's relatively expensive healthcare facilities are pushing many developing countries to invest heavily in medical tourism facilities to attract foreign patients for some economic gain. Recent economic developments and poor healthcare facilities in Bangladesh are also pushing many patients to get treatment abroad. Thus, Bangladesh loses economically as medical tourists spend abroad. To better understand how Bangladesh can promote its medical tourism both globally and domestically, this chapter evaluates factors that affect medical tourism on the basis of existing literature and theories.

Chapter 13 is written by Rabbi and Nekmahmud that covers medical tourism in Bangladesh regarding issues, opportunities, and the Strategic Marketing Plan Model for growth and development. At the initial stage of the chapter, the authors define health and medical tourism. The globalization of the health care sector can help. The authors opine that cost-effective and higher quality treatment in many developing countries has managed to attract tourists' attention from across the world. This research aims to explain the pragmatic marketing strategies for health care institutions involved in medical tourism in Bangladesh. The chapter also evaluates the current perspective of the health sector, drivers, and market structure of medical tourism in Bangladesh. It identifies major opportunities, investment, growth, and healthcare facilities development.

In Chap. 14, Alauddin, Rahman, and Hassan clarify the investment and development of agri-tourism in Bangladesh. This chapter focuses on introducing agri-tourism in Bangladesh with its scope, importance, and prospects.

Chapter 15 is contributed by Saha, Akhter, Hassan, Anas, and Shathi that addresses the investment and development perspective of agri-tourism in Bangladesh. The authors inform that emphasizing the tourism industry can divert a nation towards sustainable development, and agri-tourism can appear as an excellent performer in this regard. In most countries, sustainable development characteristics lack useful strategies to promote the investment and development situation in agri-tourism. This research reviews the data and information generated by both the public and private agencies.

In Chap. 16, Alam, Mia, and Banna discuss garment tourism development in Bangladesh. Following the authors, garments tourism, as consonant to fashion tourism, still stays in its nascent stage in a developing country like Bangladesh. Currently, the government is one of the largest exporters of apparel suppliers globally and international fashion companies. For this purpose, it has brought billions of dollars to the Bangladesh economy and created a platform for thousands of business people/companies to travel to the country. This research's key objective is to study the Bangladeshi garment industry's overall development and how this industry can be translated into tourism.

In Chap. 17, Alauddin, Kamal, and Chowdhury demonstrate the investment and development context of ecotourism in Bangladesh. The authors feature ecotourism as one of the fastest-growing forms of tourism globally. The authors then state that Bangladesh has abundant natural wonders and ecotourism resources. This chapter identifies the prospects of ecotourism products and services, major constraints for its development, and suggests measures to overcome these constraints. The authors describe Bangladesh's relevant ecotourism resources that they consider as the primary ecotourism products.

In Chap. 18, Roy assesses the tourism industry in Bangladesh through Advanced SWOT (strengths, weaknesses, opportunities, and threats) analysis and TOWS (threats, opportunities, weaknesses, and strengths) Matrix. The author opines that Bangladesh has a rich culture and various tourist spots to support its national economy through tourism. Still, the country fails to utilize both its opportunities and strengths to appear like a good tourist destination. As a result, the research searches the general concept, impacts, types, and the tourism industry of Bangladesh. The chapter also highlights the present tourism status based on the national budget, income-expenditure and contribution. The author explains marketing strategies and conducts SWOT analysis and TOWS matrix for tourism development strategies and communication in Bangladesh.

In Chap. 19, Islam and Akhtar precisely analyze the development and investment scenario in Bangladesh's tourism education. The authors believe that Bangladesh with diverse tourism resources can be a potential tourism market. The country is yet to gain deserved shares of the global tourism market. This chapter explores different development and investment concerns in tourism education in Bangladesh.

In Chap. 20, Ahmad analyses the investment and development of tourism and hospitality revenue collection in Bangladesh. The chapter presents some pieces of evidence of the tourism industry's expansion in Bangladesh. The chapter then follows employment status, investments, and revenue collection. Roles of the public organizations (i.e. BPC, BTB, NBR), and different ministries of the GoB and international organizations (i.e. World Travel & Tourism Council, World Travel & Tourism Council, etc.) and private organizations in the country are also mentioned.

In Chap. 21, Nekmahmud and Fekete-Farkas write on Green marketing, investment, and sustainable development for Green tourism. As a developing country with a rapidly expanding economy, the authors claim that Bangladesh has a tourism industry that steadily increased in the industrial index over the past two decades. The author believes that Green tourism gradually becomes popular in Bangladesh with tourists, investors, and the GoB. This chapter aims to converse the investment aspects in Green tourism and explain the important factors for the sustainable development of the tourism industry. The chapter also outlines Green marketing and promotional mix and answers how and why the GoB needs to support the actual or potential investments in Green tourism development in terms of environmental and sustainable development. Then the chapter views community participation in Green tourism, financial institution, the GoB's policy on Green tourism, and the initiatives for yielding sustainable development.

Chapter 22 is contributed by Chowdhury, Alauddin, and Uddin, which focus on different aspects of sustainable tourism in Bangladesh from an investment and development context. The authors draw the scope and importance of promoting and ensuring sustainability in Bangladesh's tourism industry. The authors also spot some challenges that the tourism industry is facing in implementing sustainable tourism in Bangladesh. The chapter then offers some propositions based on these challenges for improving the process to achieve sustainable tourism. The chapter also debates the global experiences that can help promote the tourism industry of Bangladesh.

Chapter 23 is contributed by Nowreen and Mohiuddin, which addresses sustainable tourism investments and development from the Bangladesh perspective. The authors opine that Bangladesh's tourism industry is rapidly growing, primarily benefitting from the excessive rise of domestic tourist flow. Thus, the chapter reveals sustainable tourism indicators' status as guided by the Department of International Development (DFID), and the World Tourism Organization (UNWTO).

In Chap. 24, Abir and Khan explain the challenges and future of investment and development in tourism and hospitality products and service innovations in Bangladesh. The authors believe diversities can become a pull factor to attract international tourists. With the natural scenery, temperate climate, UNESCO World Heritage Sites, and modern democratic culture, Bangladesh has some world-class mega-projects of development. The GoB's allocation for the tourism industry is significant in recent years because the tourism industry's contribution is increasing in the national economy of Bangladesh. Thus, this research aims to classify both the obstacles and opportunities of investment in Bangladesh's tourism industry.

In Chap. 25, Karim and Hassan primarily concentrate on the barriers and opportunities for investment in Bangladesh's tourism and hospitality industry. The study then outlines the Bangladeshi tourism industry's comprehensive scenario followed by different available tourism types.

As expected, this book's contents can attract tourism students in higher education institutions in Bangladesh. PhD students pursuing research in the tourism industry in Bangladesh can find this book useful. Also, policy planners, stakeholders, and industry professionals can be benefitted from reading this book.

References

Cró, S., & Martins, A. M. (2020). Foreign Direct Investment in the tourism sector: The case of France. *Tourism Management Perspectives, 33*, 100614.
Endo, K. (2006). Foreign direct investment in tourism – Flows and volumes. *Tourism Management, 27*, 600–614.
Hayes, A. (2021). Investment. *Investopedia*. Retrieved from: https://www.investopedia.com/terms/i/investment.asp. Accessed 1 Nov 2020.
Liu, C., Dou, X., Li, J., & Cai, L. A. (2020). Analyzing government role in rural tourism development: An empirical investigation from China. *Journal of Rural Studies, 79*, 177–188.
National Board of Revenue (NBR). (2020). *Budget speech 2019–20*. NBR.
The World Bank. (2020). *International tourism, number of arrivals – East Asia & Pacific*. Retrieved from: https://bit.ly/34NSYqA. Accessed 1 Nov 2020.
United Nations Conference on Trade and Development (UNCTAD). (2010). *Promoting foreign investment in tourism*. Retrieved from https://unctad.org/system/files/official-document/diaepcb200916_en.pdf. Accessed 1 June 2021.
United Nations Conference on Trade and Development (UNCTAD). (2021). *Global foreign direct investment fell by 42% in 2020, outlook remains weak*. Retrieved from https://unctad.org/news/global-foreign-direct-investment-fell-42-2020-outlook-remains-weak. Accessed 1 June 2021.
World Tourism Organization. (UNWTO). (2021). *Tourism investments trends and challenges during COVID-19*. Retrieved from https://www.unwto.org/events/tourism-investments-trends-and-challenges-during-covid-19-asia-pacific. Accessed 11 June 2021.

Part I
Theoretical and Contextual Synopsis

Chapter 2
Investment Decision-Making and Foreign Direct Investment in Tourism and Hospitality: Concepts Exposition and Implications for Bangladesh

Muhammad Shoeb-Ur- Rahman, Kamrul Hasan Bhuiyan, and Azizul Hassan

Abstract In many countries across the world, tourism is considered as the medium to promote business enterprises. Investment in the tourism and hospitality industry pertains to purchasing existing, opening, or relocating enterprises or investing in constructions or redevelopment projects. Positive decision-making for foreign public and private sector investments is important for both general economic growth and sectoral growth, for instance, the tourism and hospitality industry. Still, the knowledge in these areas is very limited in a developing country' context, such as Bangladesh. This chapter explains the theoretical aspects of investment decision-making and Foreign Direct Investment (FDI) and relates to the particular context of Bangladesh. This chapter discusses the theoretical perspective of FDI and tourism development and their link-up. It also discourses different approaches and theories regarding investment decision making and the host country's policy with empirical examples from the literature. The chapter explores both the general and the tourism and hospitality industry perspectives of investment decision-making followed by FDI and its role in the growth of the tourism and hospitality industry. The chapter then highlights the tourism and hospitality industry in Bangladesh, which reveal eventually that the country has been offering FDI benefits in this industry. Nevertheless, there are challenges, and this study puts forward some suggestions.

Keywords Tourism · Investment · Decision-making · Trends · Bangladesh

M. Shoeb-Ur- Rahman (✉)
Department of Tourism and Hospitality Management, Faculty of Business Studies, University of Dhaka, Dhaka, Bangladesh

K. H. Bhuiyan
Department of Tourism and Hospitality Management, National University, Gazipur, Bangladesh

A. Hassan
Tourism Consultants Network, The Tourism Society, London, UK

A. Hassan (ed.), *Tourism in Bangladesh: Investment and Development Perspectives*, https://doi.org/10.1007/978-981-16-1858-1_2

Introduction

The most attractive economic feature of tourism, in the context of a developing nation such as Bangladesh, is that it generates positive impacts in the forms of foreign earning and employment creation (Fauzel, 2020). To (re)create positive impacts in an economy, tourism development requires the support of capital, knowledge and expertise, infrastructure, and access to the global market (Samimi et al., 2013). Besides, tourism demands several services, including accommodation, transportation, food, entertainment, and other services (Tang et al., 2007). Inbound tourism market puts pressure on the current level of production, which necessitates foreign investments, especially for developing nations. Investment from foreign countries is considered as an important avenue for capital arrangement and contributes to developing infrastructures such as highways, airports, modern technologies, and hotels (Zhang et al., 1999).

The investment process is often linked to the personality of the investment company and its culture of planning. There are several stages in these phases for connecting to different investment ventures and companies. The development of tourism comprises some cross-cutting operations, including services and goods provision (i.e. housing, building, transport, entertainment, production of fisheries, agricultural production, etc.). The structure of the tourism industry also involves a wide variety of actors from Transnational Corporations (TNCs) to medium-sized and small businesses (Martínez-Román et al., 2015; Nunkoo & Ramkissoon, 2016; Saarinen, 2017). Tourism continues to be an activity that plays an important role in building and/or disseminating capital, infrastructure, information, and accessing marketing and distribution chains across the world (Banerjee et al., 2015; Hof & Blázquez-Salom, 2015; Stauvermann & Kumar, 2017).

Although tourism is chiefly a labor-intensive industry, the capital-intensive nature of tourism often requires the provision of development of infrastructure and superstructure such as construction of airports and hotels that ultimately bring Foreign Direct Investment (FDI) in a country for the desired level of development (Endo, 2006; Yazdi et al., 2017). Foreign direct investment is a category of international investment to acquire a permanent interest in an enterprise resident in economies other than the home country environment (Abdin, 2015). A foreign direct investor is an individual, an incorporated or unincorporated public or private corporation, a government, a group of related persons, or a group of related incorporated and/or unincorporated companies having a direct interest in a company (i.e. a subsidiary, associate or, branch, operating in a country other than that of the country of residence). Thus, it is a cross-border investment strategy that reflects the long-lasting interest and long-term relationship between the residents of two economies. FDI has become very crucial for many countries, especially for developing ones, for growth and development.

This chapter explains the theoretical underpinnings of investment decision-making and FDI. In so doing, the explanations further relate to the tourism and hospitality industry context of Bangladesh. In particular, the chapter emphasizes both the general and the tourism and hospitality industry perspectives of investment decision-making followed by FDI and its role in the tourism and hospitality industry's development with empirical evidence drawn from the literature. Finally, the tourism and hospitality industry context in Bangladesh has been synchronized with the overall discussion.

Investment Decision-Making: General Perspective

The phases or stages of the investment decision-making process have been variably identified in the literature with the most relevance to the accounting discipline (Drury, 2012; Dwyer et al., 2010; Sykianakis & Bellas, 2005; Wei & Christodoulou, 1997). However, the implication remains somewhat generic in that capital investment is primarily focused on the review of decision-making processes and their subsequent acceptance (Aharoni, 1966). While Drury (2012) emphasizes a two-stage model entailing six-phases that involve the planning stage and the management stage, Wei and Christodoulou (1997) explore four phases for investment decision-making, including initiation and preliminary thought, inquiry, assessment, and final decision-making. With a particular focus on the FDI markets, Sykianakis and Bellas (2005) identify five distinct stages of recognition, diagnosis, screening, development and design, and negotiations towards the investment decision-making process. Thus, it is implicit in various decision-making models that investment decisions should follow a four-stage process, in general, emphasizing identification of investment needs, collecting relevant information, analyzing information, and making investment decisions. It is believed that a feedback loop always accompanies the investment decision-making process, underlining the process's iterative nature (Drury, 2012; Sykianakis & Bellas, 2005).

In terms of managing the investment decisions, cash flow forecasts in capital budgeting must be given greater attention. Too much reliance on readily available information in accepting the discount rate to predict the cash flows results in ineffective investment decisions (Ferreira, 2011). In this situation, Dwyer et al. (2010) propose four key steps: estimate all project-related cash inflows and outflows, calculate the Net Present Value (NPV) of all possible cash inflows and outflows, assessment and choice of alternatives, and analysis of the investment project after implementation. This again reconfirms the iterative process by ensuring post-audit that essentially enable the project to contrast between the projected and real cash flows and the disparity between the estimated return and the actual return. When decision-makers closely observe the variations between estimates and real statistics, they appear to strengthen their decision-making to more detailed ones. Apart from the cash flow forecasts, various economic and financial indicators such as Return on Investment (ROI), Internal Rate of return (IRR), Payback Period (PP), Profitability Index (PI), etc., are used to evaluate and manage investment decisions (in tourism). Associated with the management of investment decisions, at the individual firm level, are two foci pertaining to a strategic focus that considers the overall competitive landscape within an industry and detailed profiling of potential risks (Ferreira, 2011).

Investment Decision-Making: The Tourism and Hospitality Industry Outlooks

Investment decision-making in the tourism and hospitality industry generates two viewpoints covering a destination perspective and an organization-level (individual firm with the industry) perspective. The seminal work of Butler (1980) explains how a destination evolves by potentially experiencing a series of states, including exploration, involvement, development, consolidation, stagnation, rejuvenation, and decline. At the very early stages of exploration and involvement, visitations remain minimal, and the development requirements could be managed by local- or in some cases regional- level capital investment (Keller, 1987). Nonetheless, Keller (1987) finds that a destination compromises its control from local to national or international authorities as soon as the destination further develops along the line of Tourism Area Life Cycle (TALC) through securing greater capital inputs. Perhaps, during these advanced stages of development, a destination seeks Foreign Direct Investment (FDI). The implication of TALC scales a "destination" more flexibly from a site to global label, at a national level; however, it is imperative to develop strong and appealing "country branding" statement to attract investors both from within and outside the country of reference. The branding of countries needs a strategic management approach to evaluate their strengths, vulnerabilities, challenges, and opportunities (Kotler & Gertner, 2004). When the branding plan is in place, to ensure the possible effect, governments must allocate and/or secure the requisite resources and funds. Countries must also develop a monitoring framework to ensure the desired level of efficiency (Kotler & Gertner, 2004). When investors are researching their investment targets, destination marketers need to offer reliable and accurate information (e.g. labor, tax environment, facilities, higher education, schools, legislation, energy, connectivity, and business).

At an organizational level, several factors can influence investment decision-making. For example, Newell and Seabrook (2006) have found five key factors and associated twenty-five sub-factors that influence investment decisions for a hotel. The study findings reveal financial and location factors bear the utmost importance backed by economic, diversification, and relationship factors, respectively. In terms of FDI in the global hotel chain, Santos et al. (2016) find that location factors along with economic development patterns and cultural distance of the home country versus host country remain crucial. Thus, tourism investment decisions, although chiefly dominated by the motive of profit optimization, some qualitative dimensions and factors influence the ultimate decision-making process. The decision-making to invest strongly relies on the availably of both domestic and FDI. The data, information, and literature availability on domestic investment in the tourism industry of Bangladesh is acute. This is the reason for which this research focuses on FDI and creates a link between investment decision-making and the tourism industry context of Bangladesh.

Foreign Direct Investment (FDI) and the Tourism and Hospitality Industry

FDI is a strategy to expand businesses and investment portfolios across a national boundary (Denisia, 2010). FDI facilitates an exchange process in which the transfer of resources such as monetary capital, knowledge and expertise, technology, and innovation happens between a home and host country or countries for mutual gains (Bajrami & Zeqiri, 2019). On the one hand, an investor from a home country often finds FDI rewarding by means of accessing new and promising markets, increasing investment (asset) portfolio, and minimizing risk exposures (Abdin, 2015). On the other hand, the host country is the recipient of investment benefitted from GDP growth, tax revenue, and more employment (Bajrami & Zeqiri, 2019). However, the relative attractiveness and prospect of an economic sector that receives FDI influence the extent of potential gains or losses from the investment (Denisia, 2010).

FDI plays a crucial role in the overall development of the tourism and hospitality industry (a sub-sector of the service economy) by furnishing the necessary resources (Tang et al., 2007). Tourism and hospital need financial resources for the production and maintenance of resources, infrastructure, information, and accessing the global distribution chains and marketing (Banerjee et al., 2015; Hof & Blázquez-Salom, 2015; Stauvermann & Kumar, 2017). In that, the inflow of FDI improves the tourism and hospitality industry facilities and services such as international airports, airlines, motorways, hotels, restaurants, historical places, and tourist attractions (Davidson & Sahli, 2015; Endo, 2006; Li et al., 2017; Yazdi et al., 2017). It also enables the transfer of skills, expertise, and production techniques for hosting destinations and/or countries (Markusen, 1995; Marrocu & Paci, 2011).

Although FDI for tourism is chiefly concentrated in developed nations, the investment trend is rising noticeably in developing countries with an especial focus on the global hotel chains (Jacob & Groizard, 2007; Samimi et al., 2013; Williams & Deslandes, 2008). Tecle and Schroenn (2006) argue that the ultimate effectiveness of FDI in the tourism and hospitality industry is partially dependent on the degree to which professional know-how is transferable across subsidiaries. Unlike the manufacturing sector, which is capital-intensive and considers technology transfer a key criterion of FDI effectiveness, the service sector requires mobility of expertise and skills (Caves, 1996). The tourism and hospitality industry belongs to the broader service sector, where at all stages of the development and consumption processes, customers are closely connected to service providers. Thus, the transfer of expertise and skills from TNCs to local tourism and hospitality industries and/or subsidiaries helps to improve the overall tourism experience.

United Nations Centre for Transnational Companies (UNCTC, 1982) and United Nations Conference on Trade and Development (UNCTAD, 2007) acknowledge that transfer of expertise and skills remains a core indicator of FDI to the hotel sector both by developing and developed countries. Correspondingly, Transnational Companies (TNCs) investing in the cross-border hotel industry are emphasizing human capital growth with a view to achieving positive learning curve impact and economies of scale (Salifou & Haq, 2017). In this respect, TNC hotels have developed training programs and information sharing platforms for their employees that cover structured modular education, formal face-to-face training, on-the-job training, online seminars, case study videos, and instruction manuals.

UNCTAD (2007) reports that tourism-related TNCs are capable of improving the international reputation of destinations and improving the image of the host countries. Travelers, mainly from developed countries, are accustomed in their home environment to modern infrastructure and high-quality hospitality facilities (United Nations World Tourism Organization: UNWTO, 2007). They, therefore, expect to have exactly the same level of standards and comforts while traveling in a foreign country (Cohen, 1984). The involvement of TNCs and multinational restaurant and hotel chains in some destinations leads to raising standards by means of availing more sophisticated systems and quality control, which in turn cater to quality the tourism and hospitality industry services and facilities. This eventually results in a more satisfied pool of visitors. Moreover, given the unpredictable nature of the tourism and hospitality industry, TNCs can be stronger and more resilient than local companies ensuring the stability of the economy (host country or the recipient of FDI) and retaining trust in it (UNCTAD, 2007). FDI is therefore supposed to encourage the quantum and standard of services and facilities, which can be connected later to increased foreign arrivals of tourists (Endo, 2006; Selvanathan et al., 2012; Zhang et al., 1999).

The involvement of TNCs allows foreign investors, contacts, and employees to travel to the source country on a regular basis for business purposes, such as acquiring more accurate and nuanced information that is not readily available via the private sector or government institutions (Haley & Haley, 1997). In order to ensure business success, such travel is necessary in order to understand the various economies, cultures, and political systems. Therefore, the increase of tourism in a country is directly linked to the extent of FDI (Tang et al., 2007). Furthermore, in the form of foreign real estate investment, FDI is also found to generate further tourism growth and increase tourist arrivals, in particular in Integrated Residential Estate Schemes and Resort Schemes. Real estate owners in host countries also travel with their families and friends to the destinations, which in turn creates extra travel and revenue for the destinations (Fereidouni & Al-mulali, 2014). There are pieces of empirical evidence that positive investment decision-making for FDI supports the tourism and hospitality industry's development.

FDI for the Tourism and Hospitality Industry's Development: Empirical Evidences

Haley and Haley (1997) empirically proved that FDI contributes to the growth of accommodation facilities and new tourist attractions that boost the arrival of tourists in Vietnam. Craigwell and Moore (2008) used panel causality methods in a study focused on small island developing states and found a bidirectional relationship between FDI and tourism growth. Their results showed, on the one hand, FDI provides the small island states with extra energy enabling them to extend their tourism activities. On the other hand, tourism growth attracts a higher level of FDI. Sadi and Henderson (2001) revealed similar outcomes in the case of Vietnam.

However, Tang et al. (2007) examined the relationship between FDI and tourism by using a time series econometric method on data extracted of 27 years and concluded a one-way causality relationship indicating FDI as a big contributor to the growth of tourism in China. Selvanathan et al. (2012) reconfirmed this relationship in the Indian context by applying vector autoregressive framework. The rise in tourist arrivals led to an increased demand for hotel accommodation, which in turn was met by more FDI from the foremost international hotel chains in destinations (Broadman & Sun, 1997). Sanford and Dong (2000) found a similar result in the case of the United States of America. In the African context, Othman et al. (2012) also reported a clear positive FDI-tourism partnership in which inward FDI in Africa directed to significant tourism growth.

Fereidouni and Al-mulali (2014) studied selected countries from the Organization for Economic Co-operation and Development (OECD) to investigate the long- and short- term empirical relations between FDI in international tourism flows and the real estate sector. Not only did they find the presence of a long-run relationship for FDI and tourism flows, but also the findings confirm a bidirectional causal relationship. Similarly, Jayaraman et al. (2014) reported positive long-run correlations between tourism earnings and FDI in Fiji. On this note, a rise in the ratio of 10% of FDI to Gross Domestic Product (GDP) is expected to contribute to a rise in tourism earnings of about 0.49% (Jayaraman et al., 2014). Though tourism growth and FDI are linked to a reciprocal relationship, increased tourism demand can hurt the environment of a destination. In Mexico, to explore the relationship between FDI, the environment, and tourism growth, Garcia-Flores et al. (2008) found that a positive relationship exists between FDI and the growth of tourism while the increased production of tourism brought negative environmental impacts. Thus, the upfront implications remain for the policymakers is to adhere to the environmental standards while attracting FDI in tourism at destinations.

The following section of this chapter correspondingly highlights the linkages and implications of above-mentioned concepts and empirical evidence in reference to the tourism and hospitality industry of Bangladesh.

Concepts Exposition and Implications for the Tourism and Hospitality Industry in Bangladesh

The generalized four-stage process model of investment-decision (deduced from the literature) identifies that tourism investment decisions should explore the investment needs, collect relevant information, review the collected information, and finalize investment decisions based on the review results. Thus, the process depends on the quality and range of information, including the current trends of FDI to the host country, market size, potential growth, openness, exchange rate valuation, political stability, institutional quality, IRR, PI, and so on. Bangladesh is a potential recipient of tourism investment, lacks such information in a categorized format. To elaborate on how these information influence investment decisions, in particular FDI, the recent trend of FDI inflows to service sector is noteworthy. In the last decade, Bangladesh has seen stable political conditions, which increased the net FDI inflows into the country in general. Although categorized data for the tourism and hospitality industry is not available, we found data on net FDI inflows in hotels and restaurants, a sub-sector of the broader service sector, within the scope of the industry. Figure 2.1 correspondingly plots net FDI inflows in hotels and restaurants for the last 8 years.

Data indicate that Bangladesh is still struggling for FDI, especially in the hospitality industry. The country was securing an increasing amount of FDI until a sharp decline in 2016, which was presumed a consequence of a terrorist attack at Holey Artisan Bakery (Khan et al., 2017). Nonetheless, it recovered quickly in subsequent years with an uprising trend.

FDI is a strategy to broaden the portfolio of companies and assets across national boundaries (Denisia, 2010). For the development and maintenance of resources, facilities, information, and accessing the global marketing, and chains of distribution, the tourism industry needs financial resources (Banerjee et al., 2015; Hof & Blázquez-Salom, 2015; Stauvermann & Kumar, 2017). While tourism FDI is predominantly concentrated in developed countries, the trend of investment is

Fig. 2.1 Net FDI inflows in the service sector with relevance to the tourism and hospitality industry. (Source: Bangladesh Bank, 2019)

growing significantly in developing countries with a particular emphasis on global hotel chains (Jacob & Groizard, 2007; Samimi et al., 2013; Williams & Deslandes, 2008). Bangladesh is continuously attracting investments from renowned hotel chains such as Hotel Sheraton, Holiday Inn, JW Marriot, Dusit, Swissotel, Hyatt, etc. (The Daily Star, 2018). The increasing demand for hotel rooms is perceived as an expression of the growing urban visitor economy, mostly from the individuals visiting ready-made garments and textile with a business purpose (Chopra et al., 2015).

The empirical evidence provides an idea about the causal relationship between FDI and tourism growth, which is mostly seen as positive. Some empirical evidences represent Vietnam (Haley & Haley, 1997; Sadi & Henderson, 2001); China (Broadman & Sun, 1997; Tang et al., 2007); the United States of America (Sanford and Dong, Sanford Jr & Dong, 2000); Africa (Othman et al., 2012); OECD (Fereidouni & Al-mulali, 2014); Fiji (Jayaraman et al., 2014); Mexico (Garcia-Flores et al., 2008); India (Selvanathan et al., 2012), etc. For a foreign investor and TNC, the tourism and hospitality industry ecosystem of Bangladesh can be one of the promising economies with good domestic business capabilities. Bangladesh is well-known for its competitively priced labor availability for the tourism and hospitality industry business; low utility charges; long-term taxes on seaports, air, and land ports; full or partial exemption from royalty taxes, technological know-how, and fees for technical support and repatriation services; easy access to the largest regional facilities such as India and China.

The FDI recipient countries (host countries) often welcome and invite FDI by initiating different fiscal and non-fiscal packages. For instance, to secure domestic and foreign investments in tourism, the Government of Bangladesh encourages private investments through credit facilities, tax incentives, repatriation facilities of dividend, permanent residence on investing US$ 75,000 and citizenship on investing US$ 500,000, depreciation allowances, and so on (Bangladesh Bank, n.d.; Bangladesh Planning Commission, 2015). The country categorizes tourism as an export-oriented industry being scoped within the 'Thrust Sector' for securing special facilities and venture capital support (Bangladesh Bank, n.d.). It also provides legal protection to the foreign investors for Nationalization and Expropriation based on the Foreign Private Investment Act 1980, which guarantees non-discriminatory treatment between foreign and local investments. Bangladesh has also become the signatory of several agreements and members of the relevant association to ease FDI inflows, such as ICSID (International Centre for Settlement of Investment Disputes), OPIC (Overseas Private Investment Corporation), MIGA (Multilateral Investment Guarantee Agency), and the like (Bangladesh Bank, n.d.).

Some of the major obstacles to attracting FDI to the tourism and hospitality industry in Bangladesh through positive decision-making are the absence of functional physical infrastructure; bureaucratic complexity or permission to be licensed; absence of investment-promoting organizations; absence of experts and sector-specific professionals; weak imposition of IP (Intellectual Power) regulations. The tourism and hospitality industry of Bangladesh can expect increasing inflows of FDI by means of availing the support facilities and services (e.g., electricity, coal, water) to the generating agency; planning and approaching TNCs with full project profile (PP); increasing public expenditure could be under the public-private partnership (PPP) modality) to gain foreign investor confidence. Responsible authorities in the tourism and hospitality industry in Bangladesh are seemingly less interested in the research of investment decision-making for FDI. This actually results in a lack of information situation, which is desired by the foreign investors in their decision-making process. On this note, the government should focus on preparing and disseminating categorized data on FDI in the tourism and hospitality industry.

Conclusion

This chapter aims to explain the theoretical reasoning of investment decision-making and FDI in the tourism and hospitality industry within the context of Bangladesh. The study reports investment decision-making for FDI in the tourism and hospitality industry with empirical evidences. Literature shows that investment decision-making for FDI and the development of the tourism and hospitality industry are inter-linked and affects the national economy positively. Findings in the tourism and hospitality industry context of Bangladesh reflect that the country has been offering benefits for FDI. The county has been inviting TNCs and individual foreign investors to provide fiscal and non-fiscal incentives. There are some evidences that Bangladesh has been receiving FDIs supported by positive decision-making by the investors but still, the amount and success remain questionably inadequate. This study emphasizes that positive decision-making for FDI by the TNCs is much needed for Bangladesh. Accordingly, the study offers some suggestions to increase FDI inflows in the tourism and hospitality industry. This study has been largely constrained by the unavailability of relevant data in the identified research areas and strongly suggests to eliminate the scarcity of categorized data on FDI in tourism by the responsible authorities. Future research studies can inculcate a deeply rooted view to explore the contributing factors to tourism investments or in particular, FDI in the tourism and hospitality industry.

References

Abdin, M. J. (2015). Foreign Direct Investment (FDI) in Bangladesh: Trends, challenges and recommendations. *International Journal of Economics and Management Sciences, 4*(8), 1–5.

Aharoni, Y. (1966). *The foreign investment decision process*. Harvard University Press.

Bajrami, H., & Zeqiri, N. (2019). Theories of foreign direct investment (FDI) and the significance of human capital. *International Journal of Business and Management, VII*(1), 11–24.

Banerjee, O., Cicowiez, M., & Gachot, S. (2015). A quantitative framework for assessing public investment in tourism – An application to Haiti. *Tourism Management, 51*, 157–173.

Bangladesh Bank. (2019). *Foreign Direct Investment (FDI) in Bangladesh: Survey Report, July–December, 2019*. Retrieved from: https://www.bb.org.bd/pub/halfyearly/fdisurvey/fdisurveyjuldec2019.pdf. Accessed 1 Feb 2021.

Bangladesh Bank. (n.d.). *Investment facilities*. Retrieved from: https://www.bb.org.bd/investfacility/invesfac.php. Accessed 1 Feb 2021.

Bangladesh Planning Commission. (2015). *Seventh Five Year Plan, FY2016-FY2020: Accelerating growth and empowering citizens*. General Economics Division, Planning Commission, Government of the People's Republic of Bangladesh. Retrieved from: http://www.plancomm.gov.bd/site/files/2eb8022b-d4fb-42e1-a67d-67bccabd8fe9/-. Accessed 1 Feb 2021.

Broadman, H. G., & Sun, X. (1997). The distribution of foreign direct investment in China. *The World Economy, 20*(3), 339–361.

Butler, R. W. (1980). The concept of a tourist area cycle of evolution: Implications for management of resources. *The Canadian Geographer/Le Géographe canadien, 24*(1), 5–12.

Caves, R. E. (1996). *Multinational Enterprise and economic analysis*. Cambridge University Press.

Chopra, D., Thapar, R., & Mann, S. S. (2015). *In focus: Bangladesh the dark horse*. Retrieved from https://www.hospitalitynet.org/file/152006053.pdf. Accessed 5 July 2020.

Cohen, E. (1984). The sociology of tourism: Approaches, issues, and findings. *Annual Review of Sociology, 10*(1), 373–392.

Craigwell, R., & Moore, W. (2008). Foreign direct investment and tourism in SIDS: Evidence from panel causality tests. *Tourism Analysis, 13*(4), 427–432.

Davidson, L., & Sahli, M. (2015). Foreign direct investment in tourism, poverty alleviation, and sustainable development: A review of the Gambian hotel sector. *Journal of Sustainable Tourism, 23*(2), 167–187.

Denisia, V. (2010). Foreign direct investment theories: An overview of the main FDI theories. *European Journal of Interdisciplinary Studies*, 53–59.

Drury, C. (2012). *Management and cost accounting* (p. 7). Cengage Learning.

Dwyer, L., Forsyth, P., & Dwyer, W. (2010). *Tourism economics and policy*. Channel View Publications.

Endo, K. (2006). Foreign direct investment in tourism – Flows and volumes. *Tourism Management, 27*, 600–614.

Fauzel, S. (2020). FDI and tourism futures: A dynamic investigation for a panel of small island economies. *Journal of Tourism Futures*. https://doi.org/10.1108/JTF-05-2018-0026

Fereidouni, H. G., & Al-mulali, U. (2014). The interaction between tourism and FDI in real estate in OECD countries. *Current Issues in Tourism, 17*(2), 105–113.

Ferreira, D. (2011). Corporate strategy and investment decisions. In H. K. Baker & P. English (Eds.), *Capital budgeting valuation: Financial analysis for Today's investment projects* (pp. 19–36). Wiley.

Garcia-Flores, F., Pérez-Cirera, V., Martín, P., Alanis, G., de Derecho Ambiental, C. M., & Guadarrama, L. (2008). Tourism trade and the environment: Tourism and coastal development in the Mexican portion of the Mesoamerican Reef. In *Fourth North American symposium on assessing the environmental effects of trade, commission for environmental cooperation*. The 23rd April.

Haley, U. C., & Haley, G. T. (1997). When the tourists flew in: Strategic implications of foreign direct investment in Vietnam's tourism industry. *Management Decision, 35*(8), 595–604.

Hof, A., & Blázquez-Salom, M. (2015). Changing tourism patterns, capital accumulation, and urban water consumption in Mallorca, Spain: A sustainability fix? *Journal of Sustainable Tourism, 23*(5), 770–796.

Jacob, M., & Groizard, J. L. (2007). Technology transfer and multinationals: The case of Balearic hotel chains' investments in two developing economies. *Tourism Management, 28*(4), 976–992.

Jayaraman, T. K., Chen, H., & Bhatt, M. (2014). Contribution of foreign direct investment to the tourism sector in Fiji: An empirical study. *Tourism Economics, 20*(6), 1357–1362.

Keller, C. P. (1987). Stages of peripheral tourism development—Canada's northwest territories. *Tourism Management, 8*(1), 20–32.

Khan, M., Islam, M. S., & Haque, S. (2017). Impact of holey artisan restaurant attack on restaurant business in Dhaka City. *Southeast University Journal of Arts and Social Sciences, 2*(1), 67–81.
Kotler, P., & Gertner, D. (2004). *Country* as *brand*, product, and beyond: A place marketing and brand management perspective. In N. J. Morgan, A. Pritchard & R. Pride (Eds.), *Destination branding*: *Creating* the *unique destination proposition* (pp. 40–45). *:* Butterworth-Heinemann.
Li, X., Huang, S., & Song, C. (2017). China's outward foreign direct investment in tourism. *Tourism Management, 59*, 1–6.
Markusen, J. R. (1995). The boundaries of multinational enterprises and the theory of international trade. *The Journal of Economic Perspectives, 9*(2), 169–189.
Marrocu, E., & Paci, R. (2011). They arrive with new information: Tourism flows and production efficiency in the European regions. *Tourism Management, 32*(4), 750–758.
Martínez-Román, J. A., Tamayo, J. A., Gamero, J., & Romero, J. E. (2015). Innovativeness and business performances in tourism SMEs. *Annals of Tourism Research, 54*, 118–135.
Newell, G., & Seabrook, R. (2006). Factors influencing hotel investment decision making. *Journal of Property Investment & Finance, 24*(4), 279–294.
Nunkoo, R., & Ramkissoon, H. (2016). Stakeholders' views of enclave tourism: A grounded theory approach. *Journal of Hospitality & Tourism Research, 40*(5), 557–558.
Othman, R., Salleh, N. H. M., & Sarmidi, T. (2012). Growth and foreign direct investment: An ARDL approach. *Journal of Applied Sciences, 12*(12), 1245–1254.
Saarinen, J. (2017). Enclavic tourism spaces: Territorialization and bordering in tourism destination development and planning. *Tourism Geographies, 19*(3), 425–437.
Sadi, M. A., & Henderson, J. C. (2001). Tourism and foreign direct investment in Vietnam. *International Journal of Hospitality & Tourism Administration, 2*(1), 67–90.
Salifou, C. K., & Haq, I. U. (2017). Tourism, globalization and economic growth: A panel cointegration analysis for selected West African States. *Current Issues in Tourism, 20*(6), 664–667.
Samimi, A. J., Sadeghi, S., & Sadeghi, S. (2013). The relationship between foreign direct investment and tourism development: Evidence from developing countries. *Institutions and Economics, 5*(2), 59–68.
Sanford, D. M., Jr., & Dong, H. (2000). Investment in familiar territory: Tourism and new foreign direct investment. *Tourism Economics, 6*(3), 205–219.
Santos, M., Brochado, A., & Esperança, J. (2016). Foreign direct investment patterns of global hotel chains. *Journal of Business Research, 69*(11), 5235–5240.
Selvanathan, S., Selvanathan, E. A., & Viswanathan, B. (2012). Causality between foreign direct investment and tourism: Empirical evidence from India. *Tourism Analysis, 17*(1), 91–98.
Stauvermann, P. J., & Kumar, R. R. (2017). Productivity growth and income in the tourism sector: Role of tourism demand and human capital investment. *Tourism Management, 61*, 426–433.
Sykianakis, N., & Bellas, A. (2005). The foreign direct investment decision making process. Strategy, politics and uses of management accounting information. *Managerial Auditing Journal, 20*(9), 954–969.
Tang, S., Selvanathan, E. A., & Selvanathan, S. (2007). The relationship between foreign direct investment and tourism: Empirical evidence from China. *Tourism Economics, 13*(1), 25–39.
Tecle, Y. H., & Schroenn, J. L. (2006). The contribution of HRD to tourism-led development in an African context: Economics. *South African Journal of Economic and Management Sciences, 9*(4), 444–457.
The Daily Star. (2018). *Dhaka to see six new luxury hotels*. Retrieved from: https://www.thedailystar.net/business/dhaka-see-six-new-luxury-hotels-1572655. Accessed 1 Feb 2021.
UNCTAD. (2007). *FDI in tourism: The development dimension*. Retrieved from: unctad.org/en/Docs/iteiia20075_en.pdf. Accessed 1 Dec 2020.
UNCTC. (1982). *Transnational corporation in international tourism*. UNCTC.
UNWTO. (2007). *A practical guide to tourism destination management*. UNWTO.

Wei, H. C., & Christodoulou, C. (1997). An examination of strategic foreign direct investment decision processes: The case of Taiwanese manufacturing SMEs. *Management Decisions, 35*(8), 619–630.

Williams, D. A., & Deslandes, D. (2008). Motivation for service sector foreign direct investments in emerging economies: Insights from the tourism industry in Jamaica. *The Round Table, 97*(396), 419–437.

Yazdi, S. K., Salehi, K. H., & Soheilzad, M. (2017). The relationship between tourism, foreign direct investment and economic growth: Evidence from Iran. *Current Issues in Tourism, 20*(1), 15–26.

Zhang, H. Q., Chong, K., & Ap, J. (1999). An analysis of tourism policy development in modern China. *Tourism Management, 20*(4), 471–485.

Chapter 3
Theoretical Understanding and Perspective Analysis of Investment and Development in the Tourism and Hospitality Industry in Bangladesh

A H M Salimullah

Abstract In many countries, tourism becomes the fast mounting industry and a pivotal source of foreign currencies. In comparison with the developed part of the world, Bangladesh has demonstrated the emerging trend, and technology-based approaches that have flourished trade and investment and ultimately have played a revolutionary role to make this industry a real success. Despite its growth prospects and accomplishments, certain factors such as lack of proper guidance, academic knowledge, and lower financial investment from private and public institutions or firms have contributed to keeping this industry neglected for many decades. As a result, Bangladesh has not exposed her immense potentiality to boom this industry in practice, given the fact that she upholds the granary of all kinds of natural beauties and cultural legacy. Proper maintenance of natural resources during the last several decades may come up with substantial amounts to her national budget. Therefore, this chapter investigates the tourism-led-growth and the evolving trends of the development and investment in new, adventurous, and community-based tourist places such as augmented reality eco-tourism and smart tourism.

Keywords Theory · Investment · Development · Tourism · hospitality · Bangladesh

Introduction

Over the last few decades, tourism has accomplished a great achievement in the world economy as a single largest industry. Many countries such as France, Spain, China, and the US have considered this growing industry a significant foreign income source and the prime hub for vital economic activities. Those countries having the

A H M Salimullah (✉)
Department of Economics, University of Chittagong, Chattogram, Bangladesh
e-mail: salimullah1977@gmail.com

A. Hassan (ed.), *Tourism in Bangladesh: Investment and Development Perspectives*, https://doi.org/10.1007/978-981-16-1858-1_3

largest industry earn myriad incomes in payments for domestic goods and services and bring about colossal employment opportunities in the service industry. Daily Telangana Today (2018) reported that in the world's trade of services, tourism's contribution is estimated at 30%, and the contribution in exports of goods and services stands at 6%. The emerging trend of tourism has become a sine-quo-non to those countries' immense economic development. The tourism industry has become increasingly appealing and magnificent both in the developed and developing countries due to augmenting and cutting-edge new destinations that reflect their historical, cultural, and natural domains. These focal points are increasingly treated for progression and prosperity to make those the most adventurous and striking tourist destinations. It has become a typical pattern that many of the rapidly flourishing tourist spots are positioned in the middle- and low-income countries, and, correspondingly, those countries have become the center of excellence in this growing industry.

Since the independence in 1971, Bangladesh has overlooked to equip this emerging industry with significant facilities and aptitudes necessary for its breakthrough. It is undoubtedly recognized that the country is blessed with extraordinary natural resources with the availability of colossal historical and cultural sights. Due to the lack of several determinants such as proper knowledge, capital, the focus of attention, research, and study on the tourism industry's promotion and protection, this vital industry was unexposed for a long time. Hemmati and Koehler (2000) claimed that despite its more excellent prospect and potentiality to improve the industry and contribute to the economy remarkably, this industry fell short of required conditions such as lack of proper guidance from the concerned institutions and authorities for scrambling off to a regular time interval.

Davenport and Davenport (2006) argued that countries having endowed with a couple of natural, technical and designed advantages made up this world that has witnessed tourism as the fourth largest industry for employment generation and contribution to the national GDP (Gross Domestic Product). In 2012, for example, travel and tourism accounted for 9% of global GDP (share) and supported approximately 260 million jobs (World Travel & Tourism Council, 2014). Still, this booming industry has encountered the existing challenges deriving from global climate change and economic recession. International Monetary Fund (IMF) has anticipated that in 2020 the many of the developing country's economy will maintain GDP growth as high as 4.6% as of 2019 (Global Finance Magazine, 2020). Muhanna (2007a) observed that in the IMF's measurement, they exhibited that the tourism and recreation industry of those Asian and African countries will become the leading portion to that of the fastest GDP growth rate. In earlier research, Tooman (1997) mentioned that the tourism industry's emerging trend may become an instrument for developing countries to combat poverty. This industry can assist those countries, including Bangladesh, by generating income and employment and expanding the economy with their unparalleled natural beauty.

Blessed with specific criteria necessary for promoting and protecting the tourism industry, Bangladesh can utilize her exceptional natural and cultural traits. Bangladesh is a small, densely populated South Asian country surrounded and bordered by India from three sides and by Myanmar and the Bay of Bengal from the Southeast and

Southside. This geographical location has been made this country a vital communication hub from traveling to the western part to the eastern part of this world either by air or by sea. Besides, Bangladesh is blessed with natural beauties and cultural legacies as the preconditions for the development of tourism industry. Particularly, the country has two undeniable natural resourceful areas: the Sundarbans, and the longest unbroken Cox's Bazar sea beach. Some other renowned cultural, historical, and archaeological visiting and recreation sites have been innovated and renovated at different periods under different governmental regimes throughout the century.

According to Sultana (2016), the decreasing trend of economic growth in Bangladesh reveals that the country needs to focus on the conditions precedent for the tourism industry's advancement. Data shows that the tourism industry's contribution to the GDP, and employment is only 2.1% and 1.9%, respectively. Comparing to the close neighbors, Bangladesh receives the lowest number of international tourists each year. Her tourism market size is very trivial for both domestic and international tourists. The government has enormous potential to enhance economic growth and, hence, develop by adopting new tourism-led-growth. Suppose Bangladesh makes substantial progress in this industry by adopting new technology, investment, and new establishment then within few years. In that case, tourism and recreation will be our country's primary foreign earning sources. This chapter will devise a model to resolve sustainable tourism and hospitality management for Bangladesh. This chapter aims to explore the challenges and problems associated with promoting the tourism industry. This chapter argues that the responsibility for overcoming those challenges that facilitate the process of development ultimately rests with the government and private organizations.

Overview of Tourism Industry in Bangladesh

Bangladesh is not merely reputed as an attractive tourist destination to foreigners and international communities. Instead, she is described as a country stuck with poverty, natural calamities, climatic disasters, and many other low images to the global communities. Moreover, the people themselves are hospitable, generous, and friendly, essential for any country to be a tourist spot. Hossain and Firozzaman (2003) stated that tourism in Bangladesh did not make the proper advancement up to a certain standard level because it falls below the in line of sufficient number of attractions but lacking from unreasonable and insufficient promotional activities. The country's tourism and hospitality industry is conquered mainly through the active and smart private industry. In the last two decades, private industry has flourished significantly with their tourism and hospitality products and services in the hole country, and in Dhaka, Rajshahi, Sylhet, Khulna, and Chattogram divisions, in particular.

According to the World Travel & Tourism Council (2014), the total contribution of travel and tourism to Bangladesh's GDP was TK 460.3 billion (4.4 percent of GDP) in 2013, and is expected to increase by 7.9% in 2014, and by 6.5 percent per year through 2024, to TK 935.5 billion (4.7 percent of GDP). However, in 2013,

Travel & Tourism contributed 3.8 percent of total jobs, including new jobs produced by the business indirectly (i.e. almost over 2 million jobs were created). In 2014, this is predicted to increase by 4.2 percent to 2,965,000 jobs, with annual increases of 3.0 percent to 3,974,000 jobs by 2024. (4.2 percent of total). Therefore, with this influential and more than expected statistical figure, it can easily be understood how much potentiality does our tourism industry possesses and there is no doubt about that each time this vital industry is moving forwards to build a new lighthouse to our economy. According to World Tourism Organization (WTO) tourism and hospitality management continue to to outpout as a world economic force with the contribution of almost US$5.5 trillion to the global economy in 2004 (World Tourism Organization, 2005). The increasing trend of tourist arrivals and revenues, according to Rahman et al. (2010), is continuing worldwide. According to prior estimates from 2005, global visitor arrivals increased to 808 million, with 1,561.1 million predicted by 2020. The ongoing rise of the tourism industry around the world is positive, and countries are demonstrating their desire to attract more tourists to their destinations and noteworthy locations.

Latif et al. (2015) studied that a significant number of countries are trying to endorse their tourism industry as a major source of foreign income and thus forming the opportunity to establish this important industry as an exciting means for national economic development.

Another study by Hossain and Nazmin (2006) showed that the tourism industry is often mentioned as one of the significant patrons who can provide extensive returns to the Protected Reserved Forests. The economic benefit of tourism is considered as reducing local communities' reliance on natural resources such as parks, sunset spots, and lakes, among other things. In Bangladesh, there is a scarcity of in-depth research on the effects of tourism on ecosystems and communities. No effective national tourism plan has been documented, and no effective rules are implemented to guide and control the entrepreneurs, manufacturers, merchants, and investors associated with this industry.

Rahman et al. (2010) revealed that Bangladesh has the opportunities to evolve and modernize many tourism spots and facilities across the country. Still, only a few famous tourist spots dispense service to the visitors and fulfill their magnetism at present. Such limited facilities are commonly clustered in some important places in Dhaka, Sylhet, and Chattogram divisions. In Chattogram division, visitors and foreigners' main attractions facilities are located at diverse locations of Cox's Bazar, Saint Martin's Island, Rangamati, Khagrachari and Bandarban districts. Two important archaeological and historic sites located in Mainamati at Cumilla and Lalbagh road and Islampur road in Dhaka (i.e. Lalbagh Fort and Ahsan Manzil). In Sylhet Division, most of the tourist areas are attracted mainly by the realistic scenarios of Jaflong, Madhabkunda, and Tea gardens. Other famous and attractive places are Kantaji's Temple, Swapnapuri, Ramsagor located in Rajbari, Dinajpur, and Rajshahi Division. Paharpur at Noagaon, and Mahasthangarh at Bogra are also well-known to tourists for their historic manifestation. In Khulna Division, visitors are mainly attracted by the Shatgombuj Mosque at Bagerhat, and the Sundarbans, the largest mangrove forest. These heritages are situated in the southern part of Khulna that lays alongside the Bay of Bengal.

In a statement, the Daily Star (2009) stated that our unfortunate Bangladesh had not made sufficient steps to protect and develop all of its historical, cultural, and environmental tourism assets. Various important tourism attractions, as well as the geographic distribution of some ancient sites and monuments, have lately been found. It depicts the distribution of tourist attractions around Bangladesh. Furthermore, many other lovely and lovely areas have yet to be discovered, necessitating tourist expansion. It explains the previously unknown fact that tourism facilities could be established and expanded across our country's many regions. However, the most modern and technologically advanced strategies may be able to handle a continuous expansion of tourism chances based on potentialities in various sections of this deltaic country.

Investment and Development Perspective Theories in Tourism

At the edge of the preceding century (i.e. during the decades of 1970s and 1980s), the framework, scheme, and outcomes of tourism investment and development evolved as a broad contextual basis that exerted influence on the critical mind of the then researchers and analysts of economics. Britton (1982) revealed that the two approaches were dominated by the dependency perspective and the life-cycle model. Both the case of dependency and the life-cycle approach emerged from Neo-Marxism's theoretical extractions and the modernization theory.

Any feasible nation or a region that wants to enter the tourism destinations' businesses initially faces strong global competition because the nation's emerging tourism industries are primarily characterized and supported by the local establishment. Local industry structures become characterized by foreign organizations or locally selected proprietorship as the number of visitors grows and their integration into the global tourist system growth. Although local firm owners receive economic benefits, they must carry a large portion of the expected costs if there is a rapid increase in tourist numbers. However, increased higher costs associated with any mass arrival of tourist people within a nation will supply both the local communities and neighborhoods to rely on it to be a real prospect for local communication and control. Still, there would be limited potential for achieving more sustainable regional development forms.

Nevertheless, both the evolving approaches are featured as problematic since they lack the mechanism to recognize the avenues that individuals, industries, and local government can regulate their destinies. As Preister (1989, p. 20) noted, "locally-affected people are not shaped passively by outside forces but react as well, at times even changing the conditions of the larger system". Preister (1989) contended and considered the development outcome as a negotiation product between individuals or local groups and structural forces. The structures are revealed as an upheaval that could not assess the empowering locals' strategy to compel the development plans. Hence, the declining sequence of progression might be minimized or circumvented (Drake, 1991). Priestley and Mundet (1998) specified that technological revolution and industrial organization modifications are not handled effectually. In particular, the presence of incompetence enriches the perceptions of capital's ability (or otherwise) to remedy the periodic actions of profit downturn and dilemma.

Another new approach (community approach) in the 1980s focused on the growing role of local agency, treating communities and its constituent members as an effective operational system in influencing tourism outcomes (Taylor, 1995). The community method, in contrast to the two previous models, identified locals as powerful planning instruments and participants in tourism development. The community's decision-making power works as a successful outcome when their rights are concerned in deciding tourism development. The community becomes a decisive factor in controlling industrial development's culmination to some extent. Murphy and Price (2005) argued that the host communities' presence and participation in the decision-making power facilitate the system to devise and design the types of tourism they desire to introduce according to their needs and can gradually regulate their pros and cons. Regrettably, proponents of community participation in tourism development have frequently forgotten local élites' proclivity to hijack the instruments of participation for their own gain (Brohman, 1996). Gender relations, race, and ethnicity have all influenced power structures inside communities and how they are connected to larger socioeconomic, political, and environmental institutions.

The economics of new tourism has emerged with its new perceptions– profitability no longer depends exclusively on economies of scale and numerous identical markets' exploitation. Somewhat economies of scope, cartel formation, systems expansion, price competition, and designed and customized vacations are becoming more critical for profitability and tourism competitiveness. This new model tradition's significant characteristics are facilitated by its focus on product flexibility and available variety that use advanced technology. Bangladesh, a major developing country in South Asia, holds high tourism potential. This country has the evident scope for establishing and expanding its opportunities for nature-based, culture-based, research-based tourism and eco-tourism.

Regulation and flexible specialized approaches are important steps in enabling Bangladesh's tourism and economic development to work together. Both methods emphasize the importance of understanding the broader processes of industrial accumulation in order to comprehend the economic prospects that areas and communities throughout the world face, as well as the intricacy of the tourism development process. Despite various criticisms, a regulations approach will provide Bangladesh a useful base from which the county can build an ampler and more understandable global–local relations that influence a concrete development outcome of our tourism industry.

Methodology and Objectives

Tourism can be the second most significant industry after ready-made garments in our country. To continue further advancement and development in this crucial industry of our economy, more investment and thorough sustenance is need from government and non-government industry. The major goal of this study is to define the possible scope of investment and development in Bangladesh's tourist and hospitality industry. It aims to pinpoint the industry's current issues and future potential in relation to the country's economic growth and development. This study is based on conceptual ideas, secondary information, and another empirical field of observation from both a national and international perspective. The secondary information is collected mainly from the related literature survey and relevant internet-based data sites. Consequently, this chapter's focal point is to assess the positive correlation between tourism and Bangladesh's economic growth and to find out the ambiguities (if any) of existing tourism and hospitality management in Bangladesh and provide a model about sustainable tourism and environmental protection.

Data Analysis on Present Status of Tourism in Bangladesh

Foreign tourists, in comparison to our large number of domestic visitors of all types, make up a significant portion of the entire tourist population each year. The overall number of international tourists in 1997 was 1,82,420, with a rising trend continuing in subsequent years. In 2003, there were 2, 44,509 tourists, an increase of 0.60 percent from the previous year. Bangladesh was visited by around 12.75 percent of foreign tourists in 2004 and 2005. However, in 2006, the number of tourists arriving in Bangladesh fell to 3.09 percent of the overall number of tourists (Rahman et al., 2010). From 1997 to 2017, Table 3.1 shows the number of tourists who arrived in Bangladesh by year.

However, Bangladesh's visitor arrivals dropped by almost 73% in December 2018, compared with an increase of 23.6% in the previous year. The data on visitor arrivals growth rate is updated yearly on the Ministry of Civil Aviation and Tourism website, available from December 1982 to December 2018. The data reached an all-time high of 23.7% in December 2015 and a record low of −7.9% in December 2018 (Census and Economic Information Centre Data, n.d.). The Global Economic Databases and indicators by Census and Economic Information Centre (CEIC) calculate the annual tourist arrivals growth rate, tourism revenue, and revenue growth rate from the different countries' yearly tourist arrivals. The Bangladesh Bureau of Statistics (BBS) provides data on Tourist Arrivals and arrival's growth rate.

Figure 3.1 displays a bar graph of the number of tourists who visited Bangladesh throughout time, which fluctuates dramatically across years. Foreign tourists climbed almost continually from 2006 to 2017, with a dip between 2006 (3.09 percent) and 2009 (4.11 percent). The rapid decrease in tourist arrivals was noticed in

Table 3.1 Number of foreign visitors in Bangladesh (1997–2017)

Year	Number of tourists arrival in Bangladesh	%	Year	Number of tourists arrival in Bangladesh	%
1997	182,400	2.81	2008	467,000	7.20
1998	172,000	2.65	2009	267,000	4.11
1999	172,800	2.66	2010	303,000	4.67
2000	199,300	3.07	2011	155,000	2.39
2001	207,200	3.19	2012	125,000	1.93
2002	207,300	3.19	2013	148,000	2.28
2003	244,500	3.77	2014	125,000	1.93
2004	271,300	4.18	2015	643,000	9.91
2005	253,300	3.90	2016	830,000	12.79
2006	200,400	3.09	2017	1,026,000	15.81
2007	289,000	4.45	Total	6,488,500	100

Source: Ministry of Civil Aviation and Tourism (2019a) and Census and Economic Information Centre Data (n.d.)

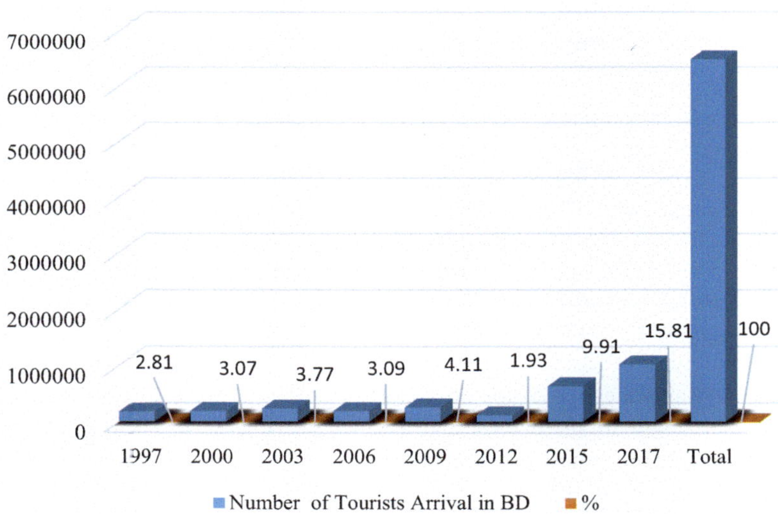

Fig. 3.1 Number of tourists' arrivals in Bangladesh from 1997–2017. (Source: Ministry of Civil Aviation and Tourism data, 2019b; Census and Economic Information Centre Data, n.d.)

2012 (1.93%) and 2014 (1.93%). Moreover, the number of tourist arrivals in this country has jumped to decrease particularly in 2011 and 2013, and the number has been fallen afterward considerably. But from 2015 and onwards, the diagram shows a skyward increase in visitors in Bangladesh. However, due to a lack of trustworthy statistics for the years after 2017, the current state of tourist volume cannot be determined (Census and Economic Information Centre Data, n.d.).

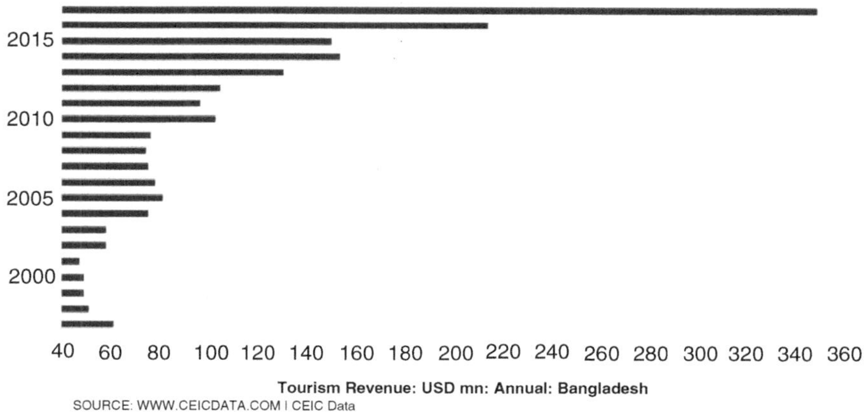

Fig. 3.2 Consolidated amount of tourist spending on domestic goods and services in Bangladesh (1997–2017). (Source: Census and Economic Information Centre Data, n.d.)

International tourism revenues, which can be defined as foreign direct expenditures by international incoming visitors, including payments to any national carriers for international transit or transportation, are one of the most important earning sources of foreign currency nowadays. Any indirect or agreed-upon payments made for goods or services provided by destination nations are likewise included in those payment receipts. The receipts from the sudden and same-day visitors or visitors for any particular purpose except when those are important enough to categorize distinctly based on the governmental invitation are also included.

Figure 3.2 shows revenue earnings from Bangladesh's tourism industry regarding spending for goods or services by foreigners from 1997 to 2017. The figure predicts that for the year 2017, this was 357,000,000, which is a 62.59% increase from 2016. But the data on tourism revenue statistics of Bangladesh has estimated the value of 367,000,000 in the year 2018, which is a 2.39% increase from 2017. All the data shown in the following charts are in current US dollars (Census and Economic Information Centre Data, n.d.). However, due to a lack of data and the ongoing pandemic scenario caused by the coronavirus spreading around the world, the current year's scenario and the current year cannot be monitored. In light of the entire scenario, it is reasonable to conclude that our country has not given sufficient attention to maximising tourism's economic potential. In Bangladesh, like in the rest of the globe, the tourist industry is growing in importance as a lucrative economic sector, and there is a significant risk of squandering this chance for export revenues and economic gains.

Existing Dodges of Bangladesh Tourism

The preconditions to promote and develop tourism are manifold such as infrastructural services, transportation facilities, effective advertisement and marketing strategy, and implementation of the holistic policy. Infrastructure facilities in many tourist spots such as Rangamati, Khagrachari, and Bandarban are not satisfactory in Bangladesh. Establishing hotels, restaurants, and rest houses, arranging picnic spots, amusement parks, theatres, camping sites, water skiing, etc., require the careful attention of the government as well as non-governmental organizations.

Despite the fact that its first national tourism policy was formulated in 1992 (Hassan & Kokkranikal, 2018), several factors such as lack of skilled manpower and technical expertise have contributed to downgrade the industry with its archaic policy guidelines for advertisement and marketing strategy. The pre-requisites of advertising tourism spots are breaking new ground for the people both in and outside the country. The BPC website lacks basic information, such as images of notable landmarks, lodging facilities, modes of transportation, distance from the capital city, information on the surrounding area, people, and weather, among other things. The political atmosphere is a condition precedent to the development of tourism in the country. Regrettably, the dearth of the congenial socio-political environment leading to political turmoil and economic strain makes such an attempt appalling and hence, challenging to sustainable tourism. There are also additional disruptions, such as a lack of good cooperation among numerous institutions, which harm the tourism business directly or indirectly. Following are the few obstacles that continue to thwart this vital industry's progression.

Shortcomings of Publicity and Marketing Activities

The strategies and maneuvering of advertisements to uphold the charms of tourism in order to mesmerize the international tourists are exposed as flawed as compared to the neighboring country's approach of advertisements in international print and electronic media. The advanced technology and innovative tactics established by the "countries of best practice" in upholding the standard of advertisement require to be practiced enhancing the constructive image of our country. The evolution of our prime tourist attractions divulging vibrant cultures, is a must to upkeep sustainable tourism. Arrangement of budgetary allocation from the government as well as private sources is necessitated to fulfill the gap. An effective brand name to flourish tourism may facilitate the process of effective publicity and marketing operations.

Backward Community

The perception and significance of tourism are not well-known to the regressive sections of society. Lack of knowledge to preserve and protect this pivotal industry is a reason behind its holistic development. Tourism is undoubtedly recognized as potential sources of employment and foreign exchange. Exploring every nook and corner of tourism demands the educational background, progressive mind, and innovative tactics of people and society to expand this business. Consciousness in discovering, devising, and extending the industry to a rural area will strengthen the business in a comprehensive way.

Lack of Training

The staff and personnel involved in the tourism industry are not proficient enough to administer the industry competently. For example, tourists guide lack enough training and knowledge to treat international tourists with sophistication and effective hands. The regular arrangement of training and coaching for the guides will help to reinvigorate the industry with its resourceful functions such as printing, folk-arts, and dying etc. Some special training programs for women to complement the functions of the staff might be needed in many cases.

Limitations in Infrastructural Development for Reinforcing Sustainable Tourism

Architectural design in infrastructure is a must to strengthen this vital industry. Embellishing an infrastructure with historical importance may beautify its overall standard that has an immense impact on developing thriving tourism industry. The amplification and introduction of necessary and primary services such as hotels and lodging, transportation and communication, safety and security, and tourism information services, among others, are all part of the process of decorating a tourist destination. Without incorporating these services in the system, this industry becomes useless, and its standard of facilities and services would be devalued.

Lack of Tourism Investments

While service investment is a well-established economic activity in rich countries, it is still in its early stages of development in developing countries. In developing nations, investment in service-oriented ventures, notably tourism, is connected with a number of obstacles, including high risk and hazardous conditions. As a result, while Bangladesh may be able to capitalize on natural tourism opportunities, securing adequate money for tourism initiatives is proving difficult. Furthermore, a lack of money may exacerbate the difficulties associated with project identification, planning, and implementation.

Lack of Consistent Tourism Policies and Strategies

In Bangladesh, National Tourism Policy 2010 works as guidelines to reduce poverty alleviation, to resolve tourism development issues, safeguard the built and natural heritage, and to achieve sustainable socio-economic development. Integrated tourism policymaking is a pre-requisite for sound and sustainable tourism development. The emerging conflicts between the government departments and the private tourism agencies give rise to discrepancies as regards the implementation of the policies to a great extent. This is conjoined in many cases with the dearth of effective administration, regulation, and institutional frameworks of touristic activity.

Lack of Tourism Safety

The safety and security of tourists play a pivotal role in making the tourism industry rewarding. The safety of tourists acts as a sine-quo-non for any successful tourism industry. One of the key attentions of tourism strategy and policy is planning and keeping a safe tourism environment. The perilous environment has harmed host countries' reputation. A negative impression of tourists formed out of an unsafe atmosphere in Bangladesh may destroy the prospects of tourism in the long run.

An intermingling of all the preconditions for a successful tourism industry is a must for its holistic growth. To promote and protect sustainable tourism, the country needs to wipe out the problems above contributed to the tourism business. Booming the tourism industry's potentials requires serious attention from both national and foreign investors. A considerable investment is looked-for. Devising a long term strategy with a well-organized plan of action may accelerate the process of development. Forming a destination wise investor's association may be a solution to this.

Investment and Development Strategies in the Tourism Industry

The tourism industry is considered a great sign of development strategy. At the primary, secondary or tertiary level, tourism's emerging development upholds a positive multiplier effect and causes growth (NumptyNerd, 2019). The positive multiplier effect has contributed to other industries generating jobs in the tertiary industry and facilitating other industries' growth. Jucan and Jucan (2013) described that the multiplier effect on the national economy is manifold. Directly, hotels' business as part of development strategy of tourism proliferates job markets, and indirectly it boosts up the national economy. The maintenance of a hotel, for instance, requires buying products from local markets or farmers who may utilize the amount in the expenditure of food or fertilizer. This chain of demand to tourists who may be eager to buy products or souvenirs from local markets contributed to secondary employment. This idea of a chain of positive multiplier effect can be identified in the following Fig. 3.3.

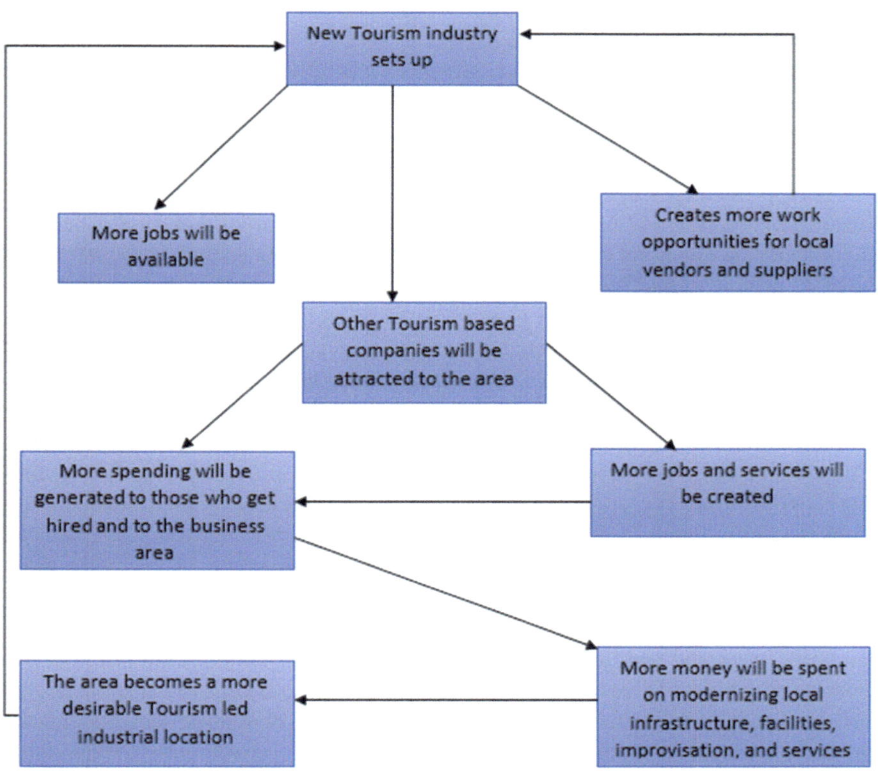

Fig. 3.3 Model of tourism led economic growth. (Source: NumptyNerd, 2019)

As a development strategy, the government of Bangladesh may promote several matters such as (a) relaxation of immigration controls, (b) easing visa requirements, (c) developing transport facilities and upgrading roads and highways, (d) devising metro rail, (e) nurturing tourist facilities such as hotels and restaurants, golf club, dive Centre, etc. for the promotion and protection of tourism industry. Introducing English courses for travel agents may facilitate the industry as a development tool. Building Tourism Information Centre to serve international tourists and reinforce international relations is another strong indicator of development strategy. Tourism encourages increased entrepreneurship for locals having the opportunity to escalate personal wealth. Strengthening new skills and technology is an added approach to invigorate tourism as a means of development strategy.

In some developing countries such as Vietnam, India, Taiwan, Barbados, Cuba, and Turkey, the tourism industry has been strengthened to strengthen local companies and entrepreneurs that contribute to devising new products and exports. As a development tool, tourism is a means to examine or test local products before the international markets, and it also urges to specific quality standards. Despite the challenges associated with global markets (e.g. to meet the adverse circumstances shortly), tourism works to motivate the system that brings about growth and gradual advancement. Maintaining labor market standards become a severe concern for tourism development.

Consequently, the burden for accessing the immigrant labor to the market has become increased. As a result, there were views of bad working conditions, career trajectories, and pay rates as compared to other industries. Re-skilling possibilities for businesses such as tourism may be available due to an ageing population. Similarly, providing adequate education and training programs is a critical area for boosting innovation and increasing productivity in the tourism business. The tourism industry faces a number of challenges in labour markets, including (a) attracting the best employees and retaining and developing them over time; (b) a lack of new skills to meet changing tourism trends; (c) effectively managing labour or skills shortages; (d) low-quality products; (e) harming the industry's competitiveness, and so on. Despite the repercussions of the current economic crisis, Muhanna (2007b) argues that the long-term prospects for tourism growth are in jeopardy.

Education and skill development are crucial for all tourism destination countries, but education and skill-related training in this field will provide a significant challenge for developing countries, particularly in those experiencing the fastest tourism boom. Asymptotically, countries with the advantages of offering low wages and an attractive payment structure for their employees will generate increasingly more challenges in the tourism industry than developed countries relying on their existing tourism enrichment. Nonetheless, there will be increased pressure on service quality standards to pursue training and skills development across countries in order to boost productivity and retain competitiveness. In the tourism industry, the prevalence of small and medium enterprises into the service providing facilities makes additional challenges to confirm the training and skills improvement plans available to small scale and infant businesses.

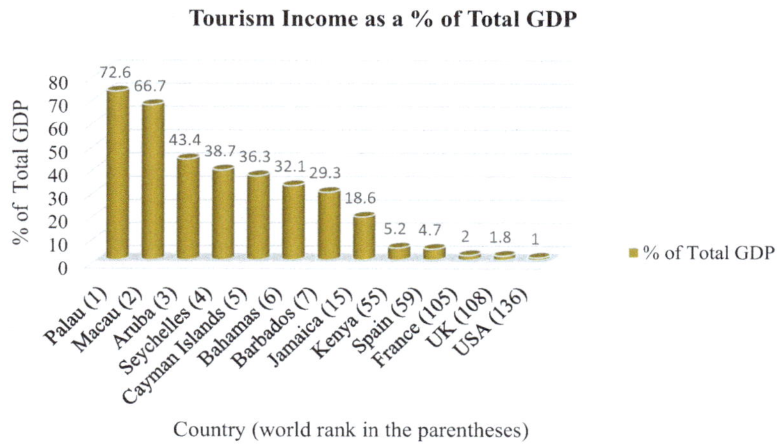

Fig. 3.4 Tourism based growth by different countries. (Source: NumptyNerd, 2019)

However, the tourism industry may be accustomed to negative aspects when applying it as a development scheme. It may increase dependency on a single industry, and other potential industries of a country could get neglected. Natural disasters and political unrest have an adverse impact on tourism industries' development. Sometimes, the government is unable to spend money on infrastructure projects due to its previous responsibility to build more schools or hospitals. Last but not least, the tourism industry may exploit workers' basic rights such as minimum wages and working hours, and overtime facilities. Figure 3.4 represents the tourism earnings as a percentage of total GDP in several tourism-promoting countries:

Any national action and supporting regional action which promotes tourism led development that will provide a guideline to any tourist destination country for negotiating specific responses that could be a part of any immediate action plan. Different methods or approaches will facilitate the accomplishment achieved by taking any supportive action if it arises. Governments, non-governmental organisations, and any regional or local groups are encouraged to be creative and original when developing strategies that utilise national and regional resources.

National Strategies

Several strategies are listed below to help the government enable to implement the suggested national action about tourism.

- A policy management plan for long-term tourism growth
- Consultative strategy between the public and commercial sectors
- Set up a committee for interpretation
- Interdepartmental teams

- Zones of sustainable development
- Demonstration projects for long-term tourism development
- A few regional strategies can be employed to support or act as a catalyst for long-term tourist development
- Information, research, and data sharing
- Comparative studies in the development of sustainable tourism
- Sub-regional and regional and collaboration

The image of any destination county is critical in drawing tourists to that location. Bangladesh, on the other hand, is having image issues as a result of some international electronic and print media. Various negative and awkward news produced by those media often highlights Bangladesh in bad shape that results in misconception among different levels of potential tourists. Thus, many foreign visitors avoid Bangladesh as a tourist destination because they lack sufficient knowledge of the country and its tourism attractions and services, or their knowledge is negative. Due to various limitations, the Bangladesh tourism industry could not spend the necessary funds to promote our country's reputation positively and, therefore, help the international media revise those negative images. This industry should place a strong emphasis on printed materials and distribute them to potential tourists via various airlines, international tour operators, foreign missions abroad, and other means. With the issues that our tourism industry is facing, the development of a sustainable tourism industry in Bangladesh will be dependent on priorities and policies concerning various touristic arrangements, technical development, and national administrative growth.

The Pursuit of Sustainable Tourism and Hospitality Management in Bangladesh

At the different part of this world, several developing countries have shown tremendous efforts of becoming rich through natural or cultural heritage assets that include national parks, safari parks, eco-parks, coral reefs, a museum for rare and uncommon species, ancient statues or monuments that were under risk or even threat. Developing or underdeveloped countries frequently lack the financial means to devote solely to the preservation of certain places, and must thus seek money for their protection in novel ways. The revenue generated from tourism could be a solution to this limitation if we could handle this productive industry in a more stylish form and thus regulate as well as manage in a responsible manner. Under current conditions, our country's tourist industry faces a number of significant issues that will influence its long-term development and performance, as well as set the agenda for public policy action. These include:

- Global affiliation and open markets policy;
- Technology-based tourism;

- Adopting the impact of climate change for sustainability;
- The knowledge economy is a new type of economy;
- Human capital formation;
- Productivity and competitiveness; and
- It improved the law and order situation.

Bangladesh is one of the adversely affected countries of climate change in this universe. Data on the last couple of decades shows that the temperature reaches to a record high in recent past years. As a result, the country found a widespread outbreak of various infectious and contagious diseases, a remarkable variation of consumers' behavior in response to climate change, possible modal shifts in transportation uses, and a lack of environmentally responsible behavior among the visitors and tourists.

Global communication, especially through the worldwide flows of transmission of information and the spread of computer, Internet-based technologies, web search engines and global positioning systems (GPS maps) have created a world with countries and communities more close to one another, and this is termed as the economy of information and communication technology under which actual competitive advantage stays in knowledge, information, and human skills.

Bangladesh's tourism destination and hospitality management need to recognize and adapt to these changes as early as possible. To better serve tourists (both domestic and foreign), the country's tourism intelligence must be effectively organized, better coordinate with local and national levels, analyze and share experience among tourism troops in order to improve the effectiveness of government activities at all levels. One area where public policy interventions might make a useful contribution is the rapid dispersion of knowledge and resources among different tourism locations through improved utilization of research-based intelligence and the construction of networking systems. Consumers have been empowered by Internet-based technologies, which have given them unprecedented access to information about travel, lodging, food, and diverse tourism items, as well as the opportunity to avoid traditional intermediaries, compare shop, and build their own product packages.

Sustainable tourism management is required in Bangladesh for the preservation and improvement of tourism resources as well as their long-term use. Diverse measures can be considered as prime concerns, and these are as follows:

- Use of friendly and innovative systems such as energy-saving technologies and usage of renewable energies etc.;
- Supporting the foreign direct and indirect investments for attaining efficacy in the world-class accommodation facilities;
- Keeping the energy usage (e.g. air cooling and heating systems) at least as possible and widening the use of renewable energy;
- Improving the quality and quantity of water and controlling air pollution; sustainable waste management;
- Protecting valuable natural regions as a possible source of long-term tourism growth.

Bangladesh must increase the quality of tourist education and training to meet labour market demands, forming a knowledge-based economy, which is critical for long-term tourist sustainability. The following measure can be taken to create more human resources in the economy:

- Use of unified and adaptive communication and information technology in the tourism industry;
- Upgrade the existing education system and launch a unique science-based education system for all and improve the staff qualifications in the tourism industry at all levels;
- Supervising and monitoring the qualification, knowledge, and access of training and research facilities for the tourism personnel/staff and preferment of mechanisms between trainers and companies;
- Coordinating the procedures for awarding work permits to overseas tour operators and tour guides.

In recent years, there has been a significant growth in consumer awareness of sustainable tourism, owing in part to substantial media coverage. Consumers all across the world are increasingly looking for ecologically friendly products and are becoming more aware of the consequences of sustainability and climate change. Bangladesh is not out of this era. The country should need to go for environment-friendly transportation as soon as possible, considering that they will be less harmful to our nature. Moreover, the Bangladesh tourist industry anticipates a significant, ground-breaking investment to eventually enhance and improve infrastructure and promotional tactics. Because the government is unable to meet the needs of heavy investment due to resource constraints, it should consider promoting (by providing tax breaks, security, and other incentives) the private sector, both domestic and foreign, in order to achieve the goal of a sustainable tourism industry in Bangladesh. The government should declare extra incentives to private investors to boost private investment.

Conclusion

To summarize, the tourism industry has been considered a profitable industry to every nation. Sustainable tourism growth in Bangladesh may be viable in the long run if the country can implement proper policies and strategies, as well as information sharing and cooperation. New locations should be investigated and conserved to produce attractive tourism sites based on existing regional and communicational contexts, in order to expand the tourist industry and pump more cash into the nation's GDP. Any illegal infringement, unsocial activity, or pollution should be avoided in certain areas. Tourism development would be successful only if diverse players in the public and private sectors work together in a coordinated and cooperative manner. Besides, tourism development and hospitality management involve different kinds of activities, knowledge-based leadership quality, productivity, and competitiveness among the national, regional, and local levels. The tourism industry has been widely dependent on the adoption of information and communications technology by various exclusive tourist destinations that will assist visitors, easing the way of online booking, advance booking, and expanding the marketing capabilities by gifts and promotions through artificial intelligence. The internet facilities, along with smartphones, tablets, etc., have fundamentally changed the way the world witnesses the international tourism industry and its interactions with customers. As Bangladesh also moves forward with the adoption of complete digitalization under the present ruling government, it is expected that the country will try to exploit every facility and potentialities of tourism in favor of its economic growth and development. The world is now using sustainable tourism for development so, Bangladesh should also start a widespread tourism business. Just like the success story of the apparel and textile industry, our country should think of the great potential it. The tourism industry can draw the investors' attention in both the local and the foreign. A huge investment is needed to develop long-term strategy for launching destination-wise tourism spots along with the exotic vision to accelerate the process of development.

References

Britton, S. G. (1982). The political economy of tourism in the Third World. *Annals of Tourism Research, 9*(3), 331–358.

Brohman, J. (1996). New directions in tourism for third world development. *Annals of Tourism Research, 23*(1), 48–70.

Census and Economic Information Centre (CEIC) Data. (n.d.). *Bangladesh BD: International tourism: Number of arrivals*. Retrieved from: https://www.ceicdata.com/en/bangladesh/tourism-statistics/bd-international-tourism-number-of-arrivals. Accessed 26 Nov 2020.

Daily Telangana Today. (2018). *Tourism a major source of income*. Retrieved from: https://telanganatoday.com/tourism-a-major-source-of-income. Accessed 28 Nov 2020.

Davenport, J., & Davenport, J. L. (2006). The impact of tourism and personal leisure transport on coastal environments: A review. *Estuarine, Coastal and Shelf Science, 67*(1–2), 280–292.

Drake, S. P. (1991). Local participation in ecotourism projects. In T. Whelan (Ed.), *Nature tourism: Managing for the environment* (pp. 132–163). Island Press.

Global Finance Magazine. (2020). *Countries with Highest GDP Growth 2020: Developing Asian and African countries lead with fastest GDP growth in the world*. Retrieved from: https://www.gfmag.com/global-data/economic-data/countries-highest-gdp-growth. Accessed 28 Nov 2020.

Hassan, A., & Kokkranikal, J. (2018). Tourism policy planning in Bangladesh: Background and some steps forward. *e-Review of Tourism Research (eRTR), 15*(1), 79–87.

Hemmati, M., & Koehler, N. (2000). Financial leakages in tourism. *Sustainable Travel and Tourism, 20*(7), 25–29.

Hossain, M. A., & Firozzaman, M. (2003). *Strategic promotion approaches to developing tourism in Bangladesh: An empirical study of some selected tour operators*. University of Dhaka.

Hossain, M. A., & Nazmin, S. (2006). *Development of tourism industry in Bangladesh – An empirical study on its problems and prospects*. University of Dhaka.

Jucan, C. N., & Jucan, M. S. (2013). Travel and tourism as a driver of economic recovery. *Procedia Economics and Finance, 6*, 81–88.

Latif, W. B., Islam, M. A., Mohamad, M., Sikder, M. A. H., & Ahmed, I. (2015). A conceptual framework of brand image on customer-based brand equity in the hospitality industry at Bangladesh: Tourism management and advertisement as moderators. *Journal of Scientific Research and Development, 2*(11), 1–16.

Ministry of Civil Aviation and Tourism (MoCAT). (2019a). *Major tourism place of Bangladesh*. MoCAT.

Ministry of Civil Aviation and Tourism (MoCAT). (2019b). *Major responsibilities of Tourism-1*. MoCAT.

Muhanna, E. (2007a). The contribution of sustainable tourism development in poverty alleviation of local communities in South Africa. *Journal of Human Resources in Hospitality & Tourism, 6*(1), 37–67.

Muhanna, E. (2007b). Tourism development strategies and poverty elimination. *Problems and Perspectives in Management, 5*(1), 37–49.

Murphy, P. E., & Price, G. G. (2005). Tourism and sustainable development. *Global Tourism, 3*, 167–193.

NumptyNerd. (2019). *Tourism as a development strategy*. Retrieved from: https://www.numptynerd.net/tourism-as-a-development-strategy.html. Accessed 26 Nov 2020.

Preister, K. (1989). The theory and management of tourism impacts. *Tourism Recreation Research, 14*(1), 15–22.

Priestley, G., & Mundet, L. (1998). The post-stagnation phase of the resort cycle. *Annals of Tourism Research, 25*(1), 85–111.

Rahman, M. L., Hossain, S. N., Miti, S. S., & Kalam, A. A. (2010). An overview of present status and future prospects of the tourism sector in Bangladesh. *Journal of Bangladesh Institute of Planners, 3*(1), 55–75.

Sultana, S. (2016). Economic contribution of tourism industry in Bangladesh. *Journal of Tourism, Hospitality and Sport, 22*, 45–54.

Taylor, G. (1995). The community approach: Does it really work? *Tourism Management, 16*(7), 487–489.

The Daily Star. (2009). *Tourism potential in Bangladesh*. Retrieved from: https://www.thedailystar.net/news-detail-81436. Accessed 20 Nov 2020.

Tooman, L. A. (1997). Tourism and development. *Journal of Travel Research, 35*(3), 33–40.

World Tourism Organization (UNWTO). (2005). *UNWTO tourism highlights*. Retrieved from: http://www.unwto.org/sdt/mission/en/mission.php. Accessed 26 Nov 2020.

World Travel & Tourism Council (WTTC). (2014). *Bangladesh annual report*. Retrieved from: http://www.wttc.travel. Accessed 26 Nov 2020.

Part II
The Present Context

Chapter 4
Development and Investment in the Tourism Industry of Bangladesh: Scenario and Prospect Analysis

Ayesha Afrin and Azizul Hassan

Abstract A country's touristy appeal majorly relies on its culture, archaeology, ancient heritage, historic palaces, natural beauty, etc. From this angle, Bangladesh can be considered as a touristic country. The present-day landmass that constitutes Bangladesh has a rich history with many tourist attractions. The Bangladesh economy is more integrated with the tourism industry. In this context, this research aims to outline the scenario and prospect of development and investment of tourism in the country. Findings of this review chapter show that adequate development and investments in the tourism industry of Bangladesh can facilitate the economy's integration with greater financial and technical strengths.

Keywords Tourism · Archaeology · Bangladesh · Development · Mosaic · Financial

Introduction

Tourism also has evolved into one of the most significant types of foreign trade. In terms of export revenue production, international tourism ranks very close to fuels, chemicals, and automotive goods. It is a key source of income and an important export category for many developing countries, provides much-needed jobs and growth opportunities. Tourism is important to many countries because it allows them to generate revenue through the purchase of goods and services by visitors and

---

A. Afrin (✉)
Department of Chemistry, University of Chittagong, Chattogram, Bangladesh

Department of Applied Chemistry and Chemical Engineering, University of Chittagong, Chattogram, Bangladesh
e-mail: aafrin.aec@cu.ac.bd

A. Hassan
Tourism Consultants Network, The Tourism Society, London, UK

A. Hassan (ed.), *Tourism in Bangladesh: Investment and Development Perspectives*, https://doi.org/10.1007/978-981-16-1858-1_4

tourists, taxes imposed on tourism companies and service offers. Transportation services (i.e. rail, air, cruise ship, boat travel etc.); lodging services (i.e. restaurants, hotels, bars, motels, entertainment venues etc.); and some other hospitality industry services (i.e. resorts, spas etc.), are all examples of these service industries (Tuhin & Majumder, 2011).

Bangladesh is a country in South Asia having a considerable population. In recent years, the country experienced steady growth in its economy and socio-cultural developments. With a population of over 170 million, the country possesses possibilities in tourism. The stronghold of a social class having affordability to spend for tourism and leisure activities, the country already placed attention for the development of its tourism industry. Arguably, the supportive roles of the government policy are favoring the development of tourism in Bangladesh (Hassan et al., 2020; Hassan & Kokkranikal, 2018; Hassan & Burns, 2014).

The market and interest in Bangladesh's tourism are expanding enormously. The expansion of tourism and hospitality enterprises in Bangladesh is an evidence. Bangladesh is believed to have numerous less explored products and services available for tourism. The contribution of tourism in the economy becomes significant in terms of creating earning and employment opportunities. However, such contribution remains uneven and improperly distributed. However, tourism as an industry has not been offered the desired importance for investment and development. On this ground, the aim of this study is to sketch out the scenario and prospects for tourism growth and investment in the region.

Tourism in Bangladesh

The Bangladeshi tourism industry's strengths are: first, Bangladesh has distinctive natural resources. There are some UNESCO world Heritages Sits that can attract tourists. Second, the country has the world's largest sea beach. Third, she has the Sundarbans World Heritage Site. She has exclusive natural beauty and greenery. Fourth, Bangladesh is prominent for its historical and archaeological places. Fifth, the nation's hospitality is well-known. Sixth, cultural heritage and religious harmony are well-off. Bangladeshi culture and history are enriched than many other countries. Seventh, despite of having some problems, Bangladesh has transportation and be reached by sea or air from any country of the world. Biman Bangladesh Airlines, the national airliner Dhaka, Chattogram, and Sylhet to major cities of the world. The national air, water, or land transportation is relatively cheaper. Eighth, Bangladesh is a democratic as well as a Muslim state also. The country is liberal democratic with a common homogeneous society. Ninth, apart from some areas like the hill tracts, there are available accommodation facilities throughout the country. However, not all of these accommodation facilities have the acceptable global standard (Tuhin & Majumder, 2011). Bangladesh has lots of tourist attractions to be a prime host of tourists. The total general scenario of tourism attractions/products can be explained under the following wider product categories:

Eco-tourism Products

A country that is filled and abundant with natural wonders, beauties, and untouched treasury and a place to a variety of exceptional and superb creatures, one of the largest deltas in the world that is Bangladesh. Bangladesh is ideal in eco-tourism for its hills, valleys, forests, beaches, lakes, rivers. However, in Bangladesh, eco-tourism is rising and still remains as a relatively newer tourism form. The Sundarbans is a part of the largest delta and the world's largest mangrove forest. This is a UNESCO has already declared Sundarbans as a World Heritage Site for preserving and protecting its bio-diversity and eco-tourism potentials. This can be a precious element for eco-tourism in Bangladesh. The Sundarbans Cox's Bazar, for example, is Bangladesh's tourist capital, with a 120-kilometer beach that is the world's longest sandy, drivable, and straight beach. In Bangladesh, Ramu is famous for its Buddhist temple, Maheskhali Island is famous for its dry fish industry and the Hindu and the Buddhist temples, Teknaf is Bangladesh's southernmost point, St. Martin Island is the country's only coral island. Also, Sonadia, Himchhari, Sylhet, Kuakata beach, Inani beach etc. are famous and can attract many tourists. Thus, the varied and rich flora and fauna simply describe the uniqueness of Bangladesh (Tuhin & Majumder, 2011).

Archeological Tourism Products

Bangladesh is a country that is not poor in archeological wealth, especially for the era of the pre-Muslims and the Muslims. In Bangladesh, under their imperial patronage, abundant efficient, independent monasteries sprang up in almost all over the country, gradually. Some major archeological sites are Paharpur, Mainamati, the Shait-Gumbad Mosque, the Bayazid Bostami Majar, the Kantanagar Temple, the Kantaji Temple etc.

Cultural Tourism Products

There are lots of cultural tourism products like Sonargaon, Dighapatiya Rajbari, Shilaidaha Kuthibari, Sagordari, Trishal, Shahjadpur Kuthibari, Ahsan Manzil museum etc.

Entertaining Tourism Products

The world's largest uninterrupted sea beach, thousands of rivers and rivulets, plentiful lakes and channels are situated in Bangladesh. These tourism products make available some recreation facilities like sea boat, paddle boat, simple boat, swimming, surfing, riding cars and kitting on the beach, squash, etc. Apart from these, there are also angling, swimming, boating, fishing and many other activities for the tourist. The country also has enriched classical arts, dances and music that can attract both the local and foreign tourists.

Exciting Activity Tourism Products

The Sundarbans and the Hill Districts in Chattogram offer exclusive and challenging opportunities for voyage tourists. They can also manage breathtaking experience to see spotted deer, the Royal Bengal tiger, crocodiles, barking deer, many type reptiles and the highest peak of Bangladesh's northern part.

Historical Tourism Products

Historical monuments in Bangladesh signify Muslim, Hindu, British, Pakistan periods, and independence spread all over the country. Among them, the famous historical attractions are Central Shahid Minar, Lalbagh Fort, Sonargaon, National Memorial, World War II Cemetery, Mujibnagar Memorial, Gandhi Ashram, Martyred Intellectual Memorial, National Poet's Grave, and Old High Court Building etc.

Religious Tourism Products

The renowned religious tourism sites are the Shrines of Hazrat Shah Jalal, Hazrat Shah Amanat, Maizbandari, Bayazid Bostami Majar, historic mosques, and Kantaji Temple, etc. (Al-Masud, 2015). The national economy and the country's performance are getting improved. The weather is relatively consistent throughout the year, but tourists prefer the colder months because they are more comfortable for travel. The inhabitants of Bangladesh speak the Bangla as the mother tongue while the English is spoken by the well-educated urban citizens. Bangladesh is a Commonwealth member with close cultural, linguistic and historical links to the United Kingdom, Pakistan and India. In general, local, regional, and the global tourism activities are growing, due to this reason, it is expected that the number of tourists in Bangladesh will be increased. Now liable tourism is a fast-growing world market recess. Following to Assignment Point (2019), when compared to the travel patterns of tourist countries, this is assumed that Bangladesh has a good travel season. Also, Bangladesh is a strategic stopover point in the country as the region's one of the largest destinations for inbound tourist, as well as India, Nepal, and Bhutan, three popular world eco-tourism destinations. Thus, following these strategies, the investment and development of the Bangladeshi tourism industry can be expanded.

Contemporary Investment and Development Scenario of Tourism in Bangladesh

Following Islamic Markets (2021), in Bangladesh, the tourism industry, like almost all other industries, relies on investment to expand. However, investment in this industry has been low, which may account for the poor output in terms of income and jobs. Capital investment, in particular, was estimated to be 41 billion in 2013, or 0.39 percent of GDP. Another noteworthy finding is that tourism spending has been steadily declining since 2009. Bangladesh is ranked 127th in the world in terms of tourism structure in the World Economic Forum's 2013 Travel & Tourism Competitiveness Report. Bangladesh is ranked 129th out of 136 countries in the World Travel and Tourism Competitiveness Report-2013, according to the World Travel and Tourism Council (2014). Bangladesh's position compares favorably to India, Nepal, Sri Lanka, and Pakistan among South Asian countries. It is necessary to measure hotel rates, airline fares, airport services, travel agents, and leisure and entertainment services by looking at the Gross Domestic Product (GDP). World Travel and Tourism Council (WTTC) report (2019) shows that, the direct contribution of travel and tourism in the GDP of Bangladesh in 2018 was USD2,750.7 billion (3.2% of GDP). This was forecasted to rise by 3.6% to USD 2849.2 billion in 2019 and by 6.5% per annum to 4.7% of GDP by 2024. In 1990-91, Bangladesh earned BDTk 267.7 million from tourism, which rose to BDTk 5562.70 million in 2010. Most of the time from 2000 to 2010, the growth rate was positive and this was also a very good indication (Reza et al., 2016). In Bangladesh, the tourism industry's rapid growth has resulted in substantial job creation. The number of direct workers in the Travel & Tourism industry is referred to as direct contribution to employment. In 2012, travel and tourism funded 1,281,500 jobs directly. Tourism is expected to provide 1,785,000 employment by 2023, according to estimates.

Bangladesh does not have a strong place in world tourism, as seen in the table above, but the pattern of earnings growth is promising. Still, all of these data present that the tourism trade in Bangladesh will have a significant rise that should have a positive influence on her tourism system and network. The tourism industry has grown in importance over time. Bangladesh's international tourism receipts accounted for USD 391,000,000 in 2019 (The World Bank, 2021). According to Roy and Roy (2015), the tourism sector received considerable allocation from the national budget in 2009-2010. In 2011-2012, this was raised to 1818. However, tourism spending has been declining in recent years. Tourism expenditures in the national budget fell in 2013-2014. This downward trend may be attributed to Bangladesh's ongoing political crisis, which has deterred tourists. The government has allotted BDTk 34 billion to the Civil Aviation (Biman) and Tourism Ministries in FY2019-20, which is more than double the previous fiscal year's allocation, demonstrating that the Bangladesh government appreciates tourism (Ministry of Finance 2019). Bangladesh jumped from 125th to 120th place among 140 nations in the

Travel and Tourism Competitiveness Report 2019, showing the biggest percentage rise.

Tourism was recognized as an industry in Bangladesh in 1999. However, it has not received much attention from the government as a thriving industry. The government established the Bangladesh Tourism Board (BTB) in 2009. BTC was established in addition to the Bangladesh Parjatan Corporation (BPC) that had been in place for over 40 years to promote tourism in the country (Majumder, 2015). The country's eye-catching attractions, magnificent natural beauty, cheap services and facilities, rich cultural heritage, and people leave tourists delighted and happy. Despite the reality, Bangladesh has taken no steps to improve its international reputation as a poor, flood-prone, and cyclone-prone country. However, both BTB and BPC tend to be less involved in their respective fields of operation and their performances are negligible in comparison to the overall earnings of the industry globally and even in Asia (The Daily Star, 2013). BTB and BPC, as well as other private tour operators, are expected to make boosting Bangladesh's image a priority by executing promotional measures to increase tourism. Tourism is a world's fast growing industry. For several countries, this is a significant source of foreign currency and a source of jobs. The government should cooperate and play leading roles to build an affirmative image that can enable this industry to become more responsible to the country's well-being (Reza et al., 2016).

Bangladesh Parjatan Corporation: Achievements, Challenges, and Limitations

According to BPC (2020a), major achievements of current years (past three years), challenges, and limitations are:

Achievements

The tourism industry is such an industry that consists of low construction, running, and salary expenses. There are dependable and plentiful labor forces. English is commonly used as an international language in global the tourism industry. The main tourist divisions in the country are divided into four main areas as: Rajshahi, Sylhet, Chattogram, and Khulna. Cox's Bazar, The Sundarbans, and the Chittagong Hill Tracts are some popular tourist destinations. The Bangabandhu Museum has printed the political life of the Father of the Nation Bangabandhu in a brochure in both Bengali and English. In the last three years, they have improved their catering facilities for passengers of Dhaka- Chattogram route, and 4500 trainees were trained by the Tourism Training Institute. As a result, they won 1st prize in Stall Pavilion Management and Exhibition Category of Bangladesh Tourism Corporation at

Dhaka International Trade Fair 2017. Newly constructed buildings at Hotel Beach in Chattogram and Rangamati district commenced as a part of commercial activities. Following the promises of the Hon'ble Prime Minister, tourist centers were constructed at Parki and Bagerhat. The documentation system through e-pin was launched. Four microbus and four coasters were bought for BPC. The eGP system has been supporting tender activities and they now have their own data center and achieved membership of Corporate Category from Pacific Asia Travel Association (PATA).

In 2019-2020, outcomes/impacts were: first, taking direct part as a key role-player in the socio-economic improvement of the country by creating service opportunities by increasing the tourism industry. Second, to provide eGRS services to make certain enhanced services. Third, to raise the consciousness of the local community, 18 meetings and seminars were organized to identify new products/attractions locally. Fourth, organized to ensure sustainable income and to achieve 46% residential occupancy. Fifth, all audit objections related to broadsheet responses were received. Sixth, to develop tourism industry provides training for 2000 young people in the human resource sector. Seventh, almost 80% of physical construction work of tourist building at Sher-e-Bangla Nagar in Agargaon established. Finally, to launch advanced current and up-to-date services like Hotel, Motel Private Hotel/Motel of BPC.

Challenges

In both rural and urban Bangladesh, a wide range of tourist accommodations are available. The investors like the most for their business in business-related tourism and the metropolitan cities. The increased salinity and climate change are another headache in Bangladesh that is concerned with safeguarding heritage damage. If the level of sustainable commercial use of such resources often stays as poor, it will have an effect on conservation standards, innovation, and international demand. Bangladesh is not old but comparatively new, can be visited by inbound tourists, and a destination to attract the global tourist market. Increasing Bangladesh's global image can be motivated by many new and sustainable tourist attractions with the help of the understanding of the country's background and history. Nonetheless, bearing in mind the global warming that can effect Bangladesh's inbound travel industry, the country is one of the ten most vulnerable countries when it comes to rising sea levels.

Limitations

There are several indicators that the current tourism industry is unsustainable (e.g. lake of safety, the condition of several attractions, amount of garbage etc.). The are some lacks as: inadequate infrastructure, underdeveloped domestic and international transportation. These have depressing effects on the convenience of tourist destinations and accommodations. Because of the current low inbound tourism demand, basic product and advertising development is critical. It is critical that tourism products represent Bangladesh's status as a modern and moderate Muslim society as well as a developing country with dense population with a strong background of custom and heritage. One such solution is a sustainable tourism product. A guide or high level of knowledge for traveling responsibly and safely (nature and culture awareness) would be helpful for tourists, particularly inbound tourists in Bangladesh (Assignment Point, 2019).

Development and Investment Tourism in Bangladesh: Analysis and Challenges

Bangladesh is one of the world's historically significant and most beautiful countries. It has the potential to develop its tourism industry, which would allow it to attract visitors. The perception of tourists is a crucial factor in the selection of many destinations; as a result, it has an effect on destination marketing strategies. Only high-quality service will keep visitors coming back (Alauddin et al., 2020). Quality service can make a good impact on tourists' minds. Most significantly, if consumers (tourists) are pleased with the quality of the goods and services, they become more interested in buying those services again. When visitors receive high-quality goods or services that meet their needs and budget, they are extremely pleased with them. As a result, they encourage others to take advantage of the privileges, and they strive to visit as soon as possible. Then they recommend it to their friends and family. Tourist satisfaction is closely related to their travel requirements, and tourism growth is influenced by these requirements. As a result, if there are high-quality performances and high satisfaction, tourism will steadily grow (Baker & Crompton, 2000). So, it is very important to pump up the tourists' motivation level, which increases the various satisfaction levels (Polas et al., 2018). Tourist satisfaction and tourism growth have a strong and meaningful relationship.

Ecotourism marketing in Bangladesh is fraught with difficulties. The poorest people in our country have the least understanding and awareness of the country's eco-tourism growth, with locals in some destinations totally unaware of the industry. Tribal people in hilly districts set fire to the jungles for "Jhoom" (slash and burn) cultivation, causing forest and species damage. Evergreen forests have been severely degraded, "Shal" forests are being converted to glamorous mangrove forests and short-rotation plantations are being lost to shrimp farming and high population pressure. Many of the animal species and plants are already extinct from the Bangladeshi protected areas due to land encroachment, commercial agriculture in forests, and violation of environmental and forest rules etc. Habitat erosion, vegetation shifts, salinity intrusion, settlement growth, and over-exploitation of fisheries resources are all threats to St. Martin's Island. In addition, the authorities at present have to face the following common challenges in exploring eco-tourism in Bangladesh: most of the world class hotels are located in Dhaka, Chattogram, Sylhet, and Cox's Bazar having insufficient number and accommodation capacities; safety and security issues; political corruption with social, political, and cultural instability; uncontrolled road transportations; sanitation issue; poor local infrastructure, unregulated urbanization, and low institutional capacity; lack of technical know-how and weak promotional activity; uncontrolled urbanization; sanitation problems; lack of hygienic foods and pure drinking water; poor local infrastructure, and low institutional capacity; lack of technical know-how and weak promotional activity; lack of coherent tourism strategies and policies; polluted environment; biodiversity loss and degradation; lack of expert human resources; lack of eco-tourism training institute; lack of business information; the cultural decay and disregard for human rights; and tensions between tribal and non-tribal citizens (Reza et al., 2016).

Tourism Vision 2020

Some of the ongoing and future projects of BPC are: first, presenting to the tourists the latest, advanced, and technology-reliant facilities of the services. Second, the establishment of tourism training institutes and star quality hotels on the territory of the longest sea beach Cox's Bazar. Third, construction of eco-friendly tourism building at Sher-e-Bangla Nagar in Agargaon; construction of tourism training center and preface of facilities in Barisal and Chattogram area; construction of a tourist center in Barisal city Expansion of tourism facilities at Durgasagar in Barisal; construction of tourist training centers in Rangpur and Sylhet districts; construction of Munshiganjo tourist center in Satkhira. To proffer digital GRS services. Finally, to endure in the spirited market competition.

Still, the investment level in travel and tourism of Bangladesh is not adequate if compared to regional and global annual investment. The tourism industry can be featured as fragile for its significant local market share and short high season. ALSO, there is a lack of qualified personnel. Bangladesh's tourism industry is still in its infancy, and its growth has been largely dictated by local market forces. There is no official marketing or development strategy, and the Bangladesh government along with the private tourism sector are both served by a number of organizations (BPC and Bangladesh Tourism Board or BTB). For that reason, the development of tourism is largely uncontrolled. Bangladesh Parjatan Corporation (BPC) (2020b) formulated a study report by considering the present tourism industry situation and declared "Tourist Vision 2020". To increase the annual GDP from 4 to 5 percent by 2020 in the tourism industry, BPC has drafted a detailed plan to achieve it. Accordingly, 0.9 million international tourists were expected to visit Bangladesh in 2020 and the World Tourism Organization predicted 0.5 million. For the construction of infrastructure, such as highways and hotels, BPC projected that domestic visitors, as well as foreign tourists, would boost the tourism industry, and the trend of local people going on pleasure tours has begun. BPC has taken up some programs in this venture, such as improving tourism goods, marketing, advertisement, and public relations, and renovating the BPC, among others as reported in the study. According to the report, proper excavation, restoration, and conservation are needed for cultural heritage, archaeological and other tourist sites that need careful management and protection from illegal encroachment.

Coercion

There are indeed some unplumbed hindrances. Bangladesh's tourism industry faces some major challenges, especially road and rail. Infrastructural and technological improvements are also needed. The climatic changes are also another issue for Bangladesh throughout the year (and in the future). Relevant policymakers should keep in mind that when traveling in Bangladesh since the country occasionally experiences several natural disasters like harsh storms, floods, and earthquakes, cyclones, etc. The image of the country all over the worldwide is also poor due to numerous natural catastrophes and corruption. The lack of social commitment and political turbulence is also negatively affecting the national tourism industry. One of the problems that Bangladeshi society faces is the protection of natural and historical attractions (Assignment Point, 2019). The summaries of weakness are: first, services with low-quality; second, lack of security, safety, and hygiene. Third, be short of proper investment. Fourth, lack of progress of infrastructure. Fifth, complicated visa requirements and visa procedures. Sixth, the dearth of sales plans and public relations activities. Seventh, lack of private sector involvement and initiatives for tourism development. Eighth, Bangladesh is not considering how to provide complete tourism goods and destination packages to both domestic and international visitors. Thus, tourists are getting a low level of satisfaction and have to go back to their homes that not the policy of tourism. The number of trained, skilled tourism industry professionals is not sufficient for building up a well-built base. Inadequate tour operators, unimportant role of travel agencies, and disorganized and inefficient national airlines. Lack of professionalism in tour guiding. Comparing to neighboring countries like India and Nepal, some of the tourism products' price like package tours, food items, standard hotel rooms etc. are relatively higher. There is no good balance, relation, or plan between the public and private sector, as a result of gap always stays in tourism promotion and marketing (Tuhin & Majumder, 2011).

Recommendations

India, Japan, Pakistan, the United States of America, the United Kingdom, China, Korea, and Germany are the main markets for Bangladesh's tourism industry, according to Al-Masud (2015). Bangladesh was visited by 69.61 percent of European and South Asian visitors combined. Like any other industry, the tourism industry also has problems in Bangladesh. The price and security are also other sensitive issues for the consumer in tourist places. So, it is needed to develop the market to set viable cost. Bangladesh can improve tourism marketing management and sales education so that trade can play a larger role in the national economy and policymakers can sell products to the right people at the right time in both local and global contexts.

The implementation of the policy of BPC regarding the better service becomes successful if they create pressure on the government more strictly. As a result, the private service providers will be able to countenance contest on an even ground. To explore new tourist spots, we should use information technology and through online or website we can convey information to the foreign tourists. Training, workshops, and other refresher courses should be held on a regular basis to keep staff up to date and proficient. BPC should plan policies and strategies to serve tourists more productively. For this, BPC can effortlessly use a variety of distribution channels. To improve Bangladesh's competitiveness in the intercontinental and global tourism destination markets, the mobilization of effective promotional actions supported by noticeable proposals and attractive tourism items can help.

Benchmarking of tourism products and services can benefit. The performance is the key factor to determine the benchmarking of the tourism industry (Shamsuddoha, 2008). However, due to the benchmarking of small organizations against one another, a negative, real effect on tourist satisfaction and destination image can occur. There are some main performance areas in which categorization and grading systems have resulted in lower standards being achieved in Bangladesh's tourism industry. The few significant examples of benchmarking that are conducted among small tourism businesses mostly conducted by third parties (Alstete, 2008; Anand & Kodali, 2008).

It can be helpful for developing tourism potentials in Bangladesh with the implementation of multilateral and interregional projects or collaboration with foreign travel companies. So, properly explored, extracted, and managed tourism is badly needed. Due to the globalization, the dissemination of information, communication, media need to widened. There are scopes to make tourist spots more eye-catching. More research and development to attract more tourists and making constructive and productive tourism policy are thus essential (Tuhin & Majumder, 2011).

Conclusion

Bangladesh is a third-world country with limited financial resources, but it can boost its GDP by putting pressure on the tourism industry for more investment. In general, BPC is the most successful tourism service provider in Bangladesh, using modern marketing concepts. They it is however, lagging in some areas. This sector offers numerous opportunities for foreign and domestic sales, but it is largely overlooked. BPC can become a better organization if it offers more attention. Bangladesh has a lot of potential tourist attractions, but it needs to figure out how to market them. Otherwise, she will be unable to get benefit from this promising market. Also, the government should enable the private sector to collaborate for Bangladesh's tourism industry's growth. To betterment of poor performance of the tourism industry compared to other South Asian countries, effective and logical framework for investment and development can help. State of the art facilities and infrastructures need to be ensured. Tourism is a rapidly growing industry around the world. Bangladesh has greater ability to draw international visitors while also boosting its domestic economy. However, through the dynamic development of tourism services and products, this potential can be realized. The tourism industry's survival and growth will be heavily reliant on its marketing strategy. Countries are competing to promote their tourism destinations. As a result, an ambitious development plan should be implemented as soon as possible to ensure the success of Bangladesh tourism. In principle, the importance of tourism is on a continuous rise in Bangladesh that in turn deserves attention from researchers and academia.

References

Al-Masud, T. (2015). Tourism marketing in Bangladesh: What, why and how. *Asian Business Review, 5*(10).

Alauddin, M., Ahsan, S. M. H., Kamal, M. A., Alam, M. M., & Hassan, A. (2020). The assessment of perceived service quality dimensions of hotel industry in Bangladesh. *Journal of Multidisciplinary Academic Tourism (JOMAT), 5*(2), 105–114.

Alstete, J. W. (2008). Measurement benchmarks or 'real' benchmarking?: An examination of current perspectives. *Benchmarking: An International Journal, 15*(2), 178–186.

Anand, G., & Kodali, R. (2008). Benchmarking the benchmarking models. *Benchmarking: An International Journal, 15*(3), 257–291.

Assignment Point. (2019). *Role-of-Bangladesh-parjatan-corporation-in-tourism-industry.* Retrieved from: https://www.assignmentpoint.com/business/management/role-of-bangladesh-parjatan-corporation-in-tourism-industry.html. Accessed 9 July 2020.

Baker, D. A., & Crompton, J. L. (2000). Quality, satisfaction and Behavioral intentions. *Annals of Tourism Research, 27*, 785–804.

Bangladesh Parjatan Corporation (BPC). (2020a). *Annual progressive report of Bangladesh Parjatan corporation, 2019*. BPC.

Bangladesh Parjatan Corporation (BPC). (2020b). *Tourist Vision 2020*. BPC.

Hassan, A., & Burns, P. (2014). Tourism policies of Bangladesh – A contextual analysis. *Tourism Planning & Development, 11*(4), 463–466.

Hassan, A., & Kokkranikal, J. (2018). Tourism policy planning in Bangladesh: Background and some steps forward. *e-Review of Tourism Research (eRTR), 15*(1), 79–87.

Hassan, A., Kennell, J., & Chaperon, S. (2020). Rhetoric and reality in Bangladesh: Elite stakeholder perceptions of the implementation of tourism policy. *Tourism Recreation Research, 45*(3), 307–322.

Islamic Markets. (2021). *Tourism investment in Bangladesh*. Retrieved from: https://islamicmarkets.com/publications/tourism-investment-in-bangladesh. Accessed 25 Apr 2021.

Majumder, D. (2015). Contributions and loopholes of tourism industry in Bangladesh. *Jagannath University Journal of Social Sciences, 3*(1–2), 1–19.

Ministry of Finance. (2019). *Budget Speech 2019-20* (p.1). Government of Bangladesh.

Polas, M. R. H., Reza, M. M. U., & Rowza, S. (2018). Role of advertising and service quality on tourism development in Bangladesh. *IOSR Journal of Business and Management (IOSR-JBM), 19*(9), 77–86.

Reza, S., Ahmed, M., & Rehunuma, M. (2016). *Opportunities and challenges of eco-tourism marketing in Bangladesh* (pp. 1–20). Retrieved from: https://bea-bd.org/site/images/pdf/new17/96.pdf. Accessed 1 Oct 2020.

Roy, S.C., & Roy, M. (2015). Tourism in Bangladesh: Present Status and Future Prospects. *International Journal of Management Science and Business Administration, 1*(8), 53–61.

Shamsuddoha, M., Hossain, M. A., Shahriar, S. E. S., & Chakraborty, T. (2008). *Development of tourism industry in Bangladesh*. Retrieved from: SSRN: https://ssrn.com/abstract=1314800. Accessed 9 July 2020.

The Daily Star. (2013). *Tourism industry in Bangladesh*. Retrieved from: https://www.thedailystar.net/news/tourism-industry-in-bangladesh. Accessed 9 July 2020.

The World Bank. (2021). *International tourism receipts (% of total exports)- East Asia & Pacific, Bangladesh*. Retrieved from: https://data.worldbank.org/indicator/ST.INT.ARVL?end=2017&locations=Z4-BD&start=2017&type=shaded&view=map&year=2017. Accessed 25 Apr 2021.

Tuhin, K. M., & Majumder, T. M. (2011). *An appraisal of tourism industry development in Bangladesh*. Retrieved from: https://www.researchgate.net/publication/288828550. Accessed 9 July 2020.

World Economic Forum. (2013). *The travel & tourism competitiveness report 2013*. Retrieved from: http://www3.weforum.org/docs/WEF_TT_Competitiveness_Report_2013.pdf. Accessed 9 July 2020.

World Travel and Tourism Council. (2014). *World travel &tourism competitiveness report – 2013*. WTTC.

World Travel and Tourism Council. (2019). *Economic impact-2019*. Retrieved from: https://bit.ly/2m7L3k1. Accessed 5 Sept 2019.

Chapter 5
Factors Influencing Capital Investment in the Bangladesh Tourism Industry

Md Yusuf Hossein Khan, Shahriar Tanjimul Islam, and Azizul Hassan

Abstract Tourism is one of the most promising and potential industries for developing countries around the world for their economic development. In developed countries like the USA, the UK, or any other European countries, the service industry has played a huge role in bringing foreign revenues. The general income and living standards of people have improved significantly over the last few decades. This indicates a need or demand for more hospitality services. A country like Bangladesh is yet to find this answer whether it has been successful in promoting growth in this industry as well other than being one of the fastest-growing economies in the world. To develop any industry, it needs investments. Thus, this paper will look into whether the tourism industry of Bangladesh is getting enough investment or not (national or from abroad) and what matters can be done to improve the situation. This study mainly consists of secondary research, with resources using different kinds of literature from research papers, articles, books, etc. The key aspect of this study is to examine the causes for low investment into this industry, like identifying probable risks that are associated with such a potential country like Bangladesh. Identifying the external factors which are related within Bangladesh and internal factors that are associated with investors with probable solutions to improve more investments is discussed. Later, a case-based discussion will be provided to make more valid arguments and suggestions. This paper hopes to create knowledge for those who are willing to invest their money in the tourism industry of Bangladesh, to government policymakers to be more aware of the situation, and to academicians who are interested in the fields of hospitality and finance.

Md Y. H. Khan (✉)
College of Tourism and Hospitality Management, IUBAT-International University of Business Agriculture and Technology, Dhaka, Bangladesh

Faculty of Economics, University of Algarve, Faro, Portugal
e-mail: ysf.khn86@gmail.com

S. T. Islam
College of Business Administration, IUBAT-International University of Business Agriculture and Technology, Dhaka, Bangladesh

A. Hassan
Tourism Consultants Network, The Tourism Society, London, UK

A. Hassan (ed.), *Tourism in Bangladesh: Investment and Development Perspectives*, https://doi.org/10.1007/978-981-16-1858-1_5

Keywords Investment · Tourism · Hospitality · Development · Bangladesh

Introduction

Tourism is one of the world' s largest economic industries that create jobs, drives exports, and increases more income all over the world. According to World Travel & Tourism Council (WTTC) (2019a), this industry comprises a wide range of industries, with the aim to serve and support domestic, international business, and leisure visitors. Companies with different size ranges and industries ranging from accommodation and transportation to food and beverages, retail and, culture, sports and recreation, all create a positive influence on this industry. This paper will try to explain the scope and areas to spend in the tourism industry of Bangladesh and what factors investors should be considered. However, it is also important to understand what this travel and tourism are contributing worldwide and why investing here can be a good idea for the investors. From the reports of WTTC (2019a), it can be noted that travel and tourism have contributed the USD 8.88 trillion to the global economy, which is equivalent to 10.4% of global Gross Domestic Product (GDP). In 2018, it was also the second-fastest-growing industry (i.e. 3.9%), only marginally behind manufacturing, which grew by 4%. Domestic visitors hold an account for 71.2% of the tourism expenditure, compared to international visitors, which is 28.8%.

Capital Investment: Some Theoretical Analysis

As per any business, investment is a strategic variable in the determination of the level and growth of income. According to UKEssays.com (2019), it refers to any act of spending with a prospective to yield. It refers to the process of capital formation whereby there is a net addition to the existing assets in the pipeline of production. According to World Tourism Organization (UNWTO) (2019), tourism is one of the most important drivers of economic and social development as it creates jobs and enterprises, export revenues, and infrastructure development. According to Parveen (2013), capital investment is an important criterion to flourish in tourism, like it is needed in any other industry. From Itic.ie (2017) report, it is suggested that capital investment a prerequisite for economic growth. Furthermore, it is stated to provide the physical and organizational building blocks for the society, where people and enterprises need to perform. It also creates connectivity between those people and enterprises, makes more smooth and easier commercial and transactional exchange among them. Nawaz and Hasan (2016) stated found from their research that investments being a part of aggregate demand as well as a source of capital formation, while the tourism industry has been given less attention.

Capital Investment: The Global Context

A study made by Dritsaki (2004) proved that tourism has a long-run economic growth effect in Greece. This statement is further enforced by Balaguer and Cantavella-Jorda (2002) from their validity of the tourism-based growth hypothesis for the long-run economic performance of Spain. Other studies conducted by Sinclair (1998) and Dieke (2004) found that tourism is an important and essential part of economic growth and development strategies as it serves as a scarce source for financial resources, job creation, foreign exchange earnings, and technical assistance in many developing countries. The global investment, according to Blomstrom (2014), the host countries gain more technological benefits from foreign direct investment, this capital investment from companies (i.e. multinationals) has also lead to employment and export generation (Fig. 5.1).

According to the report of WTTC (2019a, b), travel &tourism is expected to have attracted capital investment of USD 940.9 billion in 2018. This global increase in capital investment is expected to rise by 4.4% in 2019 and rise by 4.2% per annum over the next 10 years to USD 1489.5 billion in 2029.

Relevant case of capital investment: The Indian tourism industry, Kumaon region of Uttarakhand and Steps Taken to Promote Tourism Capital Investment

According to Khondker and Ahsan (2015), in India, the tourism industry plays an important role in the Indian economy by creating opportunities for jobs, revenue, and foreign exchange, as well as the growth or expansion of other industries such as construction, hospitality, handicrafts, and so on. Indirectly aided by the growth of

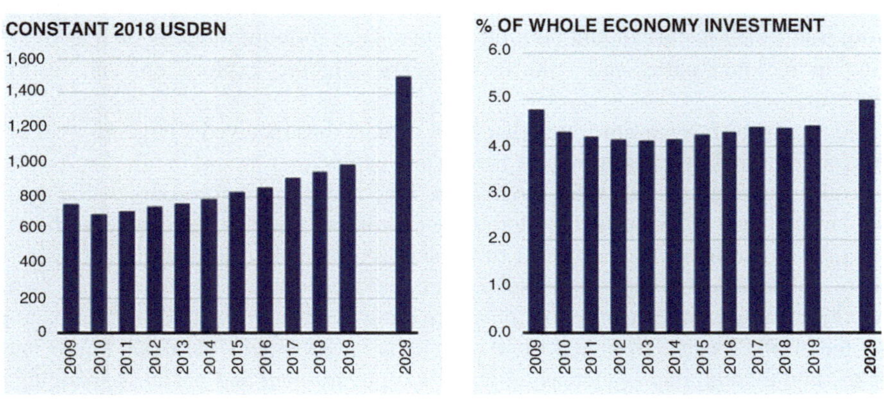

Fig. 5.1 Global capital investment in travel and tourism. (Source: WTTC, 2019b)

tourism, industries that are impacted include agriculture, construction, education, hotels, and restaurants. India serves as an excellent example of Asian tourism because of her history and culture as the primary pull for the tourists. Since investments in less useful facilities such as transportation, accommodations, and the overall infrastructuration of the economy are taking place, all the other investments must be paid off faster. Due to India's tourism policies, a new interest in traveling to Kumaon region helps growing the tourism industry.

Following Akihito (2017), with the state government's support, investment in capital equipment was provided to the project to enhance capabilities. The government place tourism as a primary pillar of its economy in the late 1980s with the belief that it would contribute significantly to India's economy. The conclusion is drawn from the research of Nakani (1996) that the policy was developed based on the belief that a good trade can be made from tourists visiting India. At that time, the Indian government viewed the country's economic opportunity in the global arena as important. Also, the Indian government thought that the foreign currency and travel markets can have significant impact on the overall economy. The Indian government has come up with five-year economic plan, and started continue this yearly rolling plans for tourism. Starting in the 1970s and continuing through the 1980s, the government placed emphasis the growth of tourism infrastructures and hotels. India undertook a lot of work to enhance transport infrastructure.

In particular, there are some selective developed regions where many foreign visitors have their key interests. Uttarakhand is rich with natural beauty and a healthy climate with plenty of playgrounds for sports, adventure, and eco-tourism. According to Jaiswal and Bisht (2017), river rafting, trekking, rock climbing, and camping etc. are among some new activities. These are presenting significant opportunities for tourism businesses. Starting in the 1990s, the government placed more focus on social infrastructure in order to encourage Indian tourism, which helped the private sector to grow (Akihito, 2017). Apart from concentrating on the construction of designated tourist sites for tourists, the government also invested on the locals. The plan started to create new tourism products in Uttarakhand and surrounding areas (Kumar, 2017). The diversification of tourism services was started from traditional historical and cultural heritage sites to more modern such as trekking in national parks and wildlife sanctuaries, nature exploration, and resort construction.

Akihito (2017) again informed that Naukuchiatal has received a large number of domestic tourists as a result of new growth and capital investment in hill resorts. As a result of Uttarakhand's close proximity to lake based tourism resources, the area saw an increase in the growth of lodging accommodations by residents and hotel capital from both inside and outside of the state. As a result, a certain influx of capital to the local economy was observed for residents and merchants in that area (working for logging accommodations, providing necessary services, and demanding material). Tourists from almost all over India have visited Uttarakhand in recent years, and this number is increasing with such government and business spending. It was obvious that many tourists came from India's north and southwest regions. A significant number came from the greater Delhi metropolitan area (Budget.uk.gov.

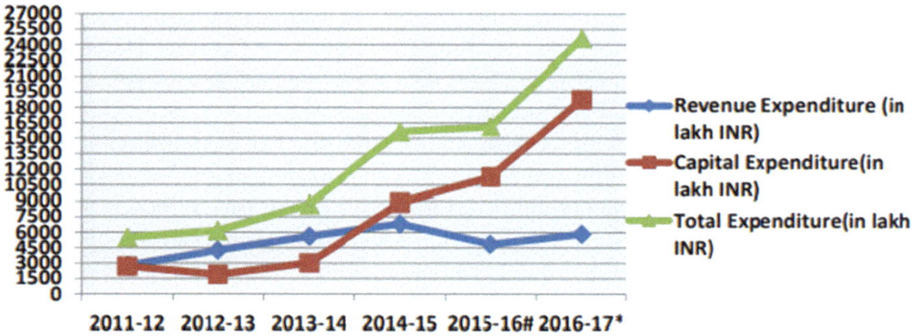

Fig. 5.2 Indian Government Expenditure in tourism. (Source: Budget.uk.gov.in, 2019)

in, 2019). Religious pilgrimage is one of the most popular reasons for locals to visit the region, followed by holiday for a growing number of visitors in hill resorts. It is thought that the majority of these visits are the outcome of middle-class peoples' leisure time. With so many tourism opportunities in surrounding lakes, Uttarakhand has seen an increase in the number of residents and hotels or lodges. The establishment of the lodging accommodation industry in Naukuchiatal accelerated quickly at the beginning of 2005 with various lodging services and room rates (Fig. 5.2).

From Fig. 5.2, it can be seen that the graph shows an overall increase in total expenditure due to a huge increase in capital expenditure and a slight increase in revenue expenditure incurred by the government. It can also be noticed that the initial trend in the year 2011–2012 revenue expenditure was greater than capital expenditure. However, the scenario changes after the year 2013–2014 and depicts a reverse picture. This indicate government has been more focusing on infrastructural development and trying to reduce more day-to-day expenses.

From their empirical findings, Selvanathan, Selvanathan, and Viswanathan (2012) concluded that Foreign Direct Investment (FDI) and tourism may be intertwined and have two effects on each other's development. Increased foreign tourists who come to India for business from companies that have invested in India visit places of tourist interest during their side trips, as seen in the case of India. As a result, it is important for India's government to draw attention of more international visitors and, as a result, it would generate more FDI. The Ministry of Tourism of India is a government body that drives the tourism industry with well-planned marketing campaigns and looks after the travel trade, such as travel agencies and tour companies.

According to Khanna (2011) and De and Chakraborty (2018), India's branding policy is known as "Incredible India". The government spent a lot of money for this strategy, and the country has taken a lot of steps to develop tourist attractions and promote them, particularly to tourists from the western countries who have a thought of India as a dangerous or boring location. The government has taken a more proactive approach to develop tourist destinations, with a particular focus on improving infrastructure in and around those areas. One of the key issues with India is her uncontrollable and fraudulent practices against international tourists, which has deterred many foreign visitors from visiting the country in the past. However, as part of the campaign, the government has taken some precautionary steps and provided opportunities for small and subsidiary businesses, such as tour operators and travel agencies, to register with the Ministry of Tourism. This registration has benefited both the initiative and the travel traders. The government encourages tourists to use services from government-approved traders or businesses to reduce the possibility of fraudulent activities. Heritage and archeological tourism are also promoted as part of the "Incredible India" campaign. Other forms of tourism are also established by the government.

According to Anholt (2006), the Indian government has taken a strategic approach to the growth of these tourism sources, like development guideline for each source as well as employment schemes for potential job seekers in each sector. According to Raj and Nayak (2019), in addition to history and culture-based tourism, India is also attempting to draw high-value tourists by promoting creative or niche products and services such as golf tourism, polo tourism, rural tourism, and cruise tourism.

Factors That Influences Capital Investment

One of the objectives of this study is to look into the main factors that affect capital investment decisions in the tourism industry. To better understand, Enoma and Mustopha (2010) clarified that at the company level, the process has usually been shown to be a multi-criterion process that takes into account a variety of variables. The country's internal (i.e. micro) and external (i.e. macro) factors have significant impacts on capital investment.

As stated by Zarnowitz (1992), from a macro viewpoint, business investment accounts for the bulk of the fluctuations in the Gross Domestic Product (GDP). From a micro viewpoint, it emphasizes on individual company growth by rising productivity and lowering unit costs. This also stresses on interest rates (the cost of borrowing), economic growth (changes in demand), confidence/expectations, technical advances (capital productivity), and the availability of finance from banks. According to Pettinger (2019), these are some other points to consider as depreciation, wage costs, inflation, infrastructure, government policy; while external/macro factors include "interest rate", "economic growth", "technological advancement", "availability of finance from banks", and "other factors"; internal factors include "trust or expectation", "firm size", "availability of capital", and the efficiency with which hotels/firms compete.

Enoma and Mustopha (2010) went on to say that, in addition to economic and risk factors, political, environmental, and government regulations are all relevant considerations. In general, the impact of these factors on individual corporations' decisions varies greatly. Bialowolski and Weziak-Bialowolska (2014) argue that investors value knowledge transparency and trustworthiness in a country or market based on their findings. Kahraman (2011) made a similar claim about investor trust, claiming that factors such as market volatility, a lack of market awareness, and a lack of investment expertise are likely to cause companies to underinvest.

Furthermore, the size of firms available in a country is an important factor for influencing capital investment. Carr et al. (2010) show the results that financial factors are more important about the investment process for small firms. Since they have very limited access to capital markets, they have to depend more on internal funds such as personal savings, funds borrowed from relatives or friends, and retained earnings (Gill et al., 2012). Furthermore, the studies of Ekanem (2005) showed that smaller firms tend to reinvest from their retained earnings to grow, avoid taking the risk of extra investment to grow more. On the other hand, large companies (i.e. limited companies and big partnership companies) have access to external funding such as better collateral and credit terms, etc. Laux (2008) claims these companies use better techniques to appraise their investment such as through capital budgeting.

According to Nelson (2005), to make the investment process work, the investment has to be implemented according to the needs of companies, which the firms can comply with and follow. However, Karim and Azmand-Sainib (2013) point out that sometimes the firms are ready to invest and use new technology, but due to barriers often created from external factors influences, investment decisions are made observing the external factors of the country.

Capital Investment in the Tourism Industry of Bangladesh

Tourism, according to the United Nations World Tourism Organization (UNWTO) (2019), is one of the most significant drivers of economic and social growth because it generates employment and businesses, as well as export revenues and infrastructure development. According to a study conducted by Dritsaki (2004), tourism has a long-term economic growth impact in Greece. Balaguer and Cantavella-Jorda (2002) support this assertion by demonstrating the validity of the tourism-based growth hypothesis for Spain's long-term economic success. Other studies by Sinclair (1998) and Dieke (2004) found that tourism is a significant and necessary component of economic growth and development strategy in many developing countries, as it provides some source of financial capital, job creation, foreign exchange earnings, and technical assistance.

According to Parveen (2013), capital investment is an important criterion to flourish in tourism, like it is needed in any other industry. From Itic.ie (2017) report, it is suggested that capital investment a prerequisite for economic growth. Furthermore, it is stated to provide the physical and organizational building blocks for the society, where people and enterprises within it need to perform. It also creates connectivity between those people and enterprises, makes more smooth and easier commercial and transactional exchange among them. Nawaz and Hasan (2016) stated found from their research that investments being a part of aggregate demand as well as a source of capital formation, while the tourism industry has been given less attention.

From TCdata360 (2019) it can be seen that Bangladesh has made steady progress and has performed much better than the world median standard. As from Fig. 5.3, it can be seen that Bangladesh has experienced a year on the average growth rate of 7.29% for the time of 1995 to 2028. The performance is good. However, when it came to regional analysis, the performance of Bangladesh was just marginally better (Knoema, 2019). In Fig. 5.3, it can be seen that, during 2000, 2008, and in 2014 Bangladesh, in terms of capital investment, has not performed poorly. Though during 2017 and 2018, the forecast for the next 10 years, Bangladesh is shown to perform much better, as per from World Bank at TCdata360 (2019) (Fig. 5.4).

With the shortage of reliable statistics on different aspects of Bangladesh tourism, it does create a big problem for academicians, researchers, and policymakers. Still, it is widely recognized that Bangladesh is by natural beauty, heritage, and panoramic views, what a tourist wants to enjoy. Many reports have been collected and prepared from the recent data. Some of the key concerns for Bangladesh is its political instability and natural calamities. As mentioned by Hasan and Islam (2014), Bangladesh could not achieve satisfactory progress in the arena of international tourism. From the information produced by WTTC (2019a, b), it can be understood that Bangladesh is far behind from global standards, and even from their neighboring countries like India, China, Bhutan, or Nepal and except Myanmar when it comes to performing in travel & tourism (Figs. 5.5, 5.6, 5.7, 5.8, 5.9, 5.10 and 5.11).

In terms of total contribution to GDP, employment, earnings from foreign tourists, and the contribution from capital invest for travel and tourism, Bangladesh has underperformed and remained at the lower part of the table.

Fig. 5.3 Capital investment in travel and tourism in USD (Billions) in Bangladesh with world median standard. (Source: TCdata360, 2019)

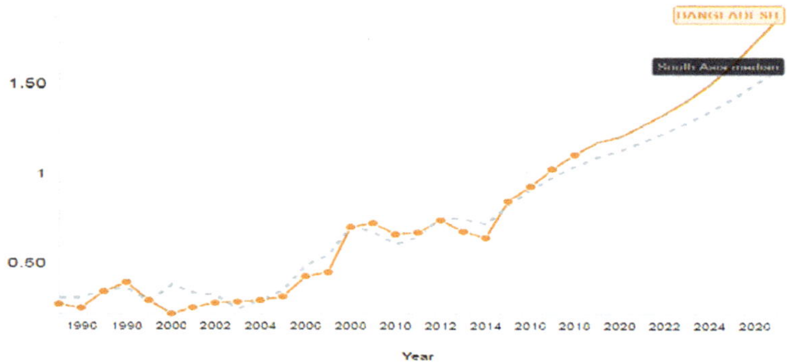

Fig. 5.4 Capital investment in travel and tourism in USD (Billions) in Bangladesh with South Asian (Regional) standard. (Source: TCdata360, 2019)

Travel & Tourism's Total Contribution to GDP		2013 % share
35	Thailand	20.2
41	Malaysia	16.1
84	Vietnam	9.6
	World	9.5
87	Sri Lanka	9.4
92	China	9.2
93	Indonesia	9.2
	Asia Pacific	8.9
108	Nepal	8.2
135	India	6.2
165	**Bangladesh**	4.4
174	Myanmar	3.7

Fig. 5.5 Travel & Tourism's total contribution to G.D.P. (Source: WTTC, 2019a)

According to WTTC (2016), there are some encouraging signs for the economy of Bangladesh in 2016, as shown below (Table 5.1):

With every passing year, Bangladesh has been performing better. WTTC (2019a, b) shows that the Asia-Pacific has experienced the most travel and tourism GDP growth in 2018. It is a good sign for Bangladesh as an Asian country. Furthermore, it is one of the top 20 countries with the most growth in travel and tourism in 2018 (see Appendix 1). The country ranked 11th and 9th for the fastest 10-year real growth from 2019 to 2029 (see Appendix 2).

From the above discussion, it could be noted that Bangladesh is having a steady progression when coming towards development in the tourism industry. However, if the performance of the industry as portrayed above is far from satisfactory, compared with neighbors, Asia-pacific countries, and as well from global averages, which does signify there are a gap and scope for Bangladesh to fill. For this, more and more capital investment is required from private and public industry and good long-term strategy from the government to Bangladesh to prosper in this area. Hence this paper will discuss the areas that influence this tourism and can promote

Travel & Tourism's Total Contribution to Employment		2013 % share
46	Thailand	15.4
50	Malaysia	14.1
	World	8.9
93	Sri Lanka	8.4
95	China	8.4
97	Indonesia	8.3
	Asia Pacific	8.2
105	Vietnam	7.9
111	India	7.7
120	Nepal	7.0
166	**Bangladesh**	3.8
176	Myanmar	3.0

Fig. 5.6 Travel & Tourism' s total contribution to employment. (Source: WTTC, 2019a)

Visitor Exports Contribution to Total Exports		2013 % share
46	Nepal	21.1
62	Thailand	16.0
67	Sri Lanka	14.3
85	Malaysia	8.5
	World	5.4
117	Myanmar	5.1
119	Indonesia	5.0
	Asia Pacific	4.9
125	Vietnam	4.7
132	India	4.1
162	China	2.1
177	**Bangladesh**	0.5

Fig. 5.7 Travel & Tourism' s total contribution to total export from visitors. (Source: WTTC, 2019a)

more investment into this industry. Lonely Planet, the largest travel guide, published and ranked Bangladesh as 7th in the research for the top ten most affordable places to visit in 2019 (Gomes, 2019). As from Abakashparjatan.com (2019), Bangladesh Parjatan Corporation classifies tourism into six different types: tourism, business, office, study, religion, service, and other purposes. Bangladesh in Travel and Tourism Competitive index is at the bottom of the table, despite having two of the unique natural attractions of the world (i.e. Cox' s Bazar and Sundarbans). It is important to learn how India has improved its tourism industry and what Bangladesh can do to attract more investors for tourism development.

Travel & Tourism Investment Contribution to Total Capital Investment		2018 % share
60	Vietnam	7.9
62	Malaysia	7.7
71	Thailand	7.0
78	India	6.2
92	Indonesia	5.3
	World	4.4
114	Nepal	4.2
117	Sri Lanka	4.0
	Asia Pacific	3.8
148	China	2.8
178	Bangladesh	1.5
182	Myanmar	1.0

Fig. 5.8 Travel & Tourism' s total contribution to total capital investment. (Source: WTTC, 2019a)

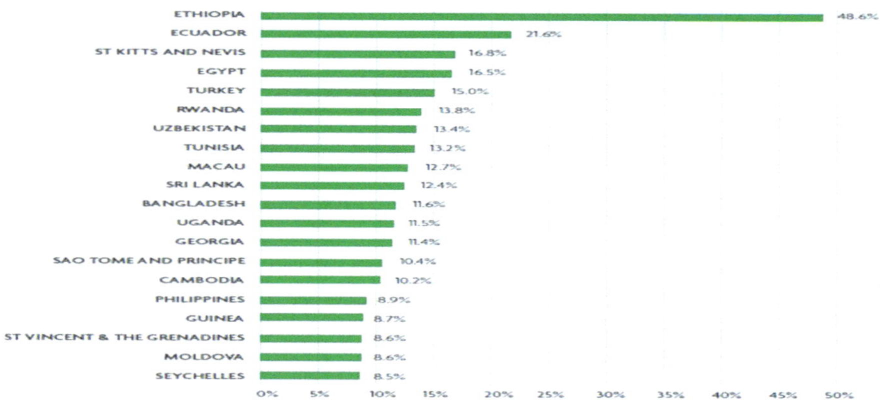

Fig. 5.9 Top 20 countries – Travel& Tourism GDP growth, 2018. (Source: WTTC, 2019a)

Suggestions to Promote Capital Investment for Tourism in Bangladesh

To promote further capital investment in Bangladesh, it needs long term plan strategies and improvement laws and regulation. Travel and Competitive Index improves the investors' confidence to invest more into a country that is ranked higher. This index is adapted from World Economic Forum that shows several areas that are identified to be critical for a sustainable tourism industry: transport infrastructure (air, port, and ground) and tourism service infrastructure. These factors to consider for Bangladesh to be more competitive to promote investment in travel and tourism, as shown in Fig. 5.6.

According to Rahman (2019), Bangladesh has been ranked 120 out of 140 countries in the Travel and Tourism Index, where other SAARC members like India, Sri

FASTEST GROWING In which countries will the contribution of Travel & Tourism to GDP grow fastest from 2019-2029?	10-YR REAL GROWTH
TRAVEL & TOURISM GDP	ANNUALISED, %
1 Qatar	7.8
2 Myanmar	6.9
3 India	6.8
4 China	6.6
5 Azerbaijan	6.4
6 Anguilla	6.3
7 Uzbekistan	6.2
8 Benin	6.1
9 Bangladesh	6.1
10 Kyrgyzstan	6.1

Fig. 5.10 Fastest Crowing (2019–2029). (Source: WTTC, 2019a)

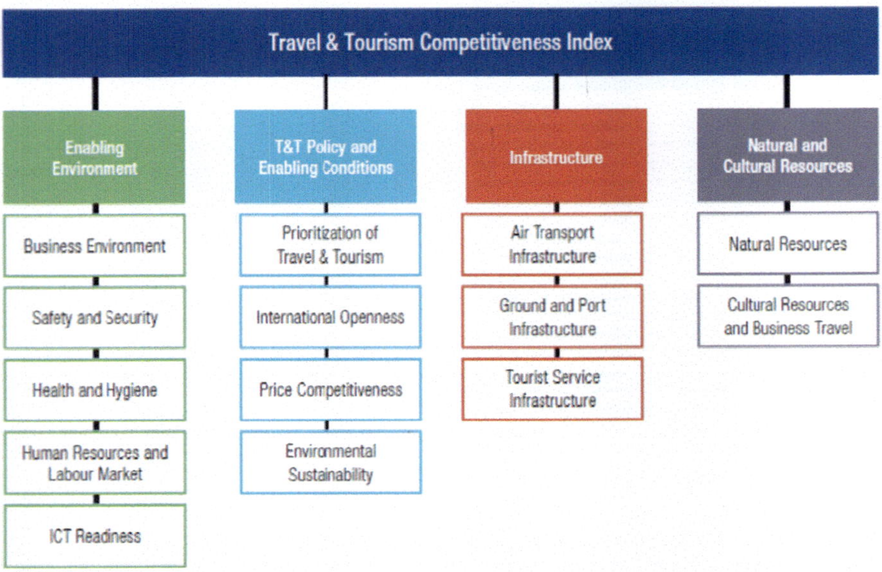

Fig. 5.11 The Travel & Tourism Competitiveness Index 2015. (Source: World Economic Forum, 2015)

Lanka, and Nepal were ranked 34, 77, and 102 and only 1 point above Pakistan, which ranked as 121. These fourteen pillars (shown above) are building blocks of this competitive index. The researchers claim that Bangladesh has performed poorly due to tourist service infrastructure, as it is one of the poorest in the world.

To focus on the natural and resources part of the index, Bangladesh can follow India' s strategy and focus on developing alternative tourism products and giving more scope capital investment in tourism. It is more rationale for Bangladesh to

Table 5.1 Impact of Bangladesh tourism 2016

Economic Indicators	Amount	% of the country total (2016)	Forecast for 2027	% of the country total
G.D.P. direct contribution	BDT 421.4bn (USD 5.3bn)	2.2	BDT806.6bn (USD10.2bn)	2.1%
G.D.P. total contribution	BDT840.2bn (USD10.6bn),	4.3%	BDT1,783.0bn (USD22.6bn)	4.7
Employment direct contribution	1,057,000 jobs	1.8%	1,138,000 jobs	1.6
Employment total contribution	2,187,000 jobs	3.8%	2,695,000 jobs	
Visitor export (out of total export)	BDT 11.1bn (USD140.0mn),	0.4%	BDT 25.6bn (USD324.2mn)	0.5%
Investment	BDT 72.5bn (USD0.9bn)	1.2%	BDT 201.8bn (USD 2.6bn)	1.8%

Source: WTTC (2016)

follow the strategies implemented by India since, in many ways, Bangladesh has many similar geographical, demographic, and historical traits as India.

Bangladesh has given significant attention towards Enabling Environment part as Khondker and Ahsan (2015) state from their report that the Bangladesh government has several active policies and legislations for the tourism following under Bangladesh Tourism Board Act, Medium Term Budget Framework (MTBF), and the Perspective Plan 2021. These policies and legislations claim to have a positive influence to flourish the tourism industry. As Islam (2013) and Parvez and Kashem (2018) assure from their findings that, the role of the Bangladesh Government has encouraged and played a positive role in promoting domestic and international tourism from the private industry.

Other things that can be considered to increase investment are: first, to increase Marketing and Promotional Campaign- As it could see from the case study that India has taken "Incredible India" campaign to promote foreign tourists, Bangladesh can do the same and thus it would attract more investors to invest into Bangladesh. Second, to ensure an Adequate supply of Qualified Human Capital- Introducing proper measures to supply proper quantity and quality of the workforce in this field is important. Bangladesh is one of the heavily populated countries in the whole world. Hence the quantity of workforce should not be a problem. However, Bangladesh still has a long way to go to have a more qualified workforce. Third, to make easier and faster Visa services for the tourists- Ensure that international tourists can come and visit the target tourist spots with ease and are not subject to the overly strict procedure. Fourth, providing more incentives to investors by allowing tax breaks or subsidies to domestic or international investors can encourage them to invest more.

If Bangladesh follows this specific macro and micro measures for tourism, along with general macro measures like low-interest rate, higher economic growth, and others, it will then definitely attract more investors (i.e. domestic and international) to invest in Bangladesh. It is also true much of Bangladesh' s potential in this tourism industry is yet to flourish poor awareness, and this is due to lack of proper development plan, promotion, and many potential tourist spots in Bangladesh is remained untapped (Dhaka Tribune, 2019).

Conclusion

From the research, it can be concluded that in Bangladesh, there is huge potential for tourism to flourish, as this offers immense scope in generating employment, poverty alleviation, and maintaining ecological balance. In recent years, the government of Bangladesh has realized the importance of tourism in its economic and social life. The economy of Bangladesh has been growing at an astonishing rate in the last decade, however much of its growth is dependent on manufacturing-based industries, like textile, pharmacy, steel, etc., but the service industry like tourism has not played any significant role in the national economy of Bangladesh yet as Das and Chakraborty (2012) point out the importance of tourism for its maintenance and growth of the economy. For this industry to grow and contribute more to the Bangladesh economy, more capital investment is required. This research has pinpointed what areas can be improved to motivate domestic and international investors to invest in the economy of Bangladesh. A case study of India on how it developed its tourism and industry was observed along with the travel & tourism industry, where it could be seen that Bangladesh lacked the infrastructure to be more competitive in tourism. If Bangladesh can give more attention to the Travel and Tourism Index and perform well, certainly more and more capital investment will flow into the economy of Bangladesh.

References

Abakash Parjatan. (2019). *Welcome to Abakash Parjatan Ltd.* Retrieved from: https://abakashparjatan.com/about_bd.php. Accessed 8 Dec 2019.

Akihito, N. (2017). Development of tourism and the tourist industry in India: A case study of Uttarakhand. *Journal of Urban and Regional Studies on Contemporary India, 3*(2), 1–12.

Anholt, S. (2006). Why brand? Some practical considerations for nation branding. *Place Branding, 2*(2), 97–107.

Balaguer, J., & Cantavella-Jordá, M. (2002). Tourism as a long-run economic growth factor: The Spanish case. *Applied Economics, 34*(7), 877–884.

Bialowolski, P., & Weziak-Bialowolska, D. (2014). External factors affecting investment decisions of companies. *Economics: The Open-Access, Open-Assessment E-Journal, 8*, 1.

Blomstrom, M. (2014). *Foreign investment and Spillovers*. Routledge.

Budget.uk.gov.in. (2019). *Home: Department of budget, government of Uttarakhand, India*. Retrieved from: http://budget.uk.gov.in/. Accessed 15 Dec 2019.

Carr, C., Kolehmainen, K., & Mitchell, F. (2010). Strategic investment decision making practices: A contextual approach. *Management Accounting Research, 21*(3), 167–184.

Das, R., & Chakraborty, J. (2012). *An evaluative study on tourism in Bangladesh*. Retrieved from: https://pdfs.semanticscholar.org/b907/e7e342f2758334703a12ce159db1e8befbae.pdf. Accessed 8 Dec 2019.

De, S., & Chakraborty, T. (2018). Foreign portfolio investment and Price discovery of Indian equity stocks. *Indian Journal of Research in Capital Markets, 5*(2), 23.

Dhaka Tribune. (2019). *World Tourism Day: Master plan to explore potential tourist spots in the offing*. Retrieved from: https://www.dhakatribune.com/bangladesh/dhaka/2019/09/27/world-tourism-day-on-friday-master-plan-to-explore-potential-tourist-spots-in-the-offing. Accessed 8 Dec 2019.

Dieke, P. U. C. (2004). Tourism in Africa' s economic development: Policy implication. *Management Decision, 41*(2), 287–295.

Dritsakis, N. (2004). Tourism as a long-run economic growth factor: An empirical investigation for Greece using causality analysis. *Tourism Economics, 10*(3), 305–316.

Ekanem, I. (2005). 'Bootstrapping': The investment decision-making process in small firms. *The British Accounting Review, 37*(3), 299–318.

Enove, A., & Mustapha, I. (2010). Factor analysis of investment decision in Nigerian insurance companies. *Interdisciplinary Journal of Contemporary Research in Business, 2*(8), 108–120.

Gill, A. S., Sharma, S. P., & Mathur, N. (2012). The impact of debt policy on the investment decision of small business owners in India. *International Research Journal of Finance and Economics, 97*, 6–13.

Gomes, E. (2019). *Bangladesh ranked 7th in best value destinations by lonely planet – SD Asia*. Retrieved from: https://sdasia.co/2018/10/24/bangladesh-ranked-7th-in-best-value-destinations-by-lonelyplanet/. Accessed 8 Dec 2019.

Hasan, D. S. R., & Islam, M. S. (2014). *Tourism in Bangladesh problems and prospects*. Retrieved from: https://www.academia.edu/9111015/Tourism_in_Bangladesh_Problems_and_Prospects. Accessed 8 Dec 2019.

Islam, S. M. N. (2013). *Tourism marketing in developing countries: A study of Bangladesh hospitality and tourism management*. University of Strathclyde.

Itic.ie. (2017). *Home*. Retrieved from: https://www.itic.ie/wp-content/uploads/2017/05/ITIC_Tourism-Submission-to-Capital-Review_Full-Report_150517.pdf. Accessed 7 Dec 2019.

Jaiswal, D. B., & Bisht, D. M. (2017). Tourism sector in Uttarakhand: A brief overview after the state formation. *International Journal of Research in Economics and Social Sciences, 7*(7), 277–295.

Kahraman, C. (2011). Investment decision making under fuzziness. *Journal of Enterprise Information Management, 24*(2), 126–129.

Karim, Z., & Azman-Saini, W. (2013). Firm-level investment and monetary policy in Malaysia: Do the interest rate and broad credit channels matter? *Journal of the Asia Pacific Economy, 18*(3), 396–412.

Khanna, D. (2011). Destination branding: Tracking brand India. *Synergy, 9*(1), 40–50.

Khondker, B., & Ahsan, T. (2015). *Background paper on tourism sector*. Retrieved from: //www.academia.edu/12098610/Final_Report_BACKGROUND_PAPER_ON_TOURISM_SECTOR. Accessed 8 Dec 2019.

Knoema. (2019). *Bangladesh contribution of travel and tourism to G.D.P. (% of G.D.P.), 1995–2018 – knoema.com*. Retrieved from: https://knoema.com/atlas/Bangladesh/topics/Tourism/Travel-and-Tourism-Total-Contribution-to-GDP/Contribution-of-travel-and-tourism-to-GDP-percent-of-GDP. Accessed 8 Dec 2019.

Kumar, A. (2017). Issues and challenges in hospitality industry: A case of Uttarakhand (India). *Tourism and Travelling, 1*(1), 51–57.

Laux, V. (2008). On the value of influence activities for capital budgeting. *Journal of Economic Behavior & Organization, 65*(3–4), 625–635.

Nakatani, T. (1996). Present condition of tourism in India. *Nara Prefectual Commerce University Studies, 7*(1), 1–9.

Nawaz, M. A., & Hassan, S. (2016). Investment and tourism: Insights from the literature. *International Journal of Economic Perspectives, 10*(4), 581–590.

Nelson, R. R. (2005). *Technology, institution, and economic growth*. Harvard University Press.

Parveen, D. J. A. (2013). Current status of tourism industry in Bangladesh: An empirical evaluation. *Scientific Research Journal, 1*(1), 41–57.

Parvez, M., & Kashem, M. J. B. (2018). Young tourists' attitude towards domestic tourism: A study on Bangladesh. *Problems and Perspectives in Management, 16*(3), 117–129.

Pettinger, T. (2019). *Factors affecting investment – Economics Help*. Retrieved from: https://www.economicshelp.org/blog/136672/economics/factors-affecting-investment/. Accessed 8 Dec 2019.

Rahman, S. (2019) *Why infrastructure is crucial to our tourism growth vision*. Retrieved from: https://www.thedailystar.net/opinion/news/why-infrastructure-crucial-our-tourism-growth-vision-1797397 Accessed 8 Dec 2019.

Rai, S., & Nayak, J. (2019). Hospitality branding in emerging economies: An Indian perspective. *Journal of Tourism Futures, 5*(1), 22–34.

Selvanathan, S., Selvanathan, E., & Viswanathan, B. (2012). Causality between foreign direct investment and tourism: Empirical evidence from India. *Tourism Analysis, 17*(1), 91–98.

Sinclair, M. (1998). Tourism and economic development: A survey. *Journal of Development Studies, 34*(5), 1–51.

TCdata360. (2019). *TCdata360: Capital investment in travel and tourism*. Retrieved from: https://tcdata360.worldbank.org/indicators/cap.inv.tat?country=BGD&indicator=24674&viz=line_chart&years=1995,2028&indicators=944. Accessed 11 Dec 2019.

UKEssays.com. (2019). *Introduction theoretical theories of investment economics essay*. Retrieved from: https://www.ukessays.com/essays/economics/introduction-theoretical-theories-of-investment-economics-essay.php. Accessed 8 Dec 2019.

World Economic Forum (WEF). (2015). *The Travel & Tourism Competitiveness Report 2015: Growth through Shocks*. Retrieved from: http://www3.weforum.org/docs/TT15/WEF_Global_Travel&Tourism_Report_2015.pdf. Accessed 7 Dec 2019.

World Tourism Organization (UNWTO). (2019). *UNWTO tourism highlights: 2017 edition*. Retrieved from: https://www.e-unwto.org/doi/pdf/10.18111/9789284419029. Accessed 7 Dec 2019.

World Travel and Tourism Council (WTTC). (2016). *Travel & tourism economic impact 2016*. Retrieved from: https://www.wttc.org/-/media/files/reports/economic%20impact%20research/regions%202015/world2016.pdf. Accessed 8 Dec 2019.

World Travel and Tourism Council (WTTC). (2019a). *Travel and tourism global impact 2019*. Retrieved from: https://bit.ly/3z2Y5Ak. Accessed 7 Dec 2019.

World Travel and Tourism Council (WTTC). (2019b). *Travel and tourism economic impact 2019 world*. Retrieved from: https://www.wttc.org/-/media/files/reports/economic-impact-research/regions-2019/world2019.pdf. Accessed 8 Dec 2019.

Zarnowitz, V. (1992). *Business Cyvles. Theory, History, Indicators, and Forecasting*. Chicago and London: The University of Chicago Press.

Chapter 6
Tourism as an Element for Economic Growth in Bangladesh: Investment Analysis for Product and Service Development

Muhammad Khalilur Rahman and Azizul Hassan

Abstract In many ways, tourism can appear as a tool for developing both the local and national economies of a country like Bangladesh. A tourism product can be identified from conceptual understanding as the total psychological and physical satisfaction provided to tourists during their traveling en-route towards the destination. On the other side, tourism services are the offers or facilities related to travel and tourism, including hotel and restaurant services, catering, tour operators, travel agents, tourist guides, visa and immigration, and other relevant services. Thus, tourism products and services are designed to meet tourist demands comprehensively. This research from a theoretical base outlines the investment scenario for tourism products and service development in Bangladesh, which is an element for gross economic growth. This is a literature review based research that relied on published and online resources for data and information. Findings show that Bangladesh has abundances in tourism product and service offers. Existing tourism products and services are mostly nature-based because of the country's geographical location and absence of required investments. However, the country's tourism service and product development are experiencing gradual growth supported by investment flows and technology application. Thus, this research suggests that investment in tourism services and products in Bangladesh can be a valid economic growth element when supported by policies.

Keywords Tourism · Products · Services · Investment · Bangladesh

M. K. Rahman (✉)
Faculty of Entrepreneurship and Business, Universiti Malaysia Kelantan,
Pengkalan Chepa, Malaysia
e-mail: mohdkhalilur@gmail.com

A. Hassan
Tourism Consultants Network, The Tourism Society, London, UK

A. Hassan (ed.), *Tourism in Bangladesh: Investment and Development Perspectives*, https://doi.org/10.1007/978-981-16-1858-1_6

Introduction

Bangladesh is a beautiful green country. The country has a favorable climate, the longest beach, hilly mountain, seven seasons, and many archaeological and historical sites that are the crucial factors and tourism element for developing economic growth (Reza et al., 2018; Sultana, 2016). The country has a local culture, hospitality of people and lifestyle are the unique selling point for attracting tourists from different countries. Tourism is a developing industry in Bangladesh in which the number of foreign tourists was 0.16 million in 2014, but it slightly decreased to 0.14 million in 2015 (The Financial Express, 2019). However, it increased again to 0.20 million in 2016 and continues to grow by 0.26 million in 2017 and around 0.27 million in 2018, respectively (The Financial Express, 2019). Sultana (2016) reported that Bangladesh's travel and tourism industry is predicted to rise by 2.9% per year to 1,785,000 jobs of total employment in 2023.

Bangladesh is a middle-income country, and many industries contribute to economic growth (Hossain, 2020; Hasan et al., 2018). The government has become competitive in the global market in many sectors (e.g. tourism, product, and services development). Remarkably, after the liberation, the country mainly focused on its agriculture sector and generated economic growth by exporting rice, tea, and jute. The agriculture sector contributes to the GDP is 14.10% in the fiscal year 2017–2018, whereas the services sector contributes to the GDP is 52.85% in 2018 (The Daily Star, 2018). The Readymade Garments (RMG) sector plays a crucial role in Bangladesh's economic growth (Roy, 2020; Hasan et al., 2016). The industry contributes to export earnings, foreign direct investment inflow, poverty alleviation, employment creation, and empowerment of women (Islam, 2020). World Bank (2019) reported that RMG exports generated more than 10% GDP and provided 84% of foreign exchange earnings in the 2018–2019 fiscal year. The industrial sector has contributed 33.71% of GDP in fiscal 2016–2017. However, despite the economy's potential growth, its international image is not a popular tourism destination (Honeck & Akhtar, 2014). The World Travel and Tourism Council (WTTC) reported that the tourism industry contributed BDT 296.6 billion (1.9% of GDP) in 2014, and it is expected to rise BDT 566.3 billion (2% of total GDP) by 2025 (Abdulkadir, 2018).

Many people probably do not consider Bangladesh to have their potential tourism destination due to overcrowding, developing countries, political violence, and suffering environmental disasters. Many people may surprise to learn it has the Sundarbans, World Heritage Site, which is part of travel circuits for cultural tourism. The Sundarbans, the Historic Mosque in the City of Bagerhat, and the Ruins of Buddhist Vihara at Paharpur are the crucial World Heritage Sites in Bangladesh (Roy & Roy, 2015). Bangladesh is full of natural beauty, for example, religious places, rivers, archaeological sites, hills, coasts, beaches, tea, waterfalls, and forests surround it. A considerable number of tourists visit these tourist attractions to see the beauty of the country's nature (Hossain & Islam, 2019).

Due to some limitations, the country still not able to introduce itself as a potential tourism destination in the world. This study aims to search the investment scenario for tourism products and service development in Bangladesh, which may contribute a lot to flourish the economic growth rate of the country. The primary concern is to emphasize the factors related to the tourism investment for product and service development. The country might be one of the potential attractive destinations around the world. Tourism is a facilitator for the development of manufacturing and construction infrastructure in Bangladesh.

Contribution of Tourism in Economic Growth

The tourism industry can play a crucial role as a driving force of economic development. Bangladesh has a vast potential to develop tourism. It has attractive natural beauty and cultural heritage, which plays a crucial role in economic growth by creating new employment opportunities, poverty alleviating, earning foreign currency through tourism, and increasing local community participation (Sultana, 2016). Bangladesh earned $421.4 billion from the travel and tourism industry, which contributes to the economic growth of Bangladesh (Mowla, 2019; Sultana, 2016). Travel and tourism have a huge contribution to the developing economy of the country. It is one of the largest and fast-growing sectors of economic development of the country. Tourism consists of small and medium-sized enterprises that attempt to be successful in a competitive and rapidly changing business environment. Tourism is also one of the profitable sectors in the economic development of Bangladesh, including the manufacturing and construction sectors. According to Patwary et al. (2020), Ferdaush and Faisal (2014), tourism plays a crucial role in the economic growth of Bangladesh. The international tourists praised this country for its natural beauty, cultural heritage, and hospitality of the people.

Foreign exchange is required for the development of the economy. The development of sustainable tourism can play a significant role in growing the national economy (Saxena et al., 2020). Bangladesh can earn huge foreign exchange through the development of the tourism industry. Readymade garments and workforce exports of Bangladesh are one of the most crucial sectors for generating foreign revenue (Rashed & Ahmadov, 2019). However, these two sectors are now at a saturated stage. The development of these two important sectors is no longer being extended to a significant level. The economic benefit of the travel industry in Bangladesh is admired regarding its contribution to foreign revenue income. Likewise, it contributed to creating job opportunity as the tourism industry in Bangladesh was created 218,000 jobs in 2016 and 2,247,000 jobs in 2017 (Hafsa, 2020).

The total contribution of the travel and tourism industry on economic growth in Bangladesh can have a wider effect directly and indirectly. The travel industry in Bangladesh directly spends money on several sectors such as transportation, accommodation, retail trade, food, and attraction. It spends indirectly through some different areas like collective government spending, security administration, flying administration, sanitary services, consumer products and services, fuel and gasoline services. Accordingly, it contributes directly and indirectly to the GDP in

Bangladesh. As per a report of World Travel and Tourism Council (WTTC), the total contribution of the tourism industry was BDTk.627.9 billion, which was 4.1% of GDP in 2014, and it is estimated that it will increase to BDTk.1, 252.8 billion by 2025 (4.4% of GDP) (Ahmed et al., 2020; Kobra et al., 2018).

The travel industry is a concentrated work industry that has enormous employment opportunities for both aptitude and unskillful employment from rural and metropolitan regions. The tourism industry can create employment from coastal to hilly zones where tourism improvement is hard to do. In certain zones, the other business industry is hesitant to contribute to economic development while the travel industry proceeds to contribute to the development of human capital towards their income (Kobra et al., 2018). Hafsa (2020) stated that tribal regions such as Chittagong Hill tracks, Coastal areas, Islands like Saint Martine, Moheskhali, Shapori, Monpura, Nijhum, and Maya are sensitive spots where the travel industry plays an important role to make job opportunities. As per WTTC, in 2014, the tourism industry made 903,500 jobs, which are 1.6% of total employment, and this is expected to increase to 1,062,000 positions by 2025 (Kobra et al., 2018). The total contribution of the travel and tourism industry to employment was 1,984,000 positions in 2014, which is 3.6% of total employment. Kobra et al. (2018) reported that by 2025, the tourism industry in Bangladesh is predicted to help 2,492,000 jobs. Bangladesh requires investment and provide quality services in the travel and tourism industry to achieve higher economic development.

Capital Investment in Tourism Products and Services

Bangladesh has huge prospects for both international and national investors for the development of the tourism industry. According to WTTC, the tourism industry is a growing sector in Bangladesh (Ahmed et al., 2020; Hafsa, 2020; Sultana, 2016). The travel and tourism-related investment were around US$612 billion in 2010, and it reached $652 billion in 2011, or 4.5% of total investment (Kobra et al., 2018). In 2018, the tourism industry in Bangladesh covered 4.4% of the total GDP (World Bank, 2017). The country has given attention to the tourism industry as the government has allocated BDTk. 34 billion in FY2019–2020 in Civil Aviation and Tourism industry (Light Castle Analytics Wing, 2020). World Bank (2017) reported that the dominant portion (97%) of earnings in the tourism industry come from local sources. It indicates that Bangladesh cannot reach a popular destination among international tourists.

Compare to the other sectors in Bangladesh; investment is the key to the development of the tourism industry (Sakib et al., 2020). The government has given low attention to investing in the tourism industry (Kobra et al., 2018). The investment scenario in the tourism industry can be visible by assessing the existing context of investment. According to Kamruzzaman (2018), Bangladesh has plenty of attractive natural places such as the lengthiest sea beach (Cox's Bazar), live coral reef (St. Martin's Island), and biggest Mangrove forest (the Sundarbans) that shows an open view of both sunrise and sunset over the Bay of Bengal (Kuakata) which can influence both local and international investors to invest in the tourism industry. In 2016, the capital investment of the tourism industry was BDTk.72.5 billion, and it increased 13.9% in 2017 and further is expected to rise 9.3% with BDTk.201.8 billion by 2027 (Kobra et al., 2018). Bangladesh has huge opportunities to invest and promote the tourism industry, but several challenges need to maintain successfully. Table 6.1 below shows the opportunities and challenges for investment in the tourism industry.

Table 6.1 Challenges and opportunities for Tourism investment

Challenges	Opportunities
Political unrest in Bangladesh is interrupting the development of the tourism industry. During the political demonstration, some arrogant people from different political parties destroy and burn the vehicles.	Bangladesh is a naturally beautiful country with its least carbon emitted place, larger sea beaches, and mountains.
The transportation systems (e.g. highways, roads, riverways, and airways) are not developed well; as a result, the tourism industry cannot growing up in Bangladesh.	The country has three World Heritage Sites as the Sundarbans, the Historic Mosque in the City of Bagerhat, and the Ruins of Buddhist Vihara at Paharpur.
Bangladesh is familiar with a poor country in the world, and it has a poor brand image worldwide.	The country's tourism board is organizing various tourism fairs, travel mart, symposiums, tradeshows, conferences, tradeshows, and exhibitions.
The country has a lengthy bureaucratic system that takes time to get approval from the government to invest in the travel and tourism industry.	The country has plenty of opportunities for rural tourism where an international and local investor can build a resort, create ethnic culture, sports and generate rural activities.
The allocation of funds in the budget is not sufficient for the growth of the tourism industry in Bangladesh.	Currently, the government has taken mega-project such as different power plants, metro rail, Padma bridge, airports, and railroads.
The country has inefficiency of human capital because they are not skilled in terms of the hospitality and tourism industry.	The various banks in Bangladesh are offering SME loans to attract small entrepreneurship.
International investors, delegates, and visitors face difficulty in visa processing issues.	The local tourism in this country is in a booming position.
International investors' confidence is below the benchmark for the unfavorable environment.	The government introduces community partnerships, private and public, to make a sustainable tourism environment.
The communication gap among the government, investors, foreigners, and local people is still a vital challenge for the country to grow up the tourism industry.	The cost of the labor force is reasonably cheap. Thus, the country can take this advantage to motivate the investor to invest.
Interest is very high for both short and long-term as financial benefits.	The government has taken special consideration for foreign direct investment.

Source: Kobra et al. (2018)

Tourism Development in Bangladesh

The travel and tourism industry has opened doors for generating revenue in a developing country. Bangladesh has huge resources in the travel industry with historic resources, old heritage, and different cultures that can influence both international and domestic tourists (Bhuiyan & Darda, 2019; Ferdaush & Faisal, 2014; Majumder & Hossain, 2011). Bangladesh Travel and Tourism Board can provide appropriate facilities to serve the tourists (Rana et al., 2020). The tourism industry of this country is growing slightly and earning some foreign revenue. Hassan and Ramkissoon (2020) reported that potential tourism development in Bangladesh could be capitalized on with the support of an effective set of policy implementation. The tourism industry is the most neglected sector in Bangladesh and is not yet reached a place that can draw a large number of international tourists. This is one of the grounds that there is an absence of effective initiatives, plans, and improper Management, and sincere efforts of the government (Jahan & Amin, 2014). However, it is known to all of the travel industry-related experts, agents, and operators that Bangladesh has natural beauty, the largest sea beach, and the Sixty-dome Mosque, Mahasthangarh, and the Sundarbans, which can satisfy both national and international tourists. The tourism industry predicted that the GDP of Bangladesh would grow 6.1% per annum with BDTk.13805.6 billion by 2027 (Roy & Roy, 2015; Ferdaush & Faisal, 2014). Tourists industry is increasingly growing that lead to social and economic advantage, and can play a crucial role in improving a nation's trade performance. The advancement of this sector should ensure the sustainability of the tourism industry with the end goal of acquiring foreign revenue for a longer timeframe.

Sustainable Tourism Development of Bangladesh

The development of sustainable tourism in Bangladesh is a new phenomenon. It is a continuous process of cooperation and improvement of products and services, amenities, and facilities that support the host communities to provide quality services for tourists and develop residents' wellbeing (Hossain & Khanal, 2020; Jahan & Amin, 2014). The tourism industry should improve tourists' needs following the trend of global sustainable development (Hassan et al., 2020). The travel industry can bring economic benefits to the country (Rani et al., 2020; Hafsa, 2020). Thus, sustainable development, effective tourism strategy, and implementation are essential for evolving the tourism industry (Bhuiya et al., 2020). These ideas may increase environmental preservation, awareness, and use of natural resources. Natural and environmental destruction can be occurred by the society, local and international tourists' consumption behavior. Therefore, the government should take proper initiative for the protection of natural beauty and environmental preservation through the development of sustainable tourism (Mowla, 2019). All levels of tourism consumption should be preserved over a long period, which can ensure a great benefit for the economic, cultural, social, and natural environments of tourist regions (Masrurul, 2019). The development of sustainable tourism requires continuous monitoring, presenting the necessary preventive and remedial measures whenever necessary. Sustainable tourism also leads to travelers' satisfaction and increase their awareness about sustainability issues and developing sustainable tourism practice. Tourism should be sustainable in three areas such as environmental, economic, and social, and control the negative impacts of tourism to make Bangladesh a sustainable tourism location.

Researchers' Assessment of the Tourism Industry in Bangladesh

Kobra et al. (2018) identified that there is a lack of research regarding tourism investment to promote the flourishing tourism industry. Thus, they empathized on sufficient investment in the travel and tourism industry in Bangladesh. Saxena et al. (2020) examined how capital contributes to community wellbeing and promote sustainable tourism in the coastal community of Cox's Bazar in Bangladesh. Hassan and Burns (2014) revealed the demand for a proper set of tourism policy in Bangladesh that leads to promote the tourism industry. Hassan et al. (2020) analyzed qualitative data using content analysis to investigate the perceptions of the national tourism policy implementation process in Bangladesh. Hafsa (2020) stated that tourism plays a crucial role in the GDP growth of Bangladesh by creating new job opportunities for both male and female, generating foreign revenue, reducing poverty, increasing local community participation, and developing the economic standard of local people. Majumder and Hossain (2011) stated that Bangladesh is a new tourism destination with and potential to promote the tourism industry as well as national economic development because of its world cultural heritage and attractive natural beauty. Reza et al. (2018) identified that the foreign direct investment flows are a crucial weapon of GDP growth for Bangladesh, capital investment, and technology development. Bangladesh government needs to create foreign investment-friendly policies, trade promotion strategy, and transfer of knowledge.

Policies for Tourism Development

The travel and tourism industry in Bangladesh was first adopted in 1992 (Hassan & Burns, 2014). The tourism policy was issued and published in 2009 (Hassan et al., 2020). The government revised the national tourism policy in 2010 to cope with global demand (Kobra et al., 2018). For implementing this policy, the Ministry of Civil Aviation, Bangladesh Parjatan Corporation, and Bangladesh Tourism Board have responsibility for implementing the national tourism policy (Hassan et al.,

2020). The revised tourism policy is commonly seen as extensive with policy instrument, implementation, and objectives. Hassan and Burns (2014) reported that the tourism policy of Bangladesh has thirty-one (31) objectives. The main objectives are (i) to increase interest in tourism with the people of Bangladesh; (ii) to create a positive image of the country to the international tourists or foreign nations; (iii) to create employment opportunities through promoting tourism industry and poverty alleviation; (iv) to make available incentive to increase private investment in the tourism industry; (v) to increase available entertainment and recreational opportunities; and (vi) to develop tourism resources, maintenance invention, and protection.

Conclusion

The tourism industry in Bangladesh has grown in recent years, but this development pattern is inconsistent because of unstructured development and lack of appropriate planning. There is required to make a comprehensive mechanism to bring all tourism players to align practical support, legal framework improvement, access to finance and investment, evolving information-sharing resources such as investment opportunity, investment website, and success stories. Local community and ecological opportunities and contribute to sustainable development are the keys to promote tourism investment. Bangladesh has a shortage of finance. However, the country can raise the GDP by promoting the tourism industry. The Parjatan Corporation in Bangladesh is practicing a modern marketing strategy for developing the tourism industry.

The country has numerous opportunities to generate foreign and domestic revenue from the tourism industry, but this sector is neglected. Bangladesh has many potential tourist spots which are required to find out the exposure. The government of this country can encourage the private sector to work together to develop the tourism industry. The country can also make a tourism portal, new TV programs for offering both local and international channels, and developing a framework for the travel industry to pull interest in inbound the tourism industry. Besides, the weak investment, lack of safety and security, poor infrastructure, improper marketing policy, inappropriate Management of tourism destinations, and lack of local community participation in promoting tourism development should be solved with priority. The managerial implications (skilled human resources, integration among stakeholders, extensive research on tourism resources, and quality of service) are essential and need to be prioritized for developing the tourism industry in Bangladesh. The government should ensure proper investment planning and policy implications to flourish in the tourism industry and dream to turn Bangladesh into a lucrative tourism destination.

References

Abdulkadir, C. M. I. (2018). Holistic approach to tourism development and destination image of Chittagong hill tracts. *NDC E-Journal, 17*(2), 161–184.

Ahmed, A., Amin, S. B., & Khan, A. M. (2020). Forecasting tourism revenue in Bangladesh using ARIMA approach: The case of Bangladesh. *International Review of Business Research Papers, 16*(1), 202–215.

Bhuiya, S., Setarnawat, S., & Viriyasuebphong, P. (2020). Factors affecting sustainable tourism destination management in Lawachara National Park, Sylhet, Bangladesh. *Research Methodology and Cognitive Science, 17*(2), 180–195.

Bhuiyan, M. A. H., & Darda, M. A. (2019). Tourists' satisfaction on heritage sites of Dhaka city in Bangladesh. *International Journal of Culture, Tourism and Hospitality Research, 14*(1), 34–43.

Ferdaush, J., & Faisal, H. (2014). Tourism potentiality & development of Bangladesh: Applicability of pragmatic governmental management policy. *Journal of Management and Science, 4*(1), 71–78.

Hafsa, S. (2020). Economic contribution of tourism industry in Bangladesh: At a glance. *Global Journal of Management and Business Research, 20*(1), 29–38.

Hasan, K. F., Mia, M. S., Rahman, M. M., Ullah, A. A., & Ullah, M. S. (2016). Role of textile and clothing industries in the growth and development of trade & business strategies of Bangladesh in the global economy. *International Journal of Textile Science, 5*(3), 39–48.

Hasan, M. M., Hossain, B. S., Alam, M. J., Chowdhury, K. A., Al Karim, A., & Chowdhury, N. M. K. (2018). The prospects of blue economy to promote Bangladesh into a middle-income country. *Open Journal of Marine Science, 8*(3), 355–369.

Hassan, A., & Burns, P. (2014). Tourism policies of Bangladesh—A contextual analysis. *Tourism Planning & Development, 11*(4), 463–466.

Hassan, A., & Ramkissoon, H. (2020). Potentials of tourism products and services in Bangladesh. In A. Hassan (Ed.), *Tourism marketing in Bangladesh*. Routledge.

Hassan, A., Kennell, J., & Chaperon, S. (2020). Rhetoric and reality in Bangladesh: Elite stakeholder perceptions of the implementation of tourism policy. *Tourism Recreation Research, 45*(3), 1–16.

Honeck, D., & Akhtar, M. S. (2014). *Achieving Bangladesh's tourism potential: Linkages to export diversification, employment generation and the "green economy"* (WTO Staff Working Paper, No. ERSD-2014-15). World Trade Organization (WTO).

Hossain, M. (2020). Macroeconomic policy, economic growth and poverty reduction in Bangladesh: An overview. In M. Hossain (Ed.), *Bangladesh's macroeconomic policy* (pp. 3–18). Palgrave Macmillan.

Hossain, M. K., & Islam, S. (2019). An analysis of destination attributes to enhance tourism competitiveness in Bangladesh. *African Journal of Hospitality, Tourism and Leisure, 8*(2), 1–17.

Hossain, M. I., & Khanal, B. P. (2020). SWOT analysis of Forest protected area based sustainable tourism development: A case study in the Lawachara National Park, Bangladesh. *Journal of Tourism and Hospitality Education, 10*, 71–89.

Islam, M. S. (2020). Ready-made garments exports earning and its contribution to economic growth in Bangladesh. *Geo Journal*. https://doi.org/10.1007/s10708-019-10131-0

Jahan, N., & Amin, M. R. (2014). Sustainable tourism development in Bangladesh: An empirical study on Sylhet. *Journal of Business, 35*(2), 239–260.

Kamruzzaman, M. (2018). Assessment of destination-specific factors of Bangladesh: A review of Saint Martin Island as an emerging tourist destination. *Journal of Business Studies, 39*(1), 113–125.

Kobra, M. K., Bhuiyan, K. H., & Zayed, N. M. (2018). Well and woes of tourism promotion in Bangladesh: Investment perspective. *Academy of Accounting and Financial Studies Journal, 22*(3), 1–8.

Light Castle Analytics Wing. (2020). *Tourism: A possible new driver for the economy of Bangladesh*. Retrieved from: https://bit.ly/3hYAS93. Accessed 14 Sept 2020.

Majumder, M., & Hossain, T. (2011). An appraisal of tourism industry development in Bangladesh. *European Journal of Business and Management, 3*(3), 287–298.
Masrurul, M. M. (2019). Investigating the compliance of tourism principles with the components of sustainable tourism: Bangladesh perspective. *Advances in Management, 12*(1), 4–7.
Mowla, M. M. (2019). Investigating the compliance of tourism principles with the components of sustainable tourism: Bangladesh perspective. *Global Journal of Management and Business Research, 19*(1), 9–13.
Patwary, A. K., Chowdury, M. M., Mohamed, A. E., & Azim, M. S. (2020). Dissemination of information and communication technology (ICT) in tourism industry: Pros and cons. *International Journal of Multidisciplinary Sciences and Advanced Technology, 1*(8), 36–42.
Rana, M. S., Rahman, M. K., Islam, M. K., & Hassan, A. (2020). Globalization effects on tourism marketing in Bangladesh. In A. Hassan (Ed.), *Tourism marketing in Bangladesh: An introduction*. Routledge.
Rani, S., Ahmed, M. K., Xiongzhi, X., Yuhuan, J., Keliang, C., & Islam, M. M. (2020). Economic valuation and conservation, restoration & management strategies of Saint Martin's coral island, Bangladesh. *Ocean & Coastal Management, 183*, 105024.
Rashed, J. S., & Ahmadov, F. S. (2019). Azerbaijan–Bangladesh: Bilateral trade opportunities and potential exporting sector for Azerbaijan. *Journal of Contemporary Economic and Business Issues, 6*(2), 83–95.
Reza, S. M., Fan, H., Reza, T., & Wang, B. (2018). The impact of foreign direct investment inflows on economic growth: Evidence from Bangladesh. *Journal of Business and Retail Management Research, 12*(2), 212–223.
Roy, R. (2020). Determinants of labor unrest in the Bangladesh readymade garments industry. *International Journal of Advanced Engineering Research and Science, 7*(5), 22–29.
Roy, S. C., & Roy, M. (2015). Tourism in Bangladesh: Present status and future prospects. *International Journal of Management Science and Business Administration, 1*(8), 53–61.
Sakib, A., Al-Kabir, F., Ainy, N., & Farhan, K. (2020). An empirical investigation between foreign direct investment (FDI) and tourism in Bangladesh. *Journal of Tourism Management Research, 7*(1), 110–121.
Saxena, G., Mowla M. M., & Chowdhury, S. (2020). Spiritual capital (Adhyatmik Shompatti) – A key driver of community well-being and sustainable tourism in Cox's Bazar, Bangladesh. *Journal of Sustainable Tourism, 28*(10), 1–27.
Sultana, S. (2016). Economic contribution of tourism industry in Bangladesh. *Journal of Tourism, Hospitality and Sports, 22*(2), 55–54.
The Daily Star. (2018). *Economy marches towards record 7.65pc growth.* Retrieved from: https://www.thedailystar.net/business/economy/gdp-growth-in-bangladesh-economy-marches-towards-record-7.65-percent-in-2018-1557706. Accessed 21 Jan 2020.
The Financial Express. (2019). *Tourist arrivals rise in five years.* Retrieved from: https://thefinancialexpress.com.bd/economy/tourist-arrivals-rise-in-five-years-1569469201. Accessed 21 Jan 2020.
World Bank. (2017). *International tourism, number of arrivals – East Asia & Pacific.* Retrieved from: //data.worldbank.org/indicator/ST.INT.ARVL?end=2017&locations=Z4&start=2017&type=shaded&view=map&year=2017. Accessed 21 Jan 2020.
World Bank. (2019). *World development indicators.* Retrieved from: https://databank.worldbank.org/data/source/world-development-indicators. Accessed 21 Jan 2020.

Part III
Institutional Involvement

Chapter 7
The Role of Public Agencies in the Tourism and Hospitality Industry Investment and Development in Bangladesh

Bapon Chandra Kuri, Sadia Afrin Ananya, Sanjoy Kumar Roy, and Azizul Hassan

Abstract Boosting economic growth is one of the major Sustainable Development Goals (SDG) that prioritized the direction of sustainable tourism investment and development for a nation. This is the absolute economic goal that Bangladesh can attain by ensuring a sustainable tourism environment through Green investment and development in the tourism industry. For this purpose, Bangladeshi tourism public agencies including Ministry of Civil Aviation and Tourism (MOCAT), Bangladesh Parjatan Corporation (BPC), Bangladesh Tourism Board (BTB), Biman Banglades h Airlines Limited update their constitutional roles, plans, and activities to boost the industry growth and GDP. These organizations are currently working on tourism-based mega projects in the coastal zone of Bangladesh to attract tourists and investors. Instead of this, these agencies should initiate a long-term sustainable incentive system that will attract both foreign and domestic investors who may invest in tourism development projects in Bangladesh. This chapter addresses how public agencies play a vanguard role in Bangladesh's tourism industry's investment and development.

Keywords Public agencies · Sustainable development · Investment · Tourism · Sustainable goals

---

B. C. Kuri (✉) · S. A. Ananya
Department of Tourism and Hospitality Management, Bangabandhu Sheikh Mujibur Rahman Science and Technology University (BSMRSTU), Gopalganj, Bangladesh
e-mail: baponchandra@gmail.com

S. K. Roy
Department of Tourism and Hospitality Management, Noakhali Science and Technology University, Noakhali, Bangladesh

A. Hassan
Tourism Consultants Network, The Tourism Society, London, UK

A. Hassan (ed.), *Tourism in Bangladesh: Investment and Development Perspectives*, https://doi.org/10.1007/978-981-16-1858-1_7

Introduction

Tourism is one of the world's major and fastest-growing economic industries, producing employment, exports, and prosperity all over the world. A report of the World Tourism and Travel Council (WTTC) shows that the impacts of the travel and tourism industry in 2017 are the causes for generating 10.4% global GDP and 313 million jobs, 9.9% of total global employment (WTTC, 2017). It is a significant concern for the governments everywhere to ensure a future with quality job opportunities and inclusive growth for their citizens. The tourism and travel industry has already contributed to the movement of employment generation by providing one job in ten jobs worldwide. And in the last decade, it has created one job opportunity out of five jobs globally. To establish the tourism and travel industry as an economy driven tool, the essential part of this arrangement is to deliver the messages to the host people, which will create an understanding of the industry's insights and prospects for the locals. To maintain this growth and establish the industry as a sustainable one for most of the countries, the requirements of investments in various developments of tourism infrastructure and superstructure seem to be significant for a long time.

Tourism institutions contribute significantly to the growth of the tourism industry by making new and increased investments, as well as providing visitors with international standardized services. Alongside the private authorities, the public stakeholders especially the established institutions, are responsible for these developments' operation. The public authorities of a country usually refer to the entities that are backed and financed by the government to meet the demands of its citizens conducting the development tasks. The continuous investment in formulating tourism services remains a crucial factor for the sustainable growth of the industry. Bangladesh has been proving to create domestic tourism demands at an increasing scale in recent years despite its inability to attract foreign tourists in significant numbers. The increasing demands of travel among local people and exploration of new destinations ask for the establishment of travel services and tourism infrastructure across the country. The governmental bodies remain the major responsible stakeholders in dealing with the investment for developing tourism facilities in various destinations. These institutions' direct and indirect involvement also establishes a new tie with the private organizations to boost the development progress. Tourism institutions contribute significantly to growth by making new and increased investments and offering international quality services to both domestic and international visitors. This chapter is designed to lay out the investment and development plans of major publicly owned tourism associations of Bangladesh towards advancing the industry.

Investment and Development in Tourism and Hospitality Industry of Bangladesh: Role of Public Agencies

The vehicles contributing to the development of an economy, the tourism and hospitality industry, are one of them. But in comparison with other sectors, it gains less attention to achieve investment and development facilities. Government, domestic and international development agencies generally contribute to seeking out the possible sources of investment.

Public agencies increase economic and sectoral growth by emerging productive capacity, the productivity of current resources, formulation of tourism plan, improvement of tourism facilities, development of communication network, controlling tourism activities, etc. (Nawaz & Hassan, 2016). The public sector makes tourism investment from governments at the national, local, and supranational levels. Public agencies of a country can manage investment bases, improve foreign exchange earnings, create jobs, and other development (Bodlender, 1984). The Ministry of Civil Aviation and Tourism (MoCAT), Bangladesh Tourism Board (BTB), Bangladesh Parjatan Corporation (BPC), and Biman Bangladesh Airlines Limited are the main public tourism agencies in Bangladesh. The role of these organizations in investment and development opportunities in the tourism and hospitality industry is described in the next section in detail (Fig. 7.1).

Investment and Development Programs of Ministry of Civil Aviation and Tourism

In Bangladesh, MoCAT is a devoted podium of government's public departments working to function internal and international flights and the country's tourism growth. The progression activities of MoCAT are operated in two different wings. Firstly, the investment plan and development in the aviation sector of the country are directly controlled and operated by the ministry itself. On the other hand, the

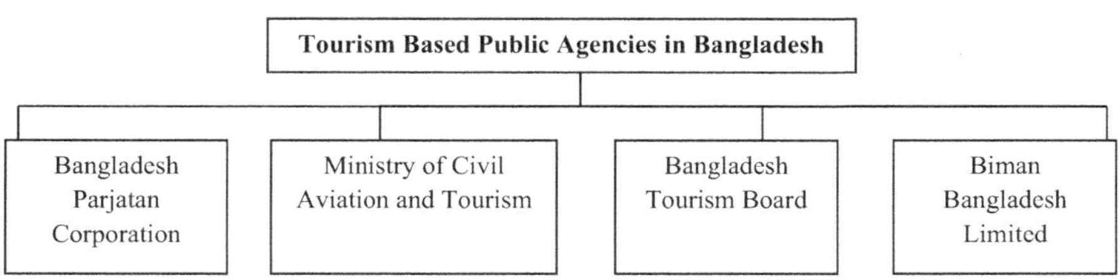

Fig. 7.1 Public organizations in Bangladesh involved in the tourism development. (Source: The authors, 2020)

Fig. 7.2 Project Development Division of MoCAT. (Source: The authors, 2020)

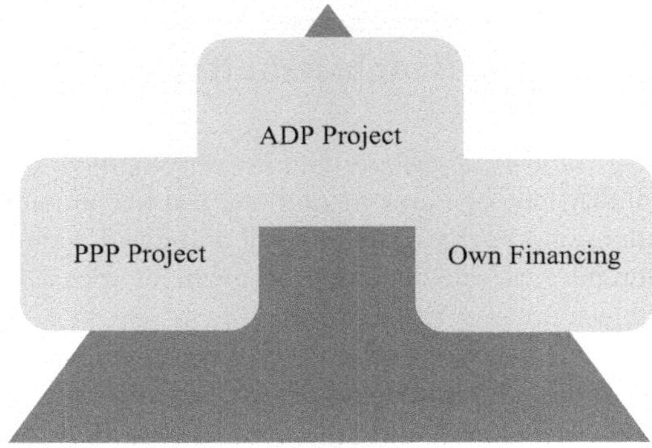

investment plan for tourism expansion is done by the authority and operated by the two governmental organizations named BTB and BPC.

It works on behalf of the People's Republic of Bangladesh to develop civil aviation regulations, improve the country's tourism industry and sponsor Bangladesh as a tourist destination. (MoCAT, 2020). Ministry of Shipping founded the Civil Aviation Division in 1972. In 1977, a new MoCAT was created, and on March 24, 1982, it became a division of the Defence Ministry. It was eventually re-established in 1986 (MoCAT, 2020).

MoCAT is working to build a major aviation pivot and a desirable tourist area. Apart from this, this ministry's mission is to ensure clean, accessible, and dependable aviation facilities to attract domestic and international tourists by facilitating comprehensive services. (MoCAT, 2020) (Fig. 7.2).

Project's Section

The projects which are planned and developed by the MoCAT for the development of tourism are under three sections (MoCAT, 2019). These three sections are running to establish a tourism-friendly environment across the country and promote the destinations to the foreign tourists by MoCAT.

Annual Development Program (ADP) projects facilitate the overall economic growth of the tourism indusrtry in Bangladesh. For example, motel development at tourist destinations such as Cox's Bazar and Chittagong, the establishment of tourism facilities at the Patenga and Parki beaches, run tourism potentiality survey at various places across Bangladesh.

The organization made extensive cooperation with the private sector to enhance the investment opportunity for the tourism industry and already completed planning for many projects jointly, and it is known as Public-Private Partnership (PPP) projects. Some of the examples are the development of existing motels' facilities at Cox's Bazar and Sylhet to turn them to international standards, the establishment of 3-star hotel at Mongla, Khulna, building a new international standard hotel with training facilities in Muzgunni of Khulna, and establishment of tourism facilities in the Payra Seaport area.

The ministry has not yet done any projects under its own financing to boost the tourism industry. Mostly the big part of the budget for this ministry goes into the aviation sector's development. Apart from these three types of developed projects, there are several projects such as the development of Parjatan Bhaban at Sher-e-Bangla Nagar in Dhaka, and renovation of different motels at tourist destinations.

There are several other projects which are already done by the MoCAT (building of Cargo apron at Shah Amanat International Airport in Chittagong, construction of Hazrat Shahjalal International Airport, Kurmitola, consulting services for Syedpur and Barisal Airports, and Sylhet Osmani International Airport development project and its supervision) whereas some of the projects are ongoing till now.

Investment and Development Perspective

Budget remains a crucial factor in establishing facilities for a booming industry. Bangladesh's tourism industry has confronted such a situation where the country could not manage to provide all the tourists' requirements at the destinations. But the scenario has started to change in recent times.

In the last ten years, the revenue of the authority has increased by 132.21%. That means the average annual growth of revenue over the last decade is 13.22%. The development expenditure of the authority has increased by 419.63% in the last ten financial years. In other words, the average annual growth of development expenditure in the last decade is 41.97% (Table 7.1).

The ministry is ensuring safe movement of all civilian aircraft, providing support for commercial development of air transport and services sector, representing of Bangladesh in the International Civil Aviation Organization, supervising all flight safety activities, formulating all rules and regulations related to aviation and transportation, maintaining installation and operation of equipment on-air, controlling the bilateral aircraft contract with various countries, establishing training areas for air traffic service in the airspace of Bangladesh, providing year traffic service and radar service for safe landing, conducting search and rescue operations, providing flight management and clearance, licensing of air traffic; controller pilot; aircrew; aircraft engineering; round crew, issuing of certificate of registration and certificate of airworthiness of aircraft crew, controlling quality and ensuring of aircraft maintenance, controlling all activities of privately owned aircraft, etc.

Investment and Development Programs of Bangladesh Parjatan Corporation

BPC, a branch of Bangladesh's National Tourism Organization, is a one-of-a-kind organization dedicated to customer service. This is a statutory board under MoCAT has been tasked with initiating an online booking system for all commercial tourism

Table 7.1 Comparative picture of the financial performance/success of MoCAT in the last 10 financial years

	Financial success[a]				
		Gives to the Government			
Fiscal year	Revenue income	NTR (Non-Tax Revenue)	DSL (Debt Service Liability)	IT (Income Tax)	Development expenditure
2009–10	642.02	30.00	–	–	121.50
2010–11	653.89	30.00	36.62	–	240.35
2011–12	731.05	35.00	36.62	–	388.28
2012–13	795.21	42.00	36.62	–	235.57
2013–14	1150.29	50.00	43.70	–	459.83
2014–15	1410.32	55.00	47.25	15.00	527.30
2015–16	1504.17	105.00	58.38	200.00	386.53
2016–17	1518.14	120.00	59.23	200.00	473.38
2017–18	1659.65	120.00	58.81	375.57	617.50
2018–19	1620.01	90.00	58.81	251.14	620.4000

Source: MoCAT (2020)
[a]The numbers are in crores of BDTk

and hospitality industry to digitalize tourism products and services. BPC is in charge of the National Hotel and Tourism Training Institute (NHTTI), which aims to improve tourist facilities across the country. Bangladesh is also attempting to achieve the vision of "Bangladesh Tourism 2020" through BPC programs. The National Tourism Organization aspires to use tourism's potential as a contributor to the country's overall development as part of this vision (Sarkar & Begum, 2013; Ishtiaque, 2013). At the end of 2020, BPC actualizes several ambitious projects with the help of foreign investment in hotel industry. BPC is working to regulate and facilitate Bangladesh's tourism industry's development; protect natural and man-made resources; construct infrastructural and superstructural development, and build strong public-private joint management (Hossain & Wadood, 2020).

BPC Initiated Policies for the Development of Tourism in Bangladesh

The government of Bangladesh supports the present policy of BPC. The first policy goal to supplying unique and diversified tourism items. Second, to take advantage of previous investments in airports, national airlines, transportation, and lodging. Third, establish a solid foundation for internal tourism. Fourth, to preserve and conserve the ecosystem. Fifth, to improve the country's universal profile. Sixth, to strengthen and preserve the country's culture. The seventh and final step is to provide tourist details in textbooks and prospectus (BPC, 2020).

BPC's duties and functions are shared with regional and local offices, as well as other tourism stakeholders for the potential investment and development prospects of Bangladesh (Hossain & Wadood, 2020). BPC has been conducting product-market research, agreeing on infrastructure and superstructure development, recruiting eligible human resources, arranging training session, marketing the tourism resources worldwide and choosing the attractions, entertainments, and other amenities to be provided at the resorts, as well as assisting in the selection of suitable organizations. BPC is in charge of foreign publicity and marketing campaigns aimed at branding and promoting the tourism resources of the country which is an important part of "Marketing Functions and Planning". The "Financial Functions and Planning" job is to collect funds from governments and international aid organizations to facilitate the economic growth for the country's crisis moment. The "Control Functions and Planning" duties and activities include: first, licensing and franchising regulations in tourism and hospitality industry; second, initiating balanced pricing strategy; third, incorporating marketing activities with private and public sectors stakeholders; fourth, organizing conference and training session; fifth, advocating socio-economic and environmental issues (Mian, 2014). Since BPC has found out new tourists spot and create to develop new facilities at important tourist spots, innovate various types of new products, and then market the product at home and abroad. Out of its commitment, the organization is striving hard to create more infrastructural facilities to provide the demand for the increasing tourist arrivals. BPC provides their dynamic efforts for the development of tourism by devising new and improved strategies and different kinds of plans to fascinate and attract tourists from around the world. There are several sources to cover up these costs to meet expenditures of development and promotional activities arranged by BPC (Fig. 7.3).

BPC is involved in product creation and offers a variety of tourism services, such as people and lifestyles, music, arts, manmade and natural resources, and so on, for local and international promotion. BPC organizes different fairs and festivals, familiarization tours, cultural events, and food festivals. It has some plans to develop some infrastructure and superstructure to attract more tourists to Bangladesh. The next section will provide some of the development programs run by BPC for the development of the tourism industry in Bangladesh.

Investment and Development Projects for Tourism by BPC

To contribute to economic development and flourish the positive image of Bangladesh abroad and enhance domestic tourism, a comprehensive work plan has been operating for a long time. The investment programs of BPC to develop tourism and other essential elements required to boost up this industry can be classified into different categories (Fig. 7.4).

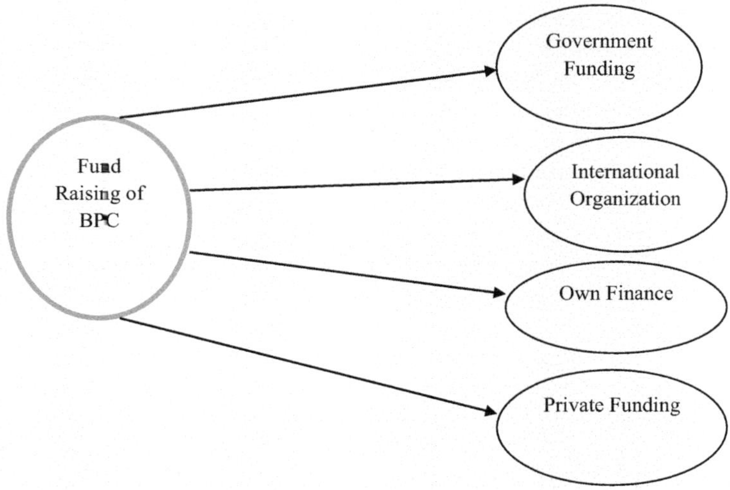

Fig. 7.3 Sources of fundraising by BPC. (Source: The authors, 2020)

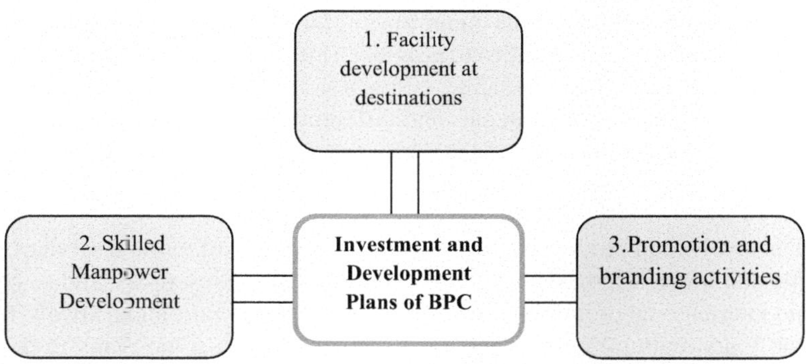

Fig. 7.4 Categorical development plans run by BPC. (Source: The authors, 2020)

Facility Development at Destinations by BPC

From the fiscal years 1995-96 to 2010-2020, BPC has built several tourism facilities. Among these ventures, the third-floor extension at Hotel Abokash at Mohakhali in Dhaka was important. In addition, the construction of the Parjatan Motels in the tourist region is implementing to meet the growing demands of the tourists.

In the previous fiscal years, some of the package projects were included in Annual Development Programs (ADP). A progressive plan was titled "The Development of Tourism Facilities in Different Places of Bangladesh" worth of BDTk 90 lacs are currently being implemented. The construction of two of the packaging project's components (Kantajee and Dinajpur) has already been accomplished. Another package project, titled "Development of Tourism Facilities at Chapainababganj, Kishoreganj, Rajshahi, and Rangpur", is currently being implemented at a cost of BDTk 1435.47 lacs. Under the Chittagong Hill Tracts Development Assistance Program, another package project named "Construction of a new Parjatan Motel at Rangamati" with a budget of BDTk 1140.01 lacs is being implemented. The government has approved another package project named "Construction of a New Parjatan Motel at Motel Shaikat Compound in Chittagong, and Horizontal Extension of Hotel Shaibal in Cox's Bazar" for BDTk 4495.85 lacs (MoCAT, 2020).

Some of the development projects were carried out by BPC and some are still implementing. For example, construction of an international standard multi-purpose tourism center in Tungipara, Gopalganj, WatchTower in Kuakata, Parjatan Bhaban in Sher-e-Bangla Nagar, Bangabandhu Chinta Nibash in Monpura, Bhola, Joy Restaurant's boundary walls in Savar, Jhinuk market in Cox's Bazar, and the beautification and erosion security of the Rangamati hanging bridge.

Repair and renovation projects are also underway in Cox's Bazar, M.L Shalook water vessel, and NHTTI building at Mohakhali. NHTTI building at Mohakhali. Some projects are creating tourism facilities near the Bangabandhu bridge of Sirajganj, the Bhairab bridge of Kishoregonj, and Shindurmati of Lalmonirhat, Panchagar, Nijhumdip of Noakhali, etc. (MoCAT, 2020).

Skilled Manpower Development by BPC

At present, several hotels, malls, restaurants, and other service providers have sprung up in the country's tourist attractions. Also, a significant number of tour operators are created. BPC founded the National Hotel and Tourism Training Institute (NHTTI) that creates skilled manpower for the tourism and hospitality industry. It is important to note that government, United Nations Development Programme and International Labour Organization provides funding to introduce professional training session within the tourism and hospitality industry of Bangladesh (NHTTI, 2020). The diploma course in Hotel and Catering Operations for management trainees is a full-time course that covers a few specializations such as front office management, housekeeping management, food and beverage production, food and beverage service, travel agency and tour operators, etc.

Promotional and Branding Activities by BPC

BPC is attempting to develop a branding strategy that will enable us to project a unique brand image on the global platform. BPC has taken the lead in launching an awareness campaign to boost and sustain the tourism and hospitality industry's Quality Services (QTS) in Bangladesh. The authority has already taken part in different international tourism fairs, and BPC also organized some pre-fair and post-fair activities like nation presentations with cultural and natural value. Recently BPC participated ITB Berlin, COTTM China, FITUR Spain, Tourism fair at Pan Pacific Sonargaon. BPC produces some unique gift items like a traditional boat, jute bag, pen drive, umbrella, coffee mug, product brochure of different destinations, tourist map of Bangladesh and Dhaka city, card-holder, keyring, etc. During the World cup 2011, a TVC, "Beautiful Bangladesh" focusing on the tourist places of Bangladesh, was prepared under the supervision of BPC, and it was broadcasted on different domestic and international TV channels, and about 2 billion people from 180 countries watched this TVC. The logo BPC was printed on the wallet, billboard, display stand, tickets during the world cup. An advertisement was printed promoting Bangladesh as a popular tourist destination in world cup cricket Souvenir published by ICC. A preparatory workshop was organized by BPC on the proposed "Visit Bangladesh Campaign" in the participation of tourism personnel, stakeholders, travel writers, creative agencies, telecom companies, journalists, tourism resource person from government/non-government organizations to gather the ideas, knowledge, experience to find out a smooth and effective way/marketing strategy for bringing Visit Bangladesh campaign into reality and make it a successful one. Every year BPC participates at various international fairs and arranges a stall and many brochures that focus on many historical and cultural places, promote Bangladeshi tourism. Many countries' people are ignorant about the beauty of Bangladesh. So, participating in these types of international fairs, Bangladeshi tourism will be promoted through supplying brochures prepared by BPC. Apart from these, BPC has produced documentaries on national pride (the glorious history of liberation war 1971, international mother language day), natural and manmade resources including forest, lake, river, sea, fountain, culture, heritage, tradition, lifestyle, the celebration of Pohela Boishakh, Mongol Sovazatra Ekushy Boi Mela, the celebration of Eid and Durgapuja broadcasting on various television channels. Artworks, product flyers, leaflets, travel guides, tourist handbooks, posters, and other advertising materials are also produced by the corporation. BPC has taken up the initiative to introduce as many as 150 Youth Clubs in different universities and institutions of the country to motivate the young generation about the prospect of the country's tourism industry. Although tourism promotion preparation has improved significantly in recent years, it still requires a more comprehensive approach. Marketing and promotional work is the responsibility of the government agencies. BPC is maintaining that type of promotional activity very effectively and efficiently.

For promoting Bangladesh BPC needs to do positive branding for this destination. By branding our attractions, it is possible to reach the mind of people easily. BPC introduced its websites (www.bangladehstourism.gov.bd) for the Digital Marketing Campaign. Taking steps for e-vamping, portal site (www.parjatan.gov.bd.com) of Digital Marketing Campaign was built especially for the World Cup Cricket in 2011 (www.visitbangladesh.gov.bd.com) where this portal acts as a gateway (Fig. 7.5).

Again, BPC participates each year in various international fairs. That is why participation in the international fair is considered another promotional tool for

Fig. 7.5 Digital marketing of BPC for tourism promotion. (Source: The authors, 2020)

Bangladeshi tourism. For the purpose of growing awareness among the country's people, BPC participates at various fairs in the country. The main aim is to promote tourism through growing awareness about Bangladeshi tourism. A remarkable domestic fair where BPC participates for knowing Bangladeshi people and growing interest making a tour in the countryside are Asian Tourism Fair (ATF), Bangladesh Travel and Tourism Fair (BTTF), 24th Scout Conference, Power Bangladesh Fair, Bangladesh International Tourism Fair (BITF), Dhaka International Trade Fair, etc.

To increase domestic tourists, BPC participates in these types of Domestic Fairs and supplies free Bangladesh tourists map, Dhaka tourists map, and many other brochures, flyers, and a large number of posters about Bangladeshi tourism. On the other hand, it has introduced some portal sites that are carried out for the promotion of Bangladeshi tourism. Many large Booking Portals in the world are used over the years to implement the promotional activities of BPC. Booking portals are- Trip Adviser, Tourism Social Networking Site Wayn.com, E-bookers.com, Sina.com, Travel Daily News International.

Presently, Bangladesh Parjatan Board (BPC) has been playing its role as the destination marketing and management organization in the public sector also. BPC has taken the following steps and strategies for making and promoting tourism in the country. The Corporation has undertaken various promotional activities for the tourism industry's growth in the country as part of the program adopted by the Government of the People's Republic of Bangladesh to make Bangladesh a middle-income country through the implementation of Vision 2021 and Development Plan 2030. Domestic and foreign tourists and guests are being provided with information about the country's tourism-related travel destinations from the Public Relations Department. Bangladesh's tourism publications are being sent to various embassies on a regular basis. Memoirs, pastors, brochures, and other publications containing information about different tourist centers, including documentaries produced by BPC are sent to Bangladesh Missions abroad (Fig. 7.6).

The country's 5-star hotels, including those in Dhaka, Chattogram, Sylhet, Khulna, and Jessore, are receiving logistical support and materials to provide tourism-related information to domestic and foreign tourists for the development and growth of the tourism industry. As part of the tourism campaign and for BPC Hotel promotion, regular motel advertisements are advertised in various papers, magazines, and memorabilia. Other exhibitions, including Bangladesh International

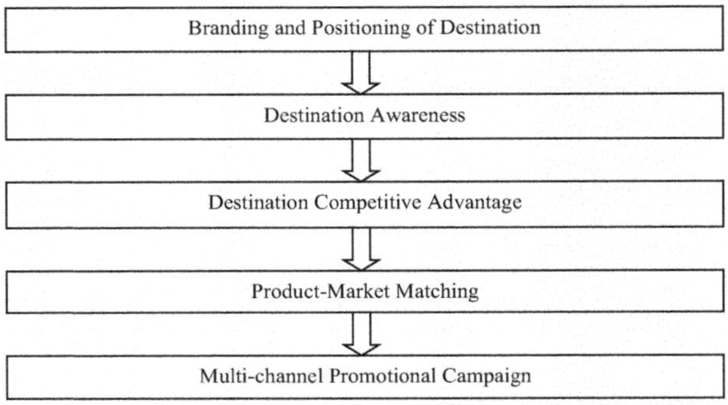

Fig. 7.6 Destination Promotional strategy of BPC. (Source: The authors, 2020)

Tourism Fair and Bangladesh Travel and Tourism Fair are supervised by taking activities and providing overall support to provide necessary promotional materials courtesy. In the intention of the development and growth of the tourism industry and to increase commercial promotion activities, a brochure has been printed on Tourism Motel-Chattogram. Printing of Visit Bangladesh brochure titled "Explore the Culture and Heritage of Bangladesh" has been completed with exciting information about BPC's commercial unit in a new updated format. This is being distributed among the visitors at the fair held at home and abroad.

Apart from these types of promotional strategies, the authority has launched the program of familiarization trips as one of their comprehensive promotional tools known as FAM tours (Fig. 7.7).

BPC has also been engaged with the private sector to establish some dedicated projects for the tourism industry. The tourism village project and international standard tourism zone are continuing at Parjatan Holiday Complex in Cox's Bazar.

BPC has already developed several regional and international organizations around the world. This Corporation is implementing the decisions taken on tourism-related issues, including participation in various meetings and seminars organized by the organization in 2018-2019. BPC is earning a lot of foreign currency by selling goods in foreign currency from duty-free shops. In duty-free shops, foreign brands of cigarettes, alcoholic beverages, cosmetics, and food items including home-made silk, traditional "Jamdani", "Katan", "Tangail" cotton sari, "Nakshi"(design) textiles, brass, bamboo, leather goods, and handicrafts. It is providing services by selling them to the general public.

BPC has a future action plan for tourism development, including the establishment of the Sheikh Hasina Tower at Khurushkul in Cox's Bazar, the formation of tourist restaurants, the establishment of tourist restaurants in different countries of the world to promote Bengali food culture abroad, a feasibility study to facilitate water rides with cable cars at various locations in the hill districts, multi-purpose Movable Tourist Installation Collection, construction of tourist resort in Maheshkhali, the introduction of the wayside tourism facility in every 100 km on the Main Highway from Teknaf to Tetulia and many others.

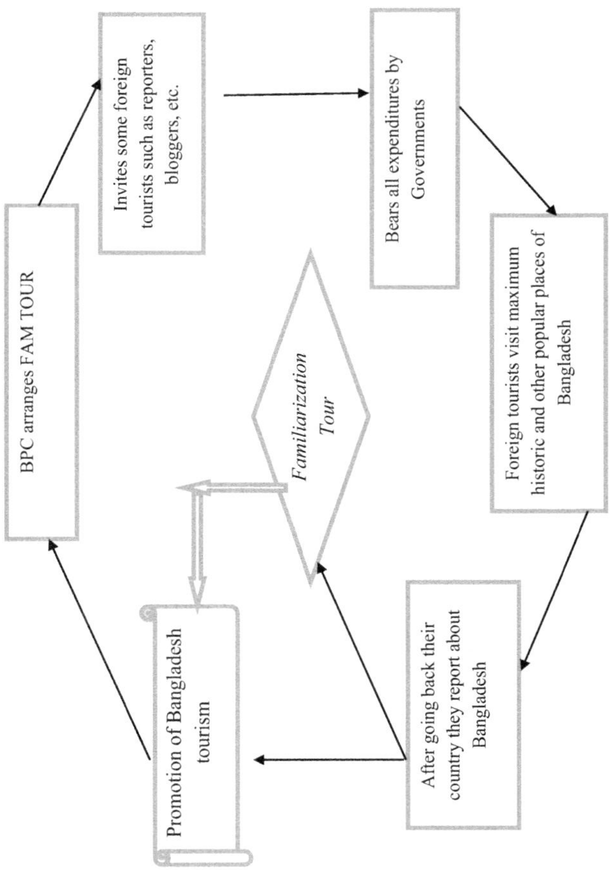

Fig. 7.7 FAM TOUR's (Familiarisation Tour) role in promoting and implementing Plan. (Source: The authors, 2020)

Investment and Development Programs of Bangladesh Tourism Board

Following BTB (2019), the current government founded BTB as the country's National Tourism Organization in response to significant demand from the tourism market and the private sector. This was one of the government's pre-election promises to the country. Over time, it was deemed necessary to establish a National Tourism Organization (NTO) in the country, which would oversee all promotion and development processes. The organization is currently funding its operations with profits raised by its hospitality industry including resorts, hotels, and motels. The Tourism Regulatory Authority of the Government was transformed into a public relations firm with significant private sector involvement. This shift in the tourism management system boosted inbound investment in destination growth and administration (Mamun, 2014). The National Tourism Organization, which is focused on public-private partnerships, has been the government's best practice for tourism management. The overall activities are undertaken and carried out by BTB in the financial year 2018–2019 are mainly focused on creating awareness throughout the entire country and abroad about the importance of the tourism industry and creating a better serving and hospitable industry.

Dedicated Activities of BTB for Tourism Developments in Bangladesh

To celebrate World Tourism Day in Bangladesh and develop awareness among the local citizens of Bangladesh about tourism significance as an essential economic key, BTB each year takes all the programs to celebrate the day in a stylish atmosphere. Colorful rallies are organized with banners, festoons, placards, and horse carts to display and promote the tourism industry. Well-equipped cars with folk music performances are there. BTB has taken the initiative to publish brochures on all the country's district tourist attractions. Branding of designated roads, including important foot over bridges of the city, the publication of particular supplements, discussion meetings, talk shows, food festivals, cultural events, etc. make the whole arrangement successful.

BTB also took part in international tourism fairs and exhibitions for Bangladesh's tourism marketing such as Travel and Tourism Fair (TTF) Kolkata, PATA Travel Mart in Malaysia, Tourism Expo Japan (JATA), International Travel Mart (ITM), Business and Luxury Travel Mart Delhi, (BLTM), International Travel Exhibition, China Outbound Travel and Tourism Mart (COTTM), Arabian Travel Mart (ATM), Seoul International Tourism Industry Fair (SITIF).

Workshops on tourism in development are organized regularly to create general awareness among people about tourism. A day-long workshop titled "Tourism in Development Thought" was organized by Bangladesh Tourism Baird on the 03rd April 2019 at CIRDAP Auditorium in Dhaka and on 25 April 2019 at Hotel Shaibal in Cox's Bazar to take public awareness activities to raise the standard of tourism facilities in Bangladesh. Bangladesh Travel and Tourism Fair (BTTF) – 2019 was organized by Bangladesh Tour Operators Association (TOAB) on the 18th–20th April of 2019 at Bangabandhu International Conference Center (TOAB, 2019). BTB has distributed tourism publications at the fair. Public awareness workshops on tourism were organized in Bogra, Sherpur, Sunamganj, Panchagarh, and Rangamati districts in January 2019 with the help of district administrations, in the presence of notable stakeholders from those districts.

To develop community-based tourism in the hospitality industry of the tourist rich areas of Saturia, Manikganj, training programs were launched in 2017, addressing various hygiene, food safety, quality control, and conservation strategies. In the end, the activities were implemented by introducing homestay facilities from the host community people. Training on Community Based Tourism Training is provided from the 14th–18th June 2019 in Delduar (Tangail), the 21st–22nd June 2019 in Sakop (Khulna), Srimangal (Sylhet), Birishiri (Netrokona), and the 29th–30th June in Paharpur (Naogaon) addressing the responsibilities and duties of the host community, preparation of travel plans for tourists, the security of tourist areas and overall measures for safe travel of tourists, protection of environment and biodiversity and waste management and the concept of sustainable and responsible Tourism.

To create tourism facilities in the country's tourist attractions in the financial year 2018–2019, a total allocation of BDTk 18,64,04,999/– was provided for the physical facilities development and other tourism facilities at 21 spots in 15 districts. With the given funds, various facilities, including water supply, public sitting areas, toilets, etc. will be created throughout the concerned districts' potential tourist areas. This will ensure the comfortable travel of tourists to the respective areas and increase the flow of tourists.

Familiarization of introductory tours has been organized by BTB on the 13th–23rd April 2019. Promising and talented tour operators and tourism journalists were selected through Bangladesh embassies abroad as a strategy to promote Bangladesh's tourism abroad: they were arranged to travel to tourist attractions in Bangladesh. The main purpose of this introductory trip was to steal the interest of the country's tourists about Bangladesh through foreign journalists. Among the Asian countries, 04 from China, 03 from Korea, 03 from Japan, 02 from Thailand, and 03 from Lebanon; total 15 and from Europe, 03 from the UK, 2 from Lenin, 04 from the Netherlands, 01 from Italy, and Belgium took part. A total of 26 tourists, including 11 people and tour operators, participated there.

Other activities for advertising purposes were undertaken such as the release of documentaries based on Buddhist Heritage in Culture and Hindus Heritage in Culture, "Unparallel Dhaka" the documentary about Dhaka city to be shown to OIC countries, taking part in the Dhaka International Trade Fair to interact with the general public, arranging and participating in several public gatherings like Asian Tourism Fair (ATF), Bangladesh International Tourism Fair (BITF), Dhaka Travel Mart (DTM), etc.

Investment and Development Programs of Biman Bangladesh Airlines Limited

According to Biman Bangladesh Airlines Limited (2019), it is a publicly owned organization working for The MoCAT. The authority has not yet established any infrastructural establishment for tourism purposes but performs some regular activities for this industry's general well-being as discussed below:

Major Activities of Biman Bangladesh Airlines Limited for Tourism Promotion of Bangladesh

The company's core business is to provide and expand safe, efficient, and integrated air transport in the heart of the country and abroad. But there are some activities done by the Biman Bangladesh Airlines Limited to participate in tourism directly or indirectly. *Providing tourism services*: One of the major roles of Biman Bangladesh Airlines Limited in contributing to tourism growth is initiated through a dedicated website named "Biman Holidays". This facility from the authority is established to ensure travel arrangements for tourists such as flight tickets, visa services, accommodations, car rental, and finally, trip booking at different destinations of Bangladesh (Biman Holidays, 2020).

Promotion through in-flight magazine: Word of mouth has been recognized as an influential medium in the promotion of the tourism industry. Most of the airlines around the world provide reading materials for their passengers during the traveling period. Bangladesh Biman also offers such materials for all kinds of guests. This magazine is known as "Bihanga," and it includes Bangladesh's popular destinations with their locations.

Training on guest services: As a public entity, the airline does not involve much in dealing with guest requirements promptly before. But there are now new facilities for the recruits to ensure a better guest impression. 68 courses were conducted from July 2018 to June 2019 with 8194 participants with a duration of 1–3 years. Keeping in mind the foreigners' first impressions, the authority regularly conducts seminars and attends a different awareness program by MoCAT to confirm valuable relationships with guests.

Ground handling services: For the guests' comfort, the authority has introduced some dedicated airport services such as arrival and departure services for any flight passengers land in Bangladeshi. There is another facility under the development stage to ensure airport services for medical tourists.

Acquiring air routes for tourists: The airline has already attained some routes formed to transport tourists from Bangladesh. The students who are going for higher studies abroad have availed of this service. Again, the flight routes to Maldives and Thailand are currently operated by Biman Bangladesh Airlines Limited to ensure transportations of staff working in travel service providing organizations and tourists from Bangladesh.

Promotion of local cuisine: Biman Bangladesh Airlines Limited has established its catering facility named Biman Flight Catering Centre (BFCC). This entity's main task is to produce foods for all the flights of Biman and provide catering facilities to the other airlines operated from the Bangladeshi national airport (BFCC, 2020). Here, the cuisine is based on local culture, and it is served to all types of passengers on board. So, the local cuisine can avail an opportunity to get promotion to the foreign people.

Apart from these services, the organization offers different facilities for the passengers during the festivals, which indirectly introduce our culture to the new buyers of travel service from abroad. The authority is also working to establish an accommodation facility, which will be operated by Biman Bangladesh Airlines Limited. All of its activities are now moving towards the confirmation of guests' satisfaction, whether locals or foreigners.

Overall Activities and Significant Investments

Biman Bangladesh Airlines Limited signed three agreements with Baying Company in 2006 for the purchase of 10 new-generation aircraft. Under the agreement and by taking a long-term loan against the Sovereign Guarantee of the Government of Bangladesh, 04 (four) T8-300ER in 2011/2014 and 02 (two) 737–800 aircraft in 2015 have been added. Among the aircraft, 2 (two) in August and November 2018, 01 (one) in July 2019, and 01 (one) in September 2019 will be combined to carry 6–7 millionaire aircraft (MoCAT, 2020).

In 2010, two 737–800 aircraft were procured from GECAS on a long-term eye lease basis, which will remain attached to the aircraft until the first half of 2021. The old F-26 and VC10–30, and A310–300 aircraft have already been phased out to modernize the fleet and increase its capacity. In April 2015, two (two) Ash7-Q400 aircraft were leased from Smart Aviation for a period of five years. On 15 May 2011, 1737–800 aircraft (S2 – AFQ) was added to the fleet, and on 29 June 2019, 1737–800 aircraft (s2 – AEW) was added to the fleet. Apart from this, in 2017, one or more aircraft were added to the fleet on a short-term or less than 12 (or) months basis for the need of transporting Hajj passengers and goods. The aircraft is equipped with a new generation electronic enabling aircraft, the 787 Dreamliner, and is equipped with the ground system hardware software required to conduct IT-related activities on the aircraft. The software is uploaded electronically to Airplane through the ground system, and the airplane system-generated data such as airplane configuration report security lock data, etc., can be downloaded. Work-related to other aircraft is done manually. Microsoft's state-of-the-art mail system Exchange 2016 has been installed, and Biman Bangladesh Airlines Limited has 1100 users till June 2019. As a result, communication with other companies inside and outside the aircraft has become easier and faster. E-documents and E-GP activities have been started. The daily activities of science have been further accelerated with the commencement of e-documents and EGP activities. Bangladesh Airlines Training Centre (BATC)s Faculty of Engineering Approves European Aviation Safety

Agency part-147 As a result of this approval, BATC conducts long-term aircraft maintenance courses that are recognized in Europe and almost all countries of the world. BATC also provides all training (simulators) for all aircraft at BATC, such as B – 737, B – 777 B – 787, Dash (2400). Domestic and international training is done commercially here without the aircraftman force. The training of people of different levels of the Bangladesh Army, Air Force, Navy, and Police is done here on the way to the UN Mission (MoCAT, 2020).

Other exceptional investments and developments are first, August/November, 2018These two 787–8 aircraft have been added to the fleet. Second, a 737–800 aircraft (S2-AEQ) was added to the fleet on the 15th May 2019. Third, on the 29th June 2019 another 737–800 aircraft (S2-AEW) was added to the fleet. Fourth, on July 23, 2019, the third 787–8 aircraft will be handed over to Biman. Fifth, the fourth and final 787–8 aircraft will be handed over to Biman from the company in September 2019. Sixth, BATC is directly involved in the human resource development of the aviation administration. Seventh, update the syllabus and notes by confirming the introductory course of all the officers and staff of the aircraft and analyzing the training needs on time. Eighth, directly participate in the development of various aircraft policies. Ninth, to organize various seminars for the development of aircraft. Finally, organizing seminars for women's participation and empowerment (MoCAT, 2020).

Future Action Plan of Biman Bangladesh Limited

First, to increase the capacity of the organization and to improve the tourism operation in the future. Second, Biman Bangladesh Airlines Limited has released the future action plan to procure cargo freighter aircraft and to establish the second hangar complex and International Aviation Institute. Third, to interline agreements with foreign airlines and with the governments to initiate 4 tier Data Centre (4tDC). This plan also includes developing Android, Ios, and Windows Phone-based apps for booking air tickets. Forth, to upgrade the current data center of the aircraft is underway and to arrange training of staff for quality improvement of work. Fifth, to deduct the number of diversions of the flight and to use automated overall operation management, etc. (MoCAT, 2020).

Conclusion

Bangladesh's government has been conducting its development plans to boost tourism and engage its people in delivering income opportunities. Several activities are operated by the public authorities of Bangladesh to set tourism for community development as well as form the tourism industry as a branding tool for the countries abroad. The establishment of tourism infrastructures such as resorts, hotels along with several promotional campaigns, remains the major activities of these public organizations. Natural and cultural values abound in Bangladesh, bearing witness to our culture, heritage, and tradition. Tourism organizations play a significant role in preserving natural resources and the dissemination of traditional values to the general public. Tourism organizations organize a variety of conferences, events, forums, and training courses regularly for the community's socio-economic development through sustainable tourism, thus bridging the distance among all parties involved in Bangladesh's tourism industry. The holistic development approach has become one of the influential strategies for the Bangladesh government to establish new places as latent tourism destinations. The government has played a crucial position in connecting with private investors and foreign organizations to develop this industry. The nature of the tourism industry remains different from the other traditional industries where the involvement of the government usually shallow. To set a competitive advantage for Bangladesh's destinations, the government needs to be more active in identifying useful plans for investments. The public authorities need to be more focused on finding opportunities to extend investment in this industry, which ultimately covers the government's objectives by ensuring viability among the economy, environment, and society. As a developing nation, it is quite obvious for Bangladesh to enact a newer form of economical driving tool where tourism can play a huge role to provide expected results for its citizens. The government should incorporate all the industry stakeholders to put the investment in place to bring the best output.

References

Bangladesh Parjatan Corporation (BPC). (2020). *Home*. Retrieved from: http://www.parjatan.gov.bd/. Accessed 15 Mar 2020.

Bangladesh Tourism Board (BTB). (2019). *Home*. Retrieved from: http://tourismboard.gov.bd/. Accessed 16 Oct 2019.

Biman Bangladesh Airlines Limited. (2019). *Home*. Retrieved from: https://www.biman-airlines.com/. Accessed 15 Oct 2019.

Biman Flight Catering Centre (BFCC). (2020). *Providing high quality food and services for airlines operating Ex-Dhaka since 1989*. Retrieved from: https://www.biman-airlines.com/service/catering. Accessed 3 June 2020.

Biman Holidays. (2020). *Home*. Retrieved from: https://www.bimanholidays.com/. Accessed 5 June 2020.

Bodlender, J. (1984). Tourism investment for the future. *Tourism Management, 5*(2), 150–152.

Hossain, B., & Wadood, S. N. (2020). Potential unexplored? Tourism and economic growth of Bangladesh. *Journal of Tourismology, 6*(1), 63–77.

Ishtiaque, A. N. A. (2013). Tourism vision 2020: A case of Bangladesh tourism with social emphasis on international tourist arrivals and tourism receipts. *Journal of Business Studies, XXXIV/2*, 13–36.

Mamun, A. (2014). *An empirical analysis on Bangladesh Tourism Board: Important activities and future roadmap to develop tourism in Bangladesh*. Unpublished Master's thesis. The University of Dhaka.

Mian, M. R. (2014). *Responsibilities of BPC and tourism marketing in developing countries: Study on Bangladesh*. Unpublished Master's thesis. The University of Dhaka.

Ministry of Civil Aviation and Tourism (MoCAT). (2019). *Home*. Retrieved from: https://mocat.gov.bd/. Accessed 15 Oct 2019.

Ministry of Civil Aviation and Tourism (MoCAT). (2020). *Annual reports from 2011–2012 to 2018–2019*. Retrieved from: https://mocat.portal.gov.bd/sites/default/files/files/mocat.portal.gov.bd/page/b0c352d4_03b2_45fd_8f2a_d0541fd6ce75/Annual%20Report%202018-19.pdf. Accessed 2 June 2020.

National Hotel & Tourism Training Institute (NHTTI). (2020). *Home*. Retrieved from: https://www.nhtti.org/. Accessed 27 June 2020.

Nawaz, M. A., & Hassan, S. (2016). Investment and tourism: Insights from the literature. *International Journal of Economic Perspectives, 10*(4), 581–590.

Sarkar, M. A. H., & Begum, S. (2013). Marketing strategies for tourism industry sector in Bangladesh: Emphasize on niche market strategy for attracting foreign tourists. *Researchers World, IV/1*, 1–5.

Tour Operators Association of Bangladesh – TOAB. (2019). *Home*. Retrieved from: http://www.toab.org/index.php?action=page&page_id=6. Accessed 18 Oct 2019.

WTTC. (2017). *Annual report of WTTC – 2017*. Retrieved from: https://wttc.org/. Accessed 2 June 2020.

Chapter 8
The Role of Banking and Financial Institutions in the Tourism and Hospitality Industry in Bangladesh

Hasanul Banna, Md Aslam Mia, and Md. Sohel Rana

Abstract The tourism industry has been contributing significantly to the national economy of any country, including Bangladesh. In recent years, beach tourism, historical tourism, natural tourism, medical tourism, apparel tourism, agricultural tourism, education tourism, entertainment tourism, and sports tourism, etc. have emerged as potential sources of income due to the many scenic and natural beauties of Bangladesh. Nevertheless, these emerging tourism niches require special attention and investment from both the public and private stakeholders. To promote the tourism industry globally, it is important to improve pertinent logistics and infrastructure, including banking and financial services. Hence, this chapter reviews and explores the role of banking and financial intermediaries to facilitate and invest in the tourism and hospitality industry so that this industry can flourish rapidly and contribute significantly to the socio-economic development process of Bangladesh. In so doing, secondary sources of data were collected and analyzed in this study. Our findings revealed that banks and financial institutions need to design travelers' friendly products and services. To date, the banking system in Bangladesh has very limited tourism-related products such as credit card facilities. Most of the lending facilities are a consumer and small and medium-sized enterprises (SME) loans. Therefore, proper banking rules and regulations should be put in place to integrate tourism-friendly products and services. Ultimately, the partnership between the bank and the tourism industry will contribute to the promotion of the tourism industry of Bangladesh.

H. Banna (✉)
Ungku Aziz Centre for Development Studies, Faculty of Economics and Administration, Universiti Malaya (UM), Kuala Lumpur, Malaysia
e-mail: banna@um.edu.my

Md A. Mia
School of Management (SOM), Universiti Sains Malaysia (USM), Penang, Malaysia

Md. S. Rana
Department of Business Policy and Strategy, Faculty of Business and Accountancy, Universiti Malaya (UM), Kuala Lumpur, Malaysia

A. Hassan (ed.), *Tourism in Bangladesh: Investment and Development Perspectives*, https://doi.org/10.1007/978-981-16-1858-1_8

Keywords Tourism · Banking and financial institutions · Hospitality · Inbound tourists · Bangladesh

Introduction

Tourism as an industry has an appeal to contribute significantly to the national economy of a developing country like Bangladesh. Tourism is, directly and indirectly, related to the development process and can play a significant role in the micro and macro level of economic functions. The tourism industry involves inland, water and air transportations, hotels, motels, homestays and rest houses, restaurants, street food shops, community-based tourism-friendly entrepreneurial activities, traditional handicrafts production and trading, employment generation, leisure industry, etc. Tourism is an industry in Bangladesh that is very much promising for its geographical placement (Rana et al., 2020). Moreover, Bangladesh has diversified seasons that involve diversified cultural and traditional activities, which are often celebrated in a festive environment. Some natural and historically significant tourism places are scattered from the north to the south and east to the west of Bangladesh.

However, in recent years, beach tourism, historical tourism, natural tourism, medical tourism, apparel tourism, agricultural, education tourism, entertainment, and sports tourism, etc., have been emerging as a potential source of economic contributors to the country. Nevertheless, these emerging tourism niches require special attention and investment from both the public and private stakeholders. To promote these tourism niches globally, it is important to improve pertinent logistics and infrastructure services in the country. For example, the attractive tourism industry requires the construction of modern hotels, improvement, and construction of airports, railroads, convenient roads, and highways, seaports, etc. (Chowdhury et al., 2013). Therefore, this industry seeks spontaneous investment from the public and private industry to improve its facilities to attract tourists from home and abroad.

Besides, tourism industries and tourists require strong financing support from the financial intermediaries. In the case of Bangladesh, both the public and private development actors can work hand in hand to make the tourism industry significantly attractive to local and international tourists. As per the public stakeholders, the government and its concerned bodies and authorities should come up with modern tourism development policies, and execution effort must be put in place to make this industry economically viable. On the other hand, along with the public stakeholders, the private industries, including financial intermediaries must come forward to support this promising industry providing adequate funding facilities so that this industry can bring benefits to the local community (Ahmed, 2000).

The economic growth in Bangladesh has been largely contributed by the export earning of the readymade garments, foreign remittances sent by the Bangladeshi immigrants from different parts of the world, and the agricultural industry as well. Such an economic growth of Bangladesh has largely been geared up by the financial industry of Bangladesh (Financial Express, 2019). The commercial, financial providers in Bangladesh, have been playing an outstanding role in the economic progression by providing investable surplus funds to both the public and private industries. The financial intermediaries, including banks, accumulate idle money for individuals and invest that surplus money mainly in three industries; the manufacturing, agriculture, and service industries (Ministry of Finance, 2010).

The tourism industry is one of the significant shareholders of the service industry. The financial intermediaries and contemporary banking system have the scope to involve themselves in making the tourism industry modern, sophisticated, smart, and convenient for both the industry players and local and international tourists. Moreover, the contemporary banking system maintains a globally recognized and compatible transactional standard that offers a variety of sophisticated financial facilities to its clients during their traveling from one location to another. Moreover, the bank can be a financial partner in promoting the tourism industry of Bangladesh throughout the world by investing more in overall infrastructural development.

In light of the above discussion, this chapter reviews and explores the role of banking and financial intermediaries to facilitate and invest in the tourism and hospitality industry so that it can flourish rapidly and contribute significantly to the socio-economic development process of Bangladesh. Thus, the chapter covers the role of banking and financial institutions in the tourism and hospitality industry highlighting the present situation of banking and financial institutions, banking products and services to promote tourism and hospitality investment and development in Bangladesh.

The remainder of the chapter is as follows: Section "Understanding tourism in Bangladesh: A snapshot" discusses the role of banking and financial institutions in promoting the tourism and hospitality industry. Section "The present situation of banking and financial institutions in Bangladesh" highlights the present overall situation of banking and financial institutions in Bangladesh. Section "Banking products and services to promote tourism and hospitality investment and development" analyzes banking products and services to better understand the services available to the tourists. Section "Conclusion" concludes the study with some policy implications to promote tourism and hospitality in Bangladesh.

Understanding Tourism in Bangladesh: A Snapshot

Over the years, tourism has become one of the fastest-growing industries around the world. This industry has been largely contributing to the national economy of any country by generating income at the micro and macro levels. The tourism industry of any country does not only significantly contribute to its economic progression, but it also upholds a country's brand image by sincerely presenting its cultural tradition, norms, natural resources, history, etc. (Rahman et al., 2019). Before discussing the perceived likely role of banking and financial institutions in the tourism and hospitality industry, it is important to understand the present status of the tourism industry in Bangladesh.

The tourism industry of Bangladesh offers both prospects and challenges. According to the World Bank (2020), the tourism industry of Bangladesh contributed 4.4% of the total national GDP in 2018. Although this share of contribution is not that significant compared to some other countries in the Southeast Asian countries, the government of Bangladesh has been focusing on this industry to be a significant economic contributor in the coming years. For example, the government of Bangladesh has allocated TK 34 billion (around US$ 0.4 billion) to the tourism industry in the fiscal years 2019 and 2020. The allocated money was distributed to the ministry of tourism to develop this industry, including civil aviation and the national air carrier of Bangladesh (Databd, 2020). Such a contribution to reshaping the tourism industry has been reflected quickly as Bangladesh has improved five positions and secured 120th position among 140 countries in the travel and tourism competitiveness report in 2019. Despite this improvement, the ranking is still disappointing as Bangladesh is only ahead of Pakistan in the south Asian region. On top of that, 97% of the earning from this industry comes from domestic tourists, which is not vibrant when it comes to attracting foreign visitors to Bangladesh (Databd, 2020).

The neighboring countries of Bangladesh like Sri Lanka, Nepal, India, Malaysia, Thailand, Indonesia, the Philippines, etc., have been performing significantly in promoting tourism. For example, it was estimated that there were 115 million tourists have visited those countries from around the world. In 2018, southeast Asian countries aggregately earned US$ 151.9 billion, whereas South Asia only earned US$ 39.4 billion as revenue from the spending of the tourists (Databd, 2020). The difference between both the region is so large. However, in recent years, both South Asia and the East Asian regions have experienced significant growth of 8% and 10%, respectively (Databd, 2020). In comparison to those countries, Bangladesh could manage to grab only a very small share of tourism dominance in these regions (World Economic Forum, 2019). According to the World Travel and Tourism Council (WTTC) (2020), the majority of the tourists travel to Bangladesh from India, which is accounted for 49% of the total visitors in Bangladesh. Besides India, the remaining 51% of visitors travel to Bangladesh from China, Myanmar, Pakistan, the United States, and the rest of the world. The share of inbound travel to Bangladesh is presented in the following Fig. 8.1.

Such a poor performance of inbound tourists' arrival in Bangladesh is attributed to many limitations and challenges. The tourism industry in Bangladesh requires special attention and necessary funding facilities to overcome these challenges so that the tourism industry and tourists of all classes can get benefits, and this industry can significantly contribute to the development process of the country. These challenges are briefly discussed in the following paragraphs.

Safety has always been a long concern for domestic and international tourists in Bangladesh. Consequently, some of the developed countries keep on imposing travel restrictions in Bangladesh from time to time. For example, the US department of state has declared a warning against traveling to the southeast parts of Bangladesh,

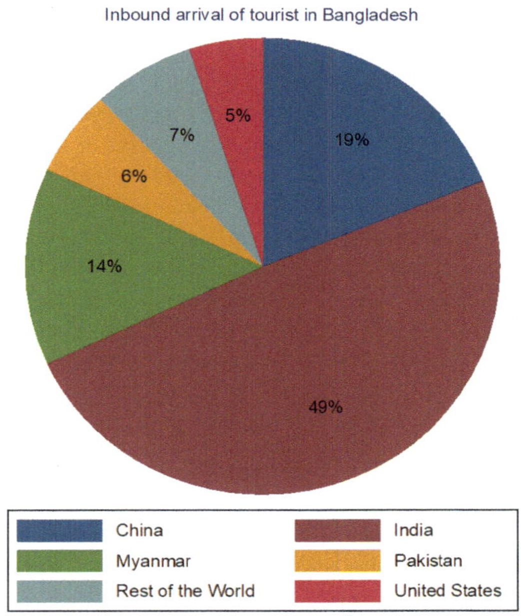

Fig. 8.1 Inbound Arrival of tourist in Bangladesh, 2018. (Source: Authors' compilation from World Travel and Tourism Council, 2020)

mentioning some reasons as crime, terrorism, kidnapping, etc. (US Department of State, 2020). Recently Bangladesh has significantly improved its safety and security index and could manage to secure the 105th rank, which is still considered as low (World Economic Forum, 2019). In such circumstance, the travel industry needs to take its measures to keep domestic and international travelers safe. To ensure security and safety, it requires specific investment by the industry stakeholders. Thus, they may require funding support from commercial financial providers.

The tourism industry in Bangladesh lacks proper infrastructural facilities. Bangladesh ranks 109th in the world as per the world's tourist infrastructure. Bangladesh is behind all other Asian countries except Nepal when it comes to tourism infrastructures. Moreover, Bangladesh is ranked 133 out of 140 countries in respect to tourists' service infrastructure, which is the worst position among Asian countries. The roads and highways, local airliners, transportations, hotels, restaurants, and other entertaining amenities are not widely present in the country.

The above challenges and limitations are not unknown to any stakeholders. The government, industry players, national and international investors, financers can play their part to develop this industry cumulatively. However, in such circumstances, the banks and other financial intermediaries can come up with diversified financial products and services that can meet the demands of the industry players and national and international tourists. The role of banking and financial institutions in Bangladesh can certainly help meet the current disadvantages encountered by the tourism industry and further improve the tourism and hospitality industry by integrating and providing diversified financial products and services. However, banks in

Bangladesh can introduce popular travel and tourism products that have been utilized globally. Moreover, some of the products like deposit schemes, especially for traveling purposes, where the bank can provide some extra benefits to the depositors. SME loans schemes are already very much popular in Bangladesh. Hence, tourism entrepreneurs can take benefits of those types of loans.

Moreover, financial institutions can play a significant role by designing tourism lending packages or bank guarantees for constructing and renovating hotels in promising tourist locations. The role of banking towards tourism development can further be extended by establishing tourism banks, which will develop and create new avenues in tourism businesses. In South Asia, Nepal is the first country to establish a tourism bank in 2010 to help develop the tourism industry. The bank can play a significant role in issuing travelers' cheques or travelers' credit cards, which will inspire many tourists to travel countrywide. In the meantime, travel and tourist insurance can be introduced by insurance companies for better security of the travelers.

The Present Situation of Banking and Financial Institutions in Bangladesh

This section outlines the overall banking situation and services in Bangladesh and makes a thorough comparison with some of its neighboring countries. This helps to examine the financial position of Bangladesh among other neighboring countries.

Since her independence in December 1971, the financial industry, especially the banking industry of Bangladesh, started with just 18 banks, including three specialized state-owned banks, six nationalized commercial banks, and nine foreign banks (Bangladesh Bank, 2019). The financial industry extended during the 1980s due to the dynamic entrance of private commercial banks. Currently, Bangladesh has two types of banks, for example (i) Scheduled Banks: those banks which get a permit to work under Bangladesh Bank Order, 1972 and Bank Company Act, 1991 (Amended up to 2013) are categorized as Scheduled Banks; (ii) Non- Scheduled Banks: the banks of this category have been established for specific and explicit goals and operated under Operating Act to meet up those objectives (Bangladesh Bank, 2019).

Total 59 scheduled banks are operating in Bangladesh under full-fledged supervision of Bangladesh Bank (BB) following the above two Acts (Bangladesh Bank, 2019) till now. Notably, scheduled banks are classified as the following four types based on ownership structure, namely: (a) State-Owned Commercial Banks (SCBs), (b) State-Owned Development Financial Institutions (DFIs), (c) Private Commercial Banks (PCBs), and (d) Foreign Commercial Banks (FCBs). Based on the operational mood (Conventional vs Islamic), Bangladesh has three types of banks: full-fledged conventional banks, full-fledged Islamic banks, and dual operational banks (both Islamic and conventional bank).

As the banking industry of Bangladesh is emerging rapidly, BB has taken various measures in terms of amendment and adaptation of various prudential policies and regulatory measures from time to time to maintain the stability and sustainability of the industry. Some of the recent prudential policies and regulatory measures are taken by the BB as follows: adaptation of guidelines for Internal Credit Risk Rating System, re-fixation of educational qualification of CEO, policy for Off-shore banking operation of banks, policy for investment in long term infrastructure projects and policy for investment in non-listed securities by banks, amendments of loan/investment write-off policy, loan classification and provisioning policy, policy to provide the incentive to good borrowers, risk management guidelines for banks and regulation regarding investment in the capital market (Bangladesh Bank, 2019). Besides, a special monitoring team is trying their best to minimize the overall non-performing loans (NPLs) of the industry by on-site inspection and supervision. With the continual effort and dedication of the regulatory and supervisory team of the BB, the banking industry shows tremendous growth over the past few years. The banking system structure, assets, and deposits of various ownership types can be seen in the following Table 8.1.

The number of bank branches increased to 10,286 at the end of December 2018 from 9,955 of December 2017, which is depicted in Table 8.1. In 2018, the PCBs held the highest percentage share of total assets (67%) and deposits (66%) of the overall industry, followed by the SCBs. The overall assets and deposits observed an increasing trend in 2018 as compared to 2017.

Deposits (74.1% in 2018) are the main sources of funds, and loans and advances (63.3% in 2018) are the main sources of total assets of the banking industry in Bangladesh (Bangladesh Bank, 2019). According to the BB, the scheduled banks in Bangladesh mainly offer two types of deposits, namely savings deposits and fixed deposits. On the other hand, banks provide various types of loans and advances to different industries such as loan to the agricultural industry, term loan and working capital to large, medium, and small-scale industry, loan to the export-import industry, trade financing loans, housing loans, consumer credit, and others. Despite having the importance, banks rarely provide the loan classification to the tourism and hospitality industry.

Table 8.1 Banking system structure, assets, and deposits (in billion BDT) in 2018

Bank types	Number of banks	Number of branches	Total assets	Share in industry assets (in percent)	Deposits	Share deposits (in percent)
SCBs	06	3746	3732.2	25.6	2868.4	25.6
DFIs	03	1412	324.0	2.2	286.0	2.6
PCBs	41	5060	9769.7	67.0	7127.2	66.0
FCBs	09	68	747.1	5.2	517.2	4.8
Total(2018)	59	10,286	14572.9	100.0	10798.7	100.0
Total(2017)	57	9955	13059.3	100.0	9874.9	100.0

Source: Bangladesh Bank (2019)

The ratio of gross NPLs to total loans and advances is the most important indicator of demonstrating the asset quality. Figure 8.1 shows the trend of the NPL ratio of the banking industry of Bangladesh from 2008 to June 2019. Figure 8.2 shows that the overall NPL ratio in June 2019 is 11.7% and SCBs had the highest (31.6%), and FCBs had the lowest (5.5%). However, the overall NPL ratio indicates a mixed trend in the banking industry during 2011–2019. The ratio started to decline from 2008 until 2011 then sharply increased in 2012, which is mainly due to the new loan classification (Bangladesh Bank, 2019). From 2013, the fluctuating drift was witnessed, and the highest peak was observed in June 2019.

The main reasons behind the low asset quality of the banking industry, more specifically the SCBs and DFIs, are the poor assessment, inadequate follow-up and supervision of the disbursed loans, poor quality of the collaterals, shadow banking practices, approval of new banks through political influence without considering the size of the economy and other concerning factors. Moreover, the structural inefficiencies and loopholes of the judicial system for the disposal of cases also significantly contributes to the overall NPL (The Dhaka Tribune, 2018). However, BB has taken several policy initiatives such as restructuring, rescheduling, recovery, one-time exit, and write-off to reduce the NPLs (The Bangladesh Bank, 2019).

The profitability of the banking industry in Bangladesh in terms of return on assets (ROA) and return on equity (ROE) indicated a low performance of the industry. In both measures, the SCBs and DFIs showed the lowest performers, and FCBs showed the best performers among the industry. Following Bangladesh Bank (2019) in June 2019, the ROA (in percent) of the different types of banks are −0.81 (SCBs), −2.68 (DFIs), 0.65(PCBs), 2.61 (FCBs) and 0.30 (Overall) as well as −16.57(SCBs),

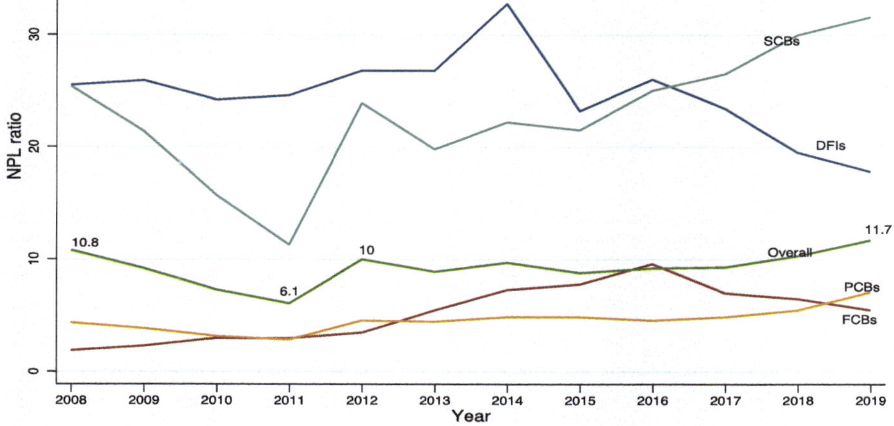

Fig. 8.2 The NPL ratio trend of the Banking industry in Bangladesh (2008–2019). (Source: authors' compilation based on secondary data)

−14.22(DFIs), 9.54(PCBs), 13.41(FCBs) and 4.68 (Overall). The capital adequacy ratio of banks is also important to look at as it highlights the bank's overall capital position, which can signal the possible losses a bank might incur due to credit, market, and operational risks during its normal business operational period (The Bangladesh Bank, 2019). Under Basel-III, banks in Bangladesh are required to have at least 10% Capital to Risk-Weighted Assets Ratio (CRAR). However, in 2018, DFIs (−31.74%) failed to meet the minimum requirement, whereas the SCBs (10.34%) just touched the mark-up. On the other hand, FCBs and PCBs maintained 25.92% and 12.80%, respectively in 2018.

The comparative banking performance of neighboring countries shows that Bangladesh is still lagged many neighboring countries like India, Pakistan, Sri Lanka, Nepal, and Bhutan in terms of financial stability (Z-score), profitability (ROA and ROE), and asset quality (NPL). Figure 8.2 shows that Nepal showed a better performance in terms of Z-score, ROA, ROE, and NPL based on 2017 data, whereas in most cases, Bangladesh showed a lower performance. It is noticeable that Nepal is a tourism-based country, and the highest economic contribution comes from the tourism industry (10.3% of total GDP contribution in 2018) (World Travel and Tourism Council, 2020). This could be one of the main reasons for better banking performance.

The banking service in terms of credit and debit card facilities to the consumers among the neighboring countries in Fig. 8.3 suggests that in 2017, banks in Sri Lanka provided credit cards to around 5% of total adult, and banks in Bangladesh provided credit card to less than 1% of total adults. On the other hand, banks in

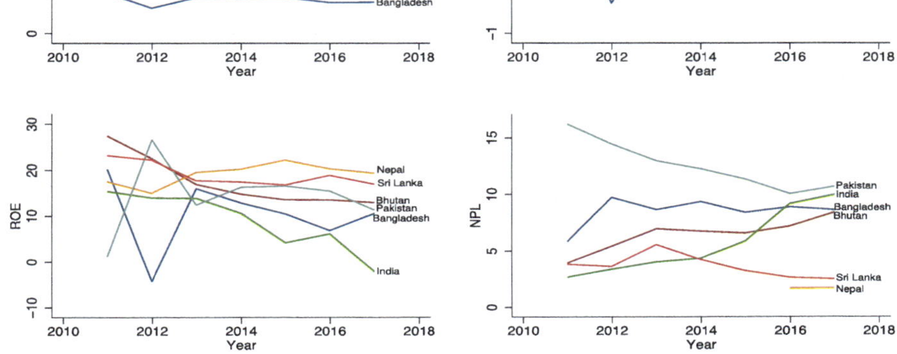

Fig. 8.3 The trend of banking performance in neighboring countries (2011–2017). (Source: World Bank, 2020)

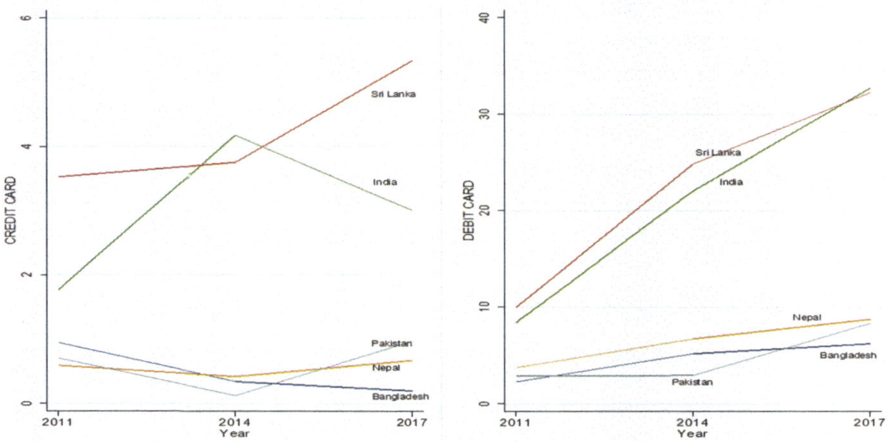

Fig. 8.4 The credit and debit card facilities to consumers among neighboring countries (2011–2017). (Source: World Bank, 2020)

India provided debit cards to 32% of the total adults, whereas banks in Bangladesh so far provided the debit card to around 6% of the total adults in 2017. Thus, banks in Bangladesh are still lagged to provide facilities and services to the consumers as compared to neighboring countries (Fig. 8.4).

Banking Products and Services to Promote Tourism and Hospitality Investment and Development

Banking institutions need to offer products and services which are accessible to all classes of people to promote tourism and hospitality investment and development in Bangladesh. More specifically, ATMs, bank branches, point of sales (POS), credit cards, debit cards should be available near to the tourist-spots and/or tourism industry so that people can have enough access to withdraw money or can get sufficient loans from banks to run their tour or tourist-based activities. Figure 8.5 shows the trend of ATMs and bank branches per 100,000 adults in Bangladesh for the period of 2008–2017.

The Figure suggests that banks in Bangladesh are dedicated to serving its clients by providing better access through increasing the number of ATMs and bank branches, which can be seen by the upward trend of ATMs (in 2008–0.81 and 2017–8.07) and bank branches (in 2008–6.97 and 2017–8.58) per 100,000 adults during 2008–2017. This suggests that the increasing number of ATMs will ease the daily transactions of the clients, and the increasing bank branches will help the entrepreneurs to apply credit/loan facilities easily to operate their businesses.

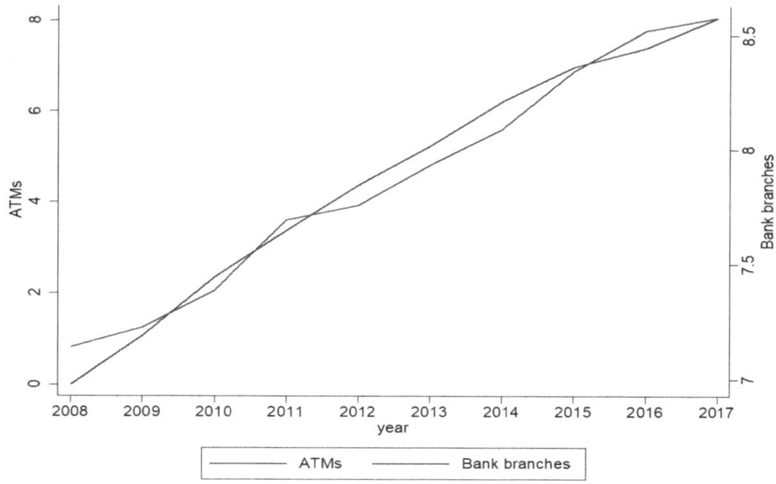

Fig. 8.5 The trend of access to ATMs and Bank Branches per 100,000 adults (2008–2017). (Source: World Bank, 2020)

It is quite difficult to see the available products and services offered by the banks in Bangladesh as banks do not specify the exact products and services for the tourists and the tourism industry as well. However, recently, many banks (mostly PCBs) in Bangladesh are offering an equated monthly installment (EMI) services with a partnership with some of the airline companies (such as Biman Bangladesh, US-Bangla, NOVOAIR, Regent) and/or hotels with up to 0% (3–12 months) interest using credit card services. Banks mainly offer this EMI to promote domestic tourism as well as foreign tourism. The following Table 8.2 shows the facilities provided by the major banks in Bangladesh to the tourists and/or tourism industry.

Conclusion

The tourism industry does not only contribute to the country's overall economy but also upholds the country's inclusive image through the presentation of sophisticated tourism environment to the visitors. Although the tourism industry in Bangladesh has been going through some challenges, it has a promising future. In recent years, the economic condition of Bangladesh and the tourism-related indexes are getting better, which pave the way for public and private industries to concentrate on this industry closely. The meticulous discussion of the banking and financial intermediaries and tourism industry in Bangladesh has exposed many spectrums. However, the above discussion suggests that banking intervention in the tourism industry has been inspired in many countries. Moreover, since tourism is a labor-intensive industry, a bank can play a significant role in the income generation of the people involved

Table 8.2 The existing facilities provided by the major banks in Bangladesh to the tourists and/or tourism industry

Bank name	Loan facilities to the Tourism industry (SME loans)	Tourist (Travelcard)		EMI with airlines and hotels	
		Credit	Debit	Domestic	International
Bank Asia Limited	No (Yes)	No	Yes	Yes	Yes
City Bank Limited	No (Yes)	Yes	No	Yes	Yes
Dhaka Bank Limited	No (Yes)	No	Yes	Yes	Yes
Eastern Bank Limited	No (Yes)	No	No	Yes	Yes
BRAC Bank Limited	No (Yes)	No	No	Yes	Yes
NCC Bank Limited	No (Yes)	No	No	Yes	Yes
South East Bank Limited	No (Yes)	No	Yes	Yes	Yes
Dutch Bangla Bank Limited	No (Yes)	No	No	Yes	Yes
Jamuna Bank Limited	No (Yes)	No	No	Yes	Yes
Midland Bank	No (Yes)	No	No	Yes	Yes
Meghna Bank	No (Yes)	No	No	Yes	Yes
United Commercial Bank Limited	No (Yes)	No	No	Yes	Yes
Mutual Trust Bank	No (Yes)	No	No	Yes	Yes
NRBCL	No (Yes)	No	No	Yes	Yes
Lanka Bangla Finance Limited	No (Yes)	No	No	Yes	Yes
Standard Chartered Bank	No (No)	No	No	Yes	Yes

Source: authors' compilation from the respective website of banks and airlines
Note: The considered banks are mostly PCBs or FCBs. No SCBs or DFIs offered any services for the tourists and/or tourism industry in terms of EMI or travel cards. Data has been collected from individual bank's website and US-Bangla and Biman Bangladesh Airline's websites

directly or indirectly with this industry by providing financial facilities to the tourism industry. However, the policy can be recommended both at the individual (tourists) and tourism industry level.

At the individual level, a bank should introduce deposit schemes, consumer travel credit facilities via travelers' cheques or travelers' credit cards, etc., to promote domestic tourism. In this regard, banks can also promote tourism-related financial products and services among their clients to inspire traveling from time to time. The travel makers, tourism agents, banks, and insurance companies can work jointly to ensure more financial guarantees during their visits to tourism locations. The credit card's EMI can be extended up to the maximum duration so that the tourists do not feel pressure to repay. Like Nepal, Bangladesh can take the initiative to introduce tourism banks to attract more foreign tourists. Moreover, banks also can offer funds, especially to the students who are inclined to continue higher studies in tourism management in the universities to build skillful human resources related to the industry.

In the tourism industry, banks can approve loans to develop tourism-related businesses. Moreover, tourism infrastructures require investment. Banks and financial intermediaries should come forward to investing in tourism development projects. Many tourism business enterprises often require working capital that can be supplied by banks and financial service providers via SME loans. Moreover, banks should invest in purchasing tourism-related tools, mechanisms, vehicles, etc.

Finally, banks and financial institutions need to design traveler friendly products and services. To date, the banking system in Bangladesh has very limited tourism-related products, such as credit card facility. Most of the lending facilities are consumer and SME loans. Therefore, proper banking rules and regulations should be taken to integrate more tourism-friendly products and services. The partnership between the bank and the tourism industry has a significant potential to contribute to the overall development of the tourism industry in Bangladesh.

References

Ahmed, M. (2000). Promoting public-private partnership in health and education: The case of Bangladesh. Paper presented at the *Seminar on Public-Private Partnerships in Health*. Ayuthya: ADB Institute and the National Economic and Social Development Board.

Bangladesh Bank. (2019). *Annual Report of Bangladesh Bank* (BB). Retrieved from: https://www.bb.org.bd/pub/annual/anreport/ar1819/full_2018_2019.pdf. Accessed 26 Dec 2020.

Chowdhury, T., Fahim, S. T., & Dooty, E. N. (2013). Promoting public private partnership for development of tourism sector of Bangladesh: An exploratory study. *European Journal of Business and Management, 5*, 194–202.

Databd. (2020). *Tourism: A possible new driver for the economy of Bangladesh*. Retrieved from: https://databd.co/stories/tourism-a-possible-new-driver-for-the-economy-of-bangladesh. Accessed 28 Dec 2020.

Dhaka Tribune. (2018). *A fair warning*. Retrieved from: http://www.dhakatribune.com/business/2018/04/09/a-fair-warning/. Accessed 26 Dec 2020.

Financial Express. (2019). *Economic growth in Bangladesh and the role of banking sector*. Retrieved from: https://www.thefinancialexpress.com.bd/views/views/economic-growth-in-bangladesh-and-the-role-of-banking-sector-1547220114. Accessed 27 Dec 2020.

Ministry of Finance (MoF). (2010). *Bangladesh economic review*. MoF.

Rahman, M., Rana, M. S., Hoque, M. N., & Rahman, M. K. (2019). Brand perception of halal tourism services and satisfaction: The mediating role of tourists' attitudes. *International Journal of Tourism Sciences, 19*(1), 18–37.

Rana, M. S., Rahman, M. K., Islam, M. F., & Hassan, A. (2020). Globalization effects on tourism marketing in Bangladesh. In A. Hassan (Ed.), *Tourism marketing in Bangladesh* (pp. 157–171). Routledge.

US Department of State. (2020). *Bangladesh travel advisory*. Retrieved from: https://travel.state.gov/content/travel/en/traveladvisories/traveladvisories/bangladesh-travel-advisory.html. Accessed 26 Dec 2020.

World Bank (WB). (2020). *International tourism, number of arrivals – East Asia & Pacific*. Retrieved from: https://data.worldbank.org/indicator/ST.INT.ARVL?end=2017&locations=Z4&start=2017&type=shaded&view=map&year=2017. Accessed 28 Dec 2020.

World Economic Forum (WEF). (2019). *The travel & tourism competitiveness report 2019*. Retrieved from: http://www3.weforum.org/docs/WEF_TTCR_2019.pdf. Accessed 28 Dec 2020.

World Travel and Tourism Council (WTTC). (2020) *Economic impact reports*. Retrieved from: https://wttc.org/Research/Economic-Impact. Accessed 28 Dec 2020.

Part IV
Products and Services

Chapter 9
Development and Investment in Core Niche Tourism Products and Services in Bangladesh

Muhammad Khalilur Rahman, Md. Sohel Rana, and Azizul Hassan

Abstract Travel and tourism have become a potential instrument in Gross Domestic Product (GDP). Bangladesh has bright prospects of niche tourism products and services in the future. But, the country's tourism development is facing challenges, including marketing policy, strategy, and investment. The government of the country may apply a niche market strategy for tourism products and services. Bangladesh can earn vast foreign currency as the country has several key tourism products such as historical places, archaeological sites, hills, islands, beaches, and forest which is most attractive to the tourists. Also, the country has tourism service facilities like travel agents, tourist guides, tour operators, hotels, restaurants, catering, visas, and immigration services, which are designed to meet tourists' demands and satisfaction. From the theoretical and statistical data sources, this study mainly outlines the development and investment scenario of core niche tourism products and services, including the detailed analysis of apparel tourism, medical tourism, beach tourism, agri-tourism, and ecotourism in Bangladesh. This study uses a review of a literature-based method that relied on publish and online resources for data and information. Findings indicate that some products and services can play significant roles in generating revenue in the tourism industry. This research suggests how niche tourism products will assist policymakers in developing the tourism market and customizing their service and investment to achieve strategic marketing objectives.

Keywords Niche tourism · Products · Services · Development · Investment · Bangladesh

M. K. Rahman (✉)
Faculty of Entrepreneurship and Business, Universiti Malaysia Kelantan,
Pengkalan Chepa, Malaysia
e-mail: mohdkhalilur@gmail.com

Md. S. Rana
Department of Business Policy and Strategy, Faculty of Business and Accountancy,
Universiti Malaya (UM), Kuala Lumpur, Malaysia

A. Hassan
Tourism Consultants Network, The Tourism Society, London, UK

A. Hassan (ed.), *Tourism in Bangladesh: Investment and Development Perspectives*, https://doi.org/10.1007/978-981-16-1858-1_9

Introduction

Bangladesh is a land of beauty and wonder. Geographically, Bangladesh is gifted with exquisite natural resources. Seven hundred rivers are occupying a total area of 24,140-kilometer crisscrossed Bangladesh from one corner to another (Banglapedia, 2020). The country is also known as a land of rivers that creates huge silt making this country's land allure for the production of any crops. Bangladesh is a delta of 148,460 square kilometers situated in South Asia (Central Intelligent Agency, 2020), where the southern part of this country is linked to the Bay of Bengal. Bangladesh contains a total land area of 130,170 square kilometers, a total water area of 18,290 square kilometers, and a total coastline of 580 kilometers (World Atlas, 2020). The northern part of Bangladesh is situated beneath the Himalayas. The scenic view of Bangladesh is accentuated by mountains, rivers, plains farm-land, sea, forest, and lakes in different parts of the country. The country is not just diverse geographically; the climate of Bangladesh is also diverse with six seasons. The country has a social custom, the hustle and bustle, pastoral livelihood, and peaceful mangrove forest, which are the potential to attract any tourists significantly from home and away. The country poses huge economic importance around the world. Consistent economic progress for more than a decade has highlighted Bangladesh on the global stage. The contribution of the manufacturing and service sector has geared up the economic development of Bangladesh over the years. The emergence and rapid expansion of the apparel sector in Bangladesh have ushered the economic progression in many folds, including employment generation, female intervention in the labor force, and uplifting standard of lives of many engaged with this sector.

The tourism industry in Bangladesh has been taken seriously by the government of Bangladesh in recent years. Since 2011, the tourism board of Bangladesh has launched "Beautiful Bangladesh" campaign to promote the country as a potential destination of tourism (Sarker, 2014). In the meantime, the government has also taken long term master plan to develop the tourism industry by preparing the tourism-friendly infrastructure gradually (The Financial Express, 2020). However, it is crucial to develop a comprehensive marketing strategy for the tourism industry in Bangladesh along with the implementation of the long-term policy. Tourism is a service industry that has huge potential to contribute to the national economy of Bangladesh. The development of the tourism service industry in Bangladesh carries immense significance since this industry is quite capable of generating direct and indirect employment, small scale entrepreneurship, and earn huge foreign currency, which would ultimately support Bangladesh to efficiently maintain its current account balance. Hence, it is time for the concerned authority, tourism stakeholders, researchers, and scholars to identify core niche tourism products and services in Bangladesh to promote the tourism industry on the global stage. However, the promotion of these identified core niche tourism products requires a comprehensive marketing strategy and active showcasing of complete tourism packages through the globally recognized channels. The examples for the development and investment of niche tourism products and services can be observed in many parts of the world. For example, in India, niche tourism products are identified as the cultural, heritage, and adventure tourism that covers diversified topology and geopolitical heritage. Apart from these, medical and wellness tourism, rural tourism, ecotourism, and wildlife tourism are being emphasized in the recent year (Travelbizmonitor, 2011). The niche tourism products in the UK are mainly the outdoor pursuit, nature-based tourism, cultural tours, screen tourism, and many other travel experiences. In Northern Ireland, it is limited to golf tourism and craft tourism (NIbusinessinfo, 2020).

The development and investment in core niche tourism products are dependent on the interest of the travelers. Tourists often grow their interest to travel a destination after analyzing the niche tourism products of any country to satisfy his or her needs. Therefore, to fill in the literature gap, this book chapter endeavors to identify core niche tourism product development and their prospects to contribute largely to the country's economy. In the subsequent section, this chapter addresses the detailed analysis of apparel tourism, medical tourism, beach tourism, agri-tourism, and ecotourism.

Apparel Tourism

Apparel means clothes, particularly formal clothes worn on an important occasion. Apparel tourism is a type of tourism when people travel to enjoy, experiment, and use clothes and fashion. The apparel industry makes ordinary clothes or mass fashion. Fashion is the style of clothing and accessories used by groups of people (Ahmed et al., 2015). Apparel tourism includes the design of clothes, designer fashion to ordinary everyday clothing, manufacturing, retailing, advertising, marketing, distribution, and promotion of all types of apparel for men, women, and children. The apparel industry is a global industry that makes and sells clothes (Buckley, 2003). The global apparel industry is facing increasing criticism related to social sustainability issues such as labor-right violations and human rights abuse in apparel manufacturing factories (Akbar and Ahsan, 2020). The apparel industry can promote the development of fashion tourism and generate economic growth. The apparel industry in Bangladesh is famous in its neighboring countries such as India, Sri Lanka, Pakistan, and Nepal. Islam et al. (2012) reported that Bangladesh has more than 5150 apparel factories, which promotes the economic growth of the country since its independence in 1971.

Apparel tourism is a new model of the tourism industry, and travelers can choose shopping for the experience of local culture, local products, and crafts. Some tourism destinations provide particular apparel or fashion clothing for tourist shopping activities. Apparel tourism is a niche market segment within the shopping-clothes that can lead to the economic growth of the country (Akbar and Ahsan, 2020). Individual brand of fashion clothes also plays a crucial role in apparel tourism marketing. The apparel industry uses sponsored athletes and specialist lifestyle entertainment media to sell clothing and accessories (Islam and Stringer, 2018). Bangladesh's apparel industry has substantial economic growth (Islam and Stringer, 2018). However, the country currently is facing numerous challenges due to the outbreak of the COVID-19 pandemic (Sen et al., 2020). Sen et al. (2020) stated that pandemic COVID-19 affects the global economy and apparel industry due to the shortage of raw materials. Apparel tourism is one of the links in the clothing, fashion, entertainment, and marketing chain. For developing apparel tourism, entrepreneur's management capability, e-commerce orientation, foreign language skills are crucial and can play an instrumental role in the internationalization process (Rana and Sørensen, 2013). Bangladesh is attempting to enter into a regional free trade agreement with its neighboring countries, workers' rights, productivity, and wage in the apparel industry, and growing economy of Bangladesh.

Medical Tourism in Bangladesh

Medical tourism refers to an individual visiting overseas to take healthcare services. Medical tourists expect to modify their travel abroad for receiving medical treatments (Hassan et al., 2015). Bangladesh is the most densely populated country (Mottaleb et al., 2019; Ali, 2012; Islam, 2009) and a key source market for medical tourism (The Daily Star, 2018) around the world. The country can help people access medical tourism services that particularly meet their needs. Hassan, Ahmed, and Rahman (2015) reported that Bangladesh could become a potential medical tourism destination if the country's development is appropriately arranged. The country has a huge investment in private and public healthcare sectors as well as the development of healthcare infrastructure. The medical travel industry of the country is offering quality and accredited services to meet patients' healthcare needs, such as cardiac care, plastic surgery, cosmetic care, surgery, and hospitals for orthopedic procedures and facilities that fit personal budgets and desires. Dhaka Tribune (2019) reported that Bangladesh ranked 133rd among 195 countries in providing access to quality medical care services.

Despite these opportunities, the increasing trend of Bangladeshi patients is visiting overseas for medical treatment leads to some dismay in healthcare policy circles. Medical care systems in Bangladesh is despondently underdeveloped (Rahman, 2019a; Mamun and Andaleeb, 2013; Ali, 2012). The major reason for the underdevelopment of the medical care industry is inefficient human resources, including absenteeism, corruption, mismanagement, and shortage of doctors and nurses (Mahmood, 2012; Ali, 2012). Besides, due to the present outbreak of the COVID-19 pandemic, the healthcare sector in Bangladesh is facing a problem and the most affected sector among the other sectors (Bodrud-Doza et al., 2020). Rahman (2020) reported that healthcare access is directly related to the healthcare sector, and the government is responsible for maintaining the health and safety of the people and areas under its intervention.

Many developing countries are generating profit from the medical tourism industry, while Bangladesh loses patients to these nations in an incessant stream. A large number of Bangladeshi patients are willingly traveling to another country for healthcare services because of the higher quality of healthcare services (Mamun and Andaleeb, 2013). The healthcare management system in Bangladesh has a problem and decreased the perception of quality medical care in the sector. The government of the country should pay attention to providing quality medical care services to retain patients within the country and promoting the local healthcare industry. There are 3976 public healthcare facilities and 975 private hospitals/clinics (Siddiqui and Khandaker, 2007), 86,800 registered medical doctors, and 56,733 registered nurses for serving the rapidly increasing population of Bangladesh (Dhaka Tribune, 2019). Dhaka Tribune (2019) also reported that out of 86,800 medical doctors, around 20,000 are currently employed with government facilities, which makes the ratio as 1.29 medical doctors for every 10,000 people.

Ahmed, Hossain, Chowdhury, and Bhuiya (2011) reported that Bangladesh has a shortage of health workers and a lack of data on human resources for the healthcare sector. According to Mamun and Andaleeb (2013) and Islam (2004), the public healthcare system in Bangladesh is four different levels of medical care delivery such as district hospitals, Upazila health complexes, community outreach, and health and family welfare centers. The majority of public hospitals are dilapidated and better equipped- private hospitals are located in cities and towns. Thus, most of the people are getting poor quality healthcare treatment from public hospitals. Andaleeb, Siddiqui, and Khandakar (2007) stated that public medical care has been declining, whereas private healthcare services have been increasing. Private hospitals can attract more people because of their quality medical care services. However, currently, certain public hospitals are providing quality services like organ transplants, cosmetic surgery, neurosurgery, heart surgery, and radiation therapy at low-cost packages. Despite this, overall public medical care services are low in Bangladesh (Mohiuddin, 2020; Rahman, 2019b; Andaleeb et al., 2007).

The medical professionals of health sectors are often prescribed unnecessary and contradictory tests particularly it is added to the diagnostic centers in Bangladesh (Mamun and Andaleeb, 2013). This inaccurate diagnostic test, misbehavior of healthcare professionals, and unsatisfactory health services provided by the hospitals have made distrust among people concerning medical care systems in Bangladesh (Ali, 2012). Thus, many Bangladeshi people are choosing to travel abroad for medical treatment. Many countries like Malaysia, Singapore, Thailand, and India are successfully providing medical treatment. Bangladesh is offering medical tourism to other nations within the region while the healthcare prospects of the country have remained unexplored.

Medical Tourism is one of the fastest-growing healthcare industries (Rahman, 2019a; Rahman et al., 2018a; Rahman and Zailani, 2017). Bangladesh can improve the medical tourism industry if the country faces challenges from developing medical tourism in neighboring countries such as Thailand, Malaysia, India, and Singapore (Rahman, 2019b; Hassan et al., 2015). The government of Bangladesh should take the initiative to promote healthcare provision to its low and middle-level people and provide quality healthcare services at a reasonable cost. Bangladesh has a great advantage of its geographical location that getting known as a potential destination for medical tourism around the globe (Hassan et al., 2015). Many tourists and emigrants, when traveling to meet with their relatives and friends, they frequently received medical tourism facilities in this country (Rahman et al., 2018b; Rahman et al., 2017). These tourists are received not only herbal or alternative medical care services but also interested in surgical and non-surgical healthcare procedures. This will assist in promoting medical tourism in Bangladesh to international tourists who can suggest Bangladesh as a potential medical tourism destination to other nations.

Beach Tourism in Bangladesh

Beach tourism refers to tourism activities including traveling, spending time, recreation, leisure activities, and businesses in the beach areas. Moreno and Amelung (2009) stated that beach tourism refers to the light activities that belong to the 3S tourism (sand, sea, and sun). Beaches are always a tourism attraction for many people at home and abroad. People usually like to enjoy their moments and vacation alone and with their friends and families in beach destinations. Beach tourism offers travelers a diversified experience, including the coastal livelihood, marine resources like marine lives and coral reefs, big waves of the sea, golden sands, and beautiful scenic views (Rani et al., 2020). Bangladesh is blessed with 580 kilometers of coastal areas, including the world's longest unbroken sandy sea beach of 120 kilometers in the southern part of the country (World Atlas, 2020). In Bangladesh, sea beach is one of the prime attractions for tourists from home and away. The tourism industry has been developed based on sea beaches in the Cox's Bazar and Patuakhali districts in Bangladesh. Both the tourism spots have been contributing a lot to the local and national economy by generating employment and entrepreneurial opportunities, especially for the local people on a large scale. However, in recent years, the infrastructural development by constructing roads and highways, marine drives, modern hotels, and resorts have significantly boosted the dynamism of tourism activities in the beach areas in Bangladesh (Tourism Board, 2020).

Many tourists like to visit sea beaches for watching or surfing on the big waves, spending a vacation with family members in the beach environment, listening to the sound of water, and enjoying the clear blue vision of the ocean. Beach tourism requires developing its infrastructures and facilities according to the needs and desires of the tourists in the sea beaches in Bangladesh. Beach tourism is one of the best core niche products of tourism because it does not only guarantee a high level of satisfaction to the visitors but ensures high income to the local community and national economy. However, the development of beaches as tourism sports started back in the mid of nineteenth century (Holloway and Humphreys, 2019). After World War two, much of the tourism activities started via beach holidays. However, before that, in the eighteenth century, beach holiday was an English culture when it was limited to the medical adaptation of popular sea-bathing (Britanica, 2020). In modern beach tourism, there are many additions to facilities like the establishment of the spa resort and gambling casinos. Nevertheless, beach tourism is significant for health enjoyment and religious rites. Beach tourism is a successful product of tourism across Europe, the Mediterranean, in the United States, republics of Oceania, South Africa, and Latin America and eventually successfully reached Asia (Britanica, 2020).

The largest sea beach in Coxes Bazar has become one of the prominent spots of attraction in the tourism world. Cox's Bazar is a district of Bangladesh situated in the southern part under Chattogram division. The name of the place became Cox's Bazar after the name of Lieutenant Cox, who died in 1798 under British rule (Parveen and Rajon, 2008). The place was established as a colony of Moth who sought shelter (due to the conquest of Arakan by Burmese) in that area, a British territory. The main attraction of tourists in Cox's Bazar are towering cliffs, surfing waves, rare conch shells, colorful pagodas, Buddhists' temples, tribal communities, coral reefs, and delicious seafood. Moreover, the water in the sea in Cox's Bazar is relatively warm and free from dangerous sharks (Parveen and Rajon, 2008). The golden sands on beaches are suitable for sunbath, and the water is clear and suitable for bathing and swimming. One can also enjoy the 80-kilometer-long marine drive from Cox's Bazar to Teknaf. In Cox's Bazar a visitor can enjoy several beaches and spots like Inani beach, Laboni beach, Himchori, Teknaf, Buddhist temples in Ramu, Sonadia islands, Moheskhali and coral island in St. Martin's, which will be a life-long experience for one. Laboni beach is very nearby in Cox's Bazar, and there are hundreds of shops from where the tourists can buy souvenirs and beach accessories. Inani beach is located 35 kilometers away in the south from Cox's Bazar. This beach contains golden sands and very famous among the tourists who usually come here for relaxing since this beach is not crowded like Laboni beach. Moheskhali is a small island of 268 square kilometers (Parveen and Rajon, 2008). This area is very attractive for the tourists for its natural beauty, surrounded by low hills, mangrove forests, temples, and pagodas. Further, Parveen and Rajon (2008) reported that Sonadia is only 9 kilometers of the sandy island where different shells are found on the beach. The main attraction of this island is during the winter when fishermen set up temporary camps and dry their catches of fishes from the sea. The most beautiful and only coral island in Cox's Bazar is St. Martin's, where tourists can find live corals (Hai and Alamgir, 2017). It is 30 kilometers away from Teknaf, and tourists can easily reach there by motorboats or tourist boats. Apart from the beaches in Cox's Bazar, there is another beach in Potuakhali district named Kuakata sea beach, which is also famous by the name of a daughter of the sea. It is a beautiful sea beach annexed to the Sundarbans. Kuakata is a perfect tourism spot from where the tourists can enjoy the sunrise and sunset on the beach. The beaches in Bangladesh are exquisite with natural resources and offer a variety of facilities. Nevertheless, the concerned authorities must concentrate on the needs of foreign tourists and develop and invest in the infrastructure accordingly so that beaches in Bangladesh can attract more and more tourists from home and abroad.

Agri-Tourism in Bangladesh

Agri-tourism or agri-tourism is a form of tourism where tourists stay with local people in rural areas overseas and stay in agricultural farms either to get experience of farm life or to explore the countryside. Agri-tourism is the act of visiting a farm-related to agriculture and agri-business operations for education, enjoyment, and active involvement in the activities of the farm (McGehee and Kim, 2004). The local and international tourists are often interested in performing a wide array of activities such as cultivating vegetables, planting trees, watching local birds, tasting honey, picking fruits, and identify women's capabilities in food manufacturing. Bangladesh is a rich place of agri-tourism resources with attractive to a tourist. Bangladesh is presently number four in vegetable production and fish production (The Independent, 2019). The local capacities concerning rice, prawn, tobacco, and flower cultivation have advanced the country's enhancement. These dynamic products of the country may maintain agri-tourism. The Independent (2019) reported that the contribution of the tourism industry to GDP was BDT 193.0 billion in 2012 and expects that it will increase BDT 384.7 billion in 2023. The travel and tourism industry directly upheld 1,281,500 jobs in 2012, and it is predicted that this tourism industry will create 1,785,000 jobs by 2023 (Reza, 2019).

Bangladesh can improve its agri-tourism promotional budget by 4–5% of its total earning from this sector (Dhaka Tribune, 2018). Agri-tourism might be a great source of income for Bangladesh if the fertile land and population of the country are actively dependent on the agriculture-based economy. Reza (2019) reported that the environment, forest, marginal farmers, indigenous people, women, and fishermen could be environmentally, socially, and economically included for sustainable rural development. Some entrepreneurs have established local farms such as goat, buffalo, cow, and poultry to eliminate poverty. The local and international agri-tourism can assist increase awareness among local farmers (Karabati et al., 2009) concerning how to decrease climate impact on health, land, water, fish, and forest resources. Tourists can exchange knowledge on farming and agriculture, cultural diversity for welfare, and distribute the benefits for the tourism industry.

Agri-tourism product in a rural area in Bangladesh is crucial as lands of Mymensingh are fertile for paddy, Satkhira for prawn, Kushtia for tobacco, Chandpur, and Barisal are known for fish, Jessore for flowers, and Dinajpur for its wheat. Accordingly, Bangladesh has huge options when it comes to supporting agri-tourism, which can contribute to its economic growth. Ahmed and Jahan (2013) stated that trekking the mountain in rural areas of Bandarban, Rangamati, and Khagrachari, wildlife safari in the Sundarbans and Chattogram division with appropriately trained tourist guide may add a new aspect in promoting agri-tourism in Bangladesh. The geographical information systems (GIS) are important for enhancing agri-tourism. Tourism operators can sell agri-tourism products and farming tools through this system. The success of agri-tourism in Bangladesh can depend on appropriate planning, organizing, coordination and implementation. For the development of agri-tourism, the country should provide comfortable access to the land, spots, and conformable residence, which can attract tourists to enjoy the natural beauty of Bangladesh. The government and agencies can work together in business and investors and can initiate a project to support the farmers for protecting land and rural development, which can lead to promoting the potential of agri-tourism and contributing economic growth of Bangladesh.

Ecotourism in Bangladesh

Ecotourism is a form of sustainable tourism that attempts to reduce its impact on the local environment and to promote global tourism development. Ecotourism refers to the individual's purposeful travel to the natural regions that make a comprehension of the social and natural history of the environment (Ahsan, 2008). Pegas and Castley (2014) stated that ecotourism protects the biological system while creating economic opportunities that make the discussion of natural resources valuable to the local community. The financial globalization of the world has developed the eco-tourism trade (Eshun and Tichaawa, 2020). The historical heritage and bio-diversification of the country can lead to ecotourism development. The development of ecotourism needs national and international tourists and host regions while ensuring opportunities for the future (Ocampo et al., 2018). It is necessary to the management of all-natural resources so that financial, social, and aesthetic needs can be satisfied while preserving cultural integrity, fundamental natural processes, biodiversity, and life support systems. Dey et al. (2020) reported that Bangladesh has the opportunity to liberate the economic backwardness by ideally using the scope of ecotourism throughout the country.

Ecotourism is mainly related to environmental preservation and community development. Local community participation is one of the key aspects of ecotourism. Community-based ecotourism is characterized by environmental consideration, the participation of local people, environmental protection, and important opportunities for the host community (Mree et al., 2020). Iqball et al., 2010) stated that the participants of the community should take a decision making, planning, and implementation by integrating their perceptions. Dey et al. (2020) indicated that Sundarbans is one of the natural-based tourism spots in Bangladesh. Mozumder, Uddin, Schneider, Islam, and Shamsuzzaman (2018) stated that fisheries-based eco-tourism has become popular in recent years and received attention as recreational fishing trips can be a significant part of the tourism business and generating economic growth in Bangladesh. Connecting fisheries to the travel industry need to make innovative ecotourism products, which may lead to discovering a new source of income for fisheries to make them profitable and sustainable business. This effort can lead Bangladesh to promote a potential fisheries-based ecotourism destination.

Ecotourism provides the protection of the environment, incorporates community involvement, makes benefits, maintains social responsibility, cultural integrity, and promotes financial well-being (Islam et al., 2011). The fundamental principle of ecotourism is sustainable tourism that might be an instrument to make public support for environmental preservation. Ecotourism can create financial advantages for the international and local community. Thus, emphasis should be on effect planning to thrive ecotourism as an incredible potential for economic growth (Dey et al., 2020). To develop ecotourism, Bangladesh should capture the global tourism market by drawing in travelers to natural zones. Ecotourism can be an apparatus for poverty reduction, economic growth, and financial emancipation of the local community in the country.

Conclusion

Bangladesh has numerous unexplored tourism products and services. These significant attractions incorporate mangrove forests, beautiful tribal culture, sea beaches, and village life. All these attractions can pull travelers from different parts of the world. These attractions are required to investigate by tourism stakeholders, government, and private sectors using the proper tourism marketing and promotional strategy. Tourism planners, operators, marketers, and stakeholders can develop the tourism market at local and global levels utilizing distinctive marketing and promotional tools, including satellite channels, websites, local TV, marketing, and communication strategies. Effective marketing strategy is required to investigate existing and potential tourism attractions that can lead to an increased tourism image of Bangladesh. Moreover, the concerned authorities and tourism stakeholders like the tourism board of Bangladesh, travel and tourism planners, tourism promoters, tourism representatives of Bangladesh in foreign countries must identify the needs, desires, and interest of the foreign tourists and prepare and design tourism products and services accordingly with full-fledged policy support.

References

Ahmed, S. M., Hossain, M. A., Chowdhury, A. M., & Bhuiya, A. U. (2011). The health workforce crisis in Bangladesh: Shortage, inappropriate skill-mix and inequitable distribution. *Human Resources for Health, 9*(3), 1–7.

Ahmed, I., & Jahan, N. (2013). Rural tourism-prospects in rustic Bengal. *European Journal of Business and Management, 5*(16), 163–172.

Ahmed, J. U., Rifat, A., Nisha, N., & Uddin, M. J. (2015). Deshi dosh: The case on integration of ten rivals in the fashion industry of Bangladesh. *Decision, 42*(1), 87–103.

Ahsan, M. N. (2008). Ecotourism in Bangladesh: A new tool for economic development. *Journal of Socioeconomic Research and Development, 5*(3), 299–304.

Akbar, S., & Ahsan, K. (2020). Investigation of the challenges of implementing social sustainability initiatives: A case study of the apparel industry. *Social Responsibility Journal*. https://doi.org/10.1108/SRJ-09-2019-0291

Ali, M. M. (2012). Outbound medical tourism: The case of Bangladesh. *World Review of Business Research, 2*(4), 50–70.

Andaleeb, S. S., Siddiqui, N., & Khandakar, S. (2007). Patient satisfaction with health services in Bangladesh. *Health Policy and Planning, 22*(4), 263–273.
Banglapedia (2020). *Rivers*. Retrieved from: http://en.banglapedia.org/index.php?title=River. Accessed 10 Sept 2020.
Bodrud-Doza, M., Shammi, M., Bahlman, L., Islam, A. R. M., & Rahman, M. (2020). Psychosocial and socio-economic crisis in Bangladesh due to COVID-19 pandemic: A perception-based assessment. *Frontiers in Public Health, 8*(341), 1–17.
Britanica (2020). *A case study: The beach holiday.* Retrieved from: https://www.britannica.com/topic/tourism/Day-trippers-and-domestic-tourism. Accessed 10 Sept 2020.
Buckley, R. (2003). Adventure tourism and the clothing, fashion and entertainment industries. *Journal of Ecotourism, 2*(2), 126–134.
Central Intelligent Agency (2020). *The world factbook*. Retrieved from: https://www.cia.gov/library/publications/the-world-factbook/geos/bg.htm. Accessed 15 Sept 2020.
Dey, T., Ahmed, S., Bachar, B. K., & Kamruzzaman, M. (2020). Prospects of community based eco-tourism in Sundarbans: A case study at Munshiganj, Satkhira, Bangladesh. *International Journal, 2*(1), 60–68.
Dhaka Tribune (2018). *Why we should invest in agro-tourism.* Retrieved from: https://www.dhakatribune.com/business/2018/02/06/invest-agro-tourism#:~:text=Foreign%20and%20domestic%20agro%2Dtourism,knowledge%20on%20farming%20and%20agriculture. Accessed 20 Sept 2020.
Dhaka Tribune (2019). *Patient, doctors, nurses ratio: Bangladesh lags far behind its neighbours.* Retrieved from: https://www.dhakatribune.com/health/2019/07/21/patient-doctors-nurses-ratio-bangladesh-lags-far-behind-its-neighbours. Acccessed: the 11 June 2020.
Eshun, G., & Tichaawa, T. M. (2020). Towards sustainable ecotourism development in Ghana: Contributions of the local communities. *Tourism: An International Interdisciplinary Journal, 68*(3), 261–277.
Hai, M. A., & Alamgir, M. B. (2017). Local community attitude and support towards tourism development at Saint Martin Island, Bangladesh: Local community attitude and support. *International Journal of Tourism and Hospitality Management in the Digital Age, 1*(2), 32–41.
Hassan, A., Ahmed, M. U., & Rahman, M.S-U-. (2015). The development, nature, and impact of medical tourism in Bangladesh. In M. Cooper, K. Vafadari, & M. Hieda (Eds.), *Current issues and emerging trends in medical tourism* (pp. 296–311). Hershey.
Holloway, J. C., & Humphreys, C. (2019). *The business of tourism*. SAGE Publications Limited.
Iqball, S., Parvin, M., Salequzzaman, Haque, E., Islam, R. and Ahmed, S. (2010). Ecotourim in the Sunderbans and its surrounding- a possible sustainable option for alternative livelihood development. Bangladesh Research Publications Journal, 4(3), pp.244–253.
Islam, A. M. (2004). *Health and health sciences. Banglapedia: National Encyclopedia of Bangladesh.* The Asiatic Society of Bangladesh.
Islam, M. A., Khadem, M., & Sayem, A. (2012). Service quality, customer satisfaction and customer loyalty analysis in Bangladesh apparel fashion retail: An empirical study. *International Journal of Fashion Design, Technology and Education, 5*(3), 213–224.
Islam, M. B. (2009). Blood transfusion services in Bangladesh. *Asian Journal of Transfusion Science, 3*(2), 108–110.
Islam, M. R., Iftekhar, M. S., & Islam, M. W. (2011). Potential of ecotourism development in Bangladesh coast: An overview. *Tourism Review International, 15*(4), 325–336.
Islam, M. T., & Stringer, C. (2018). Challenges of achieving social upgrading in Bangladesh's apparel industry. *Society and Business Review, 15*(2), 77–94.
Karabati, S., Dogan, E., Pinar, M., & Celik, L. M. (2009). Socio-economic effects of Agri-tourism on local communities in Turkey: The case of Aglasun. *International Journal of Hospitality & Tourism Administration, 10*(2), 129–142.
Mahmood, S. A. I. (2012). Health systems in Bangladesh. *Health Systems and Policy Research, 1*(1), 1–4.

Mamun, M. Z., & Andaleeb, S. S. (2013). Prospects and problems of medical tourism in Bangladesh. *International Journal of Health Services, 43*(1), 123–141.

McGehee, N. G., & Kim, K. (2004). Motivation for Agri-tourism entrepreneurship. *Journal of Travel Research 43*(2), 161–170.

Mohiuddin, A. K. (2020). An extensive review of patient health-care service satisfaction in Bangladesh. *Adesh University Journal of Medical Sciences & Research, 2*(1), 5–16.

Moreno, A., & Amelung, B. (2009). Climate change and tourist comfort on Europe's beaches in summer: A reassessment. *Coastal Management, 37*(6), 550–568.

Mottaleb, K. A., Krupnik, T. J., Keil, A., & Erenstein, O. (2019). Understanding clients, providers and the institutional dimensions of irrigation services in developing countries: A study of water markets in Bangladesh. *Agricultural Water Management, 222*, 242–253.

Mozumder, M. M. H., Uddin, M. M., Schneider, P., Islam, M. M., & Shamsuzzaman, M. (2018). Fisheries-based ecotourism in Bangladesh: Potentials and challenges. *Resources, 7*(4), 1–18.

Mree, C. L., Das, S., Ray, T. K., Chowdhury, P., & Saha, N. (2020). Residents' perception of ecotourism in Ratargul freshwater swamp Forest of Bangladesh. *Asian Journal of Research in Agriculture and Forestry, 5*(3), 1–11.

NIbusinessinfo (2020). *Niche tourism opportunities*. Retrieved from: https://www.nibusinessinfo.co.uk/content/niche-tourism-opportunities. Accessed 11 Sept 2020.

Ocampo, L., Ebisa, J. A., Ombe, J., & Escoto, M. G. (2018). Sustainable ecotourism indicators with fuzzy Delphi method–a Philippine perspective. *Ecological Indicators, 93*, 874–888.

Parveen, J. A. and Rajon, M. A. (2008). The Cox's Bazar of Bangladesh -A Rising Tourist Spot in The World: Its Economic Prospects and Constraints. *2nd International Conference on Built Environment in Developing Countries*. Penang: the 3rd to 4th December.

Pegas, F. D. V., & Castley, J. G. (2014). Ecotourism as a conservation tool and its adoption by private protected areas in Brazil. *Journal of Sustainable Tourism, 22*(4), 604–625.

Rahman, M. K. (2019a). Medical tourism: Tourists' perceived services and satisfaction lessons from Malaysian hospitals. *Tourism Review, 4*(3), 739–758.

Rahman, M. K., & Zailani, S. (2017). The effectiveness and outcomes of the Muslim-friendly medical tourism supply chain. *Journal of Islamic Marketing, 8*(4), 732–752.

Rahman, M. K., Zailani, S., & Musa, G. (2017). Tapping into the emerging Muslim-friendly medical tourism market: Evidence from Malaysia. *Journal of Islamic Marketing, 8*(4), 514–532.

Rahman, M. K., Zailani, S., & Musa, G. (2018a). The perceived role of Islamic medical care practice in hospital: The medical doctor's perspective. *Journal of Islamic Marketing, 9*(1), 2–18.

Rahman, M. K., Zailani, S., & Musa, G. (2018b). Tourists' satisfaction and loyalty intention at Shariah compliant private hospitals in Malaysia. *International Journal of Tourism Sciences, 18*(4), 295–311.

Rahman, R. (2019b). The privatisation of healthcare system in Bangladesh. *International Journal of Health Care Quality Assurance, 32*(1), 97–107.

Rahman, R. (2020). Shrinking the state: The rise of private sector healthcare in Bangladesh. *Journal of International Development, 32*(5), 717–726.

Rana, M. B., & Sørensen, O. J. (2013). Exploring management and entrepreneurial factors in the internationalisation of SMEs: Evidence from the Bangladeshi apparel industry. *International Journal of Entrepreneurship and Small Business, 19*(4), 517–542.

Rani, S., Ahmed, M. K., Xiongzhi, X., Yuhuan, J., Keliang, C., & Islam, M. M. (2020). Economic valuation and conservation, restoration & management strategies of Saint Martin's coral island, Bangladesh. *Ocean & Coastal Management, 183*, 105024.

Reza, S. (2019). *Economics of Home-grown Agro- tourism Products in Bangladesh*. Retrieved from: https://dhakacourier.com.bd/news/Society/Economics-of-Home-grown-Agro%2D%2Dtourism-Products-in-Bangladesh/944. Accessed 15 Aug 2020.

Sarker, S. (2014). Competitive marketing strategies for tourism industry in the light of "vision 2021"of Bangladesh. *European Journal of Business and Management, 6*(4), 210–220.

Sen, S., Antara, N., Sen, S., & Chowdhury, S. (2020). The apparel workers are in the highest vulnerability due to COVID-19: A study on the Bangladesh apparel industry. *Asia Pacific Journal Multidisciplinary Research, 8*(3), 1–7.

Siddiqui, N., & Khandaker, S. A. (2007). Comparison of services of public, private and foreign hospitals from the perspective of Bangladeshi patients. *Journal of Health, Population, and Nutrition, 25*(2), 221–230.

The Daily Star (2018). *Bangladesh a key source market for medical tourism.* Retrieved from: https://www.thedailystar.net/business/tourism/bangladesh-key-source-market-medical-tourism-1571314. Accessed 24 May 2020.

The Financial Express (2020). *Govt mulls formulating master plan to boost tourism.* Retrieved from: https://thefinancialexpress.com.bd/national/govt-mulls-formulating-master-plan-to-boost-tourism-1582197881. Accessed 10 Sept 2020.

The Independent (2019). *Economics of agro-based tourism in Bangladesh.* Retrieved from: http://www.theindependentbd.com/printversion/details/217073#:~:text=The%20direct%20contribution%20of%20Travel%20Tourism%20to%20GDP%20was%20BDT193,7%20billion%20in%202023. Accessed 13 Sept 2020.

Tourism Board (2020). *Bangladesh tourism board.* Retrieved from: http://www.tourismboard.gov.bd/. Accessed 25 Sept 2020.

Travelbizmonitor (2011). *Niche tourism opportunities for India.* Retrieved from: http://www.travelbizmonitor.com/Trend-Watch/niche-tourism%2D%2Dopportunities-for-india-14794#:~:text=India's%20niche%20is%20built%20around,tourism%2C%20Wildlife%20Tourism%2C%20etc. Accessed 13 Sept 2020.

World Atlas (2020). *Bangladesh.* Retrieved from: https://www.worldatlas.com/maps/bangladesh. Accessed 11 Sept 2020.

Chapter 10
Investment and Development in Nature-Based Tourism in Bangladesh

Md. Wasiul Islam

Abstract Nature-based tourism (NBT) is known as a tool for sustainable development. Bangladesh is bestowed with various natural resources and attractions of NBT. However, there is a lack of research on investment and development in NBT, particularly in the context of Bangladesh. Therefore, the research aims to study the overall investment and development status of NBT in Bangladesh. The study followed content analysis based on reviewing various secondary information. In recent years, the NBT destinations of Bangladesh are getting more popular with domestic visitors. There are significant contributions of NBT in the tourism industry of Bangladesh by contributing GDP and generating new job opportunities. Considering the present status of NBT, this sub-industry has more potentials to develop; hence it can facilitate the sustainable development of Bangladesh. It warrants to solve various challenges in NBT to harvest more from this sub-industry. It requires more political commitments, integrated and coordinated efforts from various ministries of the government, more active and responsible participation of private actors, promotion of environmental education and awareness, efficient marketing strategies to attract more international tourists, management of these destinations, and above all practicing tourism governance to improve NBT development and its investment.

Keywords Nature-based tourism · Investment · Development · Institutions · Sustainability · Governance · Bangladesh

---

Md. W. Islam (✉)
Forestry and Wood Technology Discipline, Khulna University, Khulna, Bangladesh
e-mail: wislam@fwt.ku.ac.bd

A. Hassan (ed.), *Tourism in Bangladesh: Investment and Development Perspectives*, https://doi.org/10.1007/978-981-16-1858-1_10

Introduction

How big is the tourism industry? Generally, it may seem like a small to medium industry that deals with travel and recreation! In fact, the tourism industry is recognized as one of the biggest industries in the world considering its massive economic contributions, which is the third-highest world category in export earnings in 2019 (after fuels and chemicals), representing 10.3% of the world's Gross domestic product (GDP) (considering growth 3.5% from 2018), 30% of services exports and one out of every ten jobs in the world (World Travel & Tourism Council, 2020a). Tourism has experienced persistent growth and intensifying diversification to become one of the fastest-growing economic industries in the world over the decades. Tourism has been a key driver for international commerce and socio-economic development in many developing countries.

Tourism industry is crucial for many developing and developed countries as it generates revenues from it's different branches like consumption of goods and services by tourists, taxes, and employment opportunities. Besides its (direct, indirect, and induced) economic contributions, there are immense impacts of tourism considering its socio-cultural, environmental, and institutional perspectives, which frame sustainable tourism to foster the sustainable development of a destination.

Tourism is a kind of dynamic and competitive industry involving the ability to constantly adapt to customers' varying needs and desires since the customers' satisfaction, safety, and enjoyment are particularly the heart of tourism businesses (goHR2, 2011). The urban settings and lifestyles in those settings motivate us to come closure to the natural environment to refresh and recharge ourselves. Such motivation is one of the main factors for nature-based tourism (NBT) by which people try to find their roots.

The tragedy of mass tourism and the emerging global market for ethnic and unique experiences have promoted the growth of NBT related enterprises in various corners of the globe (Ashoka Trust for Research in Ecology and the Environment, 2006). NBT is globally considered the fastest-growing type of tourism (Pacific Asia Travel Association, 2015) often includes visiting national parks and other wilderness areas that are rich in various tourist attractions that are mostly natural. Currently, it shares about 20% of the world's total travel market (Kuenzi & McNeely, 2008).

NBT embraces a wide range of activities ranging from hunting to sunbathe on the beach, which is dependent on various ecosystem services like clean air and water, pristine scenery, and unique biodiversity (Scholes & Biggs, 2004). Therefore, NBT destinations are generally rich in biodiversity and cultural diversity. Generally, these diverse destinations are in the tropical and subtropical regions, particularly in developing economies. Bangladesh is located at such a geographical position that developing an economy known as green and rich in culture and biodiversity results in a potential destination for NBT for domestic and international tourists.

There are significant contributions of NBT in the tourism industry of Bangladesh by contributing her GDP as well as generating new job opportunities. More investments are noticed in establishing various academic and training institutes that are generating trained and competent manpower to develop and promote NBT. However, there is a lack of studies particularly focusing on the development and investment of NBT in Bangladesh. Therefore, the content of this chapter aims to discuss the overall development and investment of NBT in the context of Bangladesh. The rest of the chapter consists of various perspectives of NBT development and investment in Bangladesh, like the importance of NBT, forms of NBT, NBT development, investment in NBT, stakeholders and institutional arrangements in NBT, prospects of NBT, challenges in NBT, and finally concluding remarks.

Importance of Nature-Based Tourism in Bangladesh

Tourism is considered one of the world's largest and fastest-growing industries (United Nations, 2020; World Travel & Tourism Council, 2020b). It is not only a massive generator of jobs, wealth, and economic growth but also a substantial driver of the global economic recovery, which are already discussed in the above section. Despite its increasing relevance and proven contribution to GDP, jobs, and exports, tourism still lacks proper political and economic recognition, particularly in the context of the developing economy. However, this industry is also rising in Bangladesh (World Travel & Tourism Council, 2020b) and is considered as a thrust industry improving the rural, regional, and national economies.

NBT has been suggested to practice in various forests, protected areas (both forest-based and wetland-based), islands, ecologically critical areas, and other natural environments of Bangladesh considering its multi-faced benefits (Climate-Resilient Ecosystems and Livelihoods, 2014). However, researchers also warn of the negative impacts of NBT that cause socio-cultural, economic, and environmental hazards (Elands et al., 2015; Hassan, 2012; Khandakar, 2014; Khanom et al., 2011). Therefore, controlled NBT with proper management of tourists and destinations has been prescribed for these NBT destinations (Ahmed, 2016; Alam et al., 2009; Haque et al., 2016; Islam, 2010).

NBT depends on gathering experiences that are directly related to natural attractions. Bangladesh is bestowed with a diverse range of terrestrial and aquatic ecosystems along with their unique flora, fauna, picturesque landscapes, and diversified cultural heritage. These natural settings and fascinations make Bangladesh one of the potential NBT destinations in the world. The Bay of Bengal, rivers and other water bodies, mountains, forests including national parks, wildlife sanctuaries, and other types of protected areas are the foundation for gathering NBT experiences in Bangladesh.

Generally, nature-based tourists are interested in being experienced with a high level of service and product directly concerned with natural attractions. In this connection, they are willing to pay for gathering such experiences. It is recognized that these tourists produce more economic benefits than other types of tourists as they like to spend more and stay longer at these destinations (Pacific Asia Travel Association, 2015). Therefore, NBT contributes more to GDP and new job creation than other types of tourism. Bangladesh is not an exception. It is very clear from the mass media regarding the crowd of tourists in NBT destinations, particularly in the winter season as well as during the long holidays like two Eid festivals, Durga puja festivals, Gregorian calendar year ending, and so on.

Currently, the people of Bangladesh are visiting national parks and other protected areas more than at any time. This number is increasing and expected to increase as the population of Bangladesh is increasing, and their disposable income and time have also increased due to improved socio-economic conditions. Such protected areas and other nature-based destinations offer a wide array of recreation facilities such as rest and relaxation, appreciation of nature, walking/trekking, swimming, scenic landscapes, canoeing, and so on are some of the examples by which nature-based tourists can get rid of boredom by doing their routine jobs and duties starting from household activities to any type of services. These recreational activities facilitate the tourists to change the air and recharge their energy or mental condition or ability to return to their routine jobs and duties in a better way. They can relax in an environment that is outside of their usual environment. This also helps us to strengthen our relationship among family members/friends/colleagues and so on.

Generally, the people of Bangladesh are not environmentally aware, even educated and urban people. Environment pollution is simply observed here. NBT creates a unique opportunity through nature travel to aware and educates people about the environment, its various components, ecosystem, biodiversity, conservation, and the local community, which facilitate to protect of the values of these nature-based destinations and assist the destination managers in managing and protecting the natural and cultural resources (i.e. the destination). Guided tours by experienced tour guides, exhibitions, documentaries, information boards, leaflets, booklets, etc., can aid this process to know our environment better and behave friendly. Such aids also help visitors know nature and appreciate it, which is considered the root. The young generation and students may be benefited much through visiting these NBT destinations. Generally, students of different levels are taken to these destinations for their education/sessional/practical/study tour/recreational purposes. They can learn and enjoy a lot from such visitations. Moreover, they can make other people aware of various environmental issues to make our environment safe and sound.

Through the appreciation of nature, visitors will be able to understand the conservation and protection values of these NBT destinations. Various functions and contributions of NBT may be realized by visiting and learning from these destinations, which may eventually promote the support and network to conserve and protect these destinations. From the above discussions, it is evident that NBT can promote socio-cultural, economic, and environmental benefits for the host and guest populations as well as the conservation of the destination. Therefore, NBT can contribute to people, profit, and planet (3Ps) to promote sustainable development and Sustainable Development Goals (SDGs).

Forms of Nature-Based Tourism in Bangladesh

NBT products and services are very diversified in Bangladesh due to their diverse people, nature, culture, and heritage. NBT in Bangladesh is basically focused on various types of forests (tropical moist deciduous forest, tropical evergreen, and semi-evergreen forest, mangrove forest and freshwater swamp forest including "Sal" forest, hill forest, mangrove forest, and swamp forest respectively) and various water bodies including Bay of Bengal (coastal environment). These NBT destinations are located countrywide. However, destinations located in the Chittagong Hill Tracts, Cox's Bazar, greater Khulna, and greater Sylhet districts are more popular than others.

NBT in Bangladesh is known as various names based on the characteristics of the products and services like ecotourism/responsible tourism, community-based (eco) tourism, agri-tourism, rural tourism, urban tourism, green tourism, adventure tourism, forest-based tourism, protected area-based tourism, tea garden tourism, water-based tourism (i.e. riverine tourism, wetland tourism, beach tourism, coastal and maritime tourism, cruise tourism, "haor-baor" tourism), cultural/heritage tourism, wildlife tourism, mountain tourism, orchard tourism, floriculture-based tourism, religious tourism at nature-based destinations, and so on. However, there is insufficient scientific literature available on each of these types of NBT. An accurate record of tourists' information is also lacking.

The nature of Cox's Bazar is mainly dominated by beach tourism, whereas the intertwined mountain, greenery, and sky at the Chittagong Hill Tracts; greenery, mountain, and water at Sylhet; mangrove forests, wildlife (mainly Bengal tiger and spotted deer), and water in the Sundarbans attract most nature lovers both of domestic as well as foreign origins. It has been noticed that the demand for such NBT products and services is increasing significantly, particularly among domestic tourists in recent years. Considering the present status of NBT, this sub-industry has more potentials to develop. Therefore, it is expected that NBT will facilitate the development of Bangladesh if it is managed properly. It warrants to solve various challenges in NBT to harvest most from this sub-industry.

Nature-Based Tourism Development in Bangladesh

Bangladesh is bestowed with various resources and attractions of NBT. The country is sanctified with greenery, various water bodies, mountains, wildlife-rich nature along with hospitable people, comfortable weather, diversified and co-existence of cultural, political, religious, and archaeological heritage.

A major segment of the tourism in Bangladesh is based on the domestic market where around 97% of the tourists are domestic tourists. The impacts of international tourists are still considered very low in Bangladesh (spent US$ 333.5 million), which is only 0.7% of total exports in 2019. The total spending on travel and tourism in Bangladesh is contributed only 4% by international visitors. Nevertheless, it is expected to grow 7.1% per year or 4.7% of the GDP by 2027 (World Travel & Tourism Council, 2020a). Therefore, there are immense opportunities for inbound tourism growth in Bangladesh, which will ultimately increase the overall impacts of NBT in Bangladesh.

Protected Area-Based Tourism

Since the last few decades of the twentieth century, there are public pressures for creating more protected areas to conserve nature and wilderness both in developed and developing countries (Kirkpatrick, 2001). The establishment and expansion of protected areas are recognized as a prime strategy to offset biodiversity loss (Liu et al., 2010). These protected areas have significant importance to mankind by providing various direct and indirect ecological, socio-economic and cultural benefits (World Wide Fund for Nature, 2014). Most countries have various institutional arrangements to reserve a portion of their land and water territories to protect, conserve and/or preserve their natural environments, including biodiversity, which also serve for recreational and tourism purposes (Alpízar, 2006). This resulted in increased demand for NBT worldwide.

There are a total of 2.58 m ha (17.48% of total land area) of forest land in Bangladesh. There are a total of 59 protected areas (0.67 m ha) in Bangladesh (see Table 10.1). Of them, 58 are forest-based protected areas (except Marine protected area, which is located at the Bay of Bengal). All these protected areas are popular NBT destinations. Various tourist facilities have been developed there to ensure enhanced tourists' satisfaction. Some of these protected areas are more popular than others due to their attractions, location, accessibility, safety, and security.

Three protected areas (Wildlife Sanctuaries) of the Sundarbans were declared as the World Heritage Site (area: 1400 km^2) by UNESCO in 1997. The Sundarbans (area: 6017 km^2) is the single largest mangrove forest in the world, which is an excellent NBT destination in the country very popular among both domestic and

Table 10.1 Protected area of Bangladesh

Types of protected areas	Number	Area (ha)
Wildlife sanctuaries	23	437087.28
National parks	18	52831.69
Botanical gardens	2	85.58
Eco-parks	11	8672.85
Safari parks	2	2093.93
Special biodiversity conservation area	2	221.59
Marine protected area	1	173,800
Total	59	674792.92

Source: Author's compilation

international tourists. The Sundarbans supports a unique accumulation of flora and fauna, including fascinating megafauna like the Bengal tiger, spotted deer, estuarine crocodile, and the Ganges river dolphin. There are 334 floral species and more than 425 faunal species in the Sundarbans. The other forest-based protected areas of Bangladesh are also very rich in biodiversity, which attracts a significant number of tourists.

Additionally, there are 56 wetland-based protected areas (area: 173 ha) in Bangladesh, which are mainly established to conserve the fisheries resources of the country (Sherwood, 2009). All these forest and wetland-based protected areas are government land and managed by the guidance of the Forest Department, Fisheries Department, and Directorate of Environment, being the legal custodians of these areas. Many of these protected areas are currently managed through a collaborative management approach, popularly known as the co-management approach, to ensure biodiversity conservation through improved governance by the active participation of multi-stakeholder, including the grass-root level direct resource users.

NBT (controlled ecotourism or community-based ecotourism) has been prescribed by different development projects (Sundarbans Biodiversity Conservation Project, Management of Aquatic Ecosystems through Community Husbandry Project, Nishorgo Support Project, Integrated Protected Area Co-management Project, Climate Resilient and Environment Resilient Project, Sustainable Forests and Livelihoods Project, and so on) and researchers in these protected areas to provide alternative income generation opportunities for the local poor people. Therefore, it is noticed that NBT interventions are basically implemented through various development project-based initiatives, which are mostly foreign donor-funded. Moreover, there is no big project which is dedicated solely to focus NBT development. Rather NBT has been focused on various development projects as one of the components or means to biodiversity conservation and eco-friendly alternative income generation opportunities.

The main objective of these NBT interventions was to reduce the traditional dependency of the local poor people on the natural resources of these protected areas as well as to enhance their environmental awareness and education to facilitate biodiversity conservation and protection of these areas. It is evident that the co-management approach has facilitated to improvement their environmental awareness and education as well as income generation opportunities through NBT interventions (Butler et al., 2015; Elands et al., 2015; Islam et al., 2011a, 2018). Several management plans of protected areas of Bangladesh emphasized NBT since 2004.

After the advent of several co-management approach-based projects under the financial assistance of USAID, various NBT interventions have been done at the project sites with the direct involvement of different stakeholders for developing and promoting NBT. The NBT interventions may have both direct and indirect impacts. Moreover it may also have positive and negative socio-cultural, economic, environmental, and institutional impacts on the local communities and other key stakeholders. These impacts have immense importance for the sustainable management of these protected areas. The successful implementation of these NBT interventions is judged as one of the indicators of upholding the co-management approach as well as promoting a resilient community. The success of this new approach may lead to its replication to other forests and tourism destinations of Bangladesh. The effectiveness of this approach will certainly reflect in other new projects (s) as policy guidelines.

Previously visitors used to act rampantly and like to do a picnic at these protected areas, which was simply destructive to the ecosystem and biodiversity. However, nowadays, visitors have become more environmentally conscious due to better-protected area and visitor management. Therefore, the visitors' mentality and attitude towards the protected area environment are gradually changing, which contribute to the conservation efforts of these protected areas. Trained tour guides, different co-management organizations, and Forest Department are playing positive roles in this regard.

The protected area governance has also been improved through such NBT by improving multi-stakeholder participation, accountability, transparency, rules of law, balancing power, social learning, and so on, which ultimately enhanced the living standards of the local people and improved the recreational and tourism facilities for their guests. Local-based institutional developments are imparting significant roles in such development. Several research studies highlighted these issues considering Bangladesh's protected areas (e.g. Islam et al., 2017, 2018; Elands et al., 2015).

Moreover, various research studies on NBT in these protected areas show the impacts of tourism activities, effects of climate change, governance, conservation, livelihoods, challenges, etc. However, some issues like gender issues, equality, equity, empowerment, community resilience, use of technologies, and so on are still to explore in these protected areas to better understand these destinations, which will improve the management of these destinations and tourist satisfaction.

NBT tourism also has negative impacts on the environment if it is not managed properly. Recently, the Bangladesh government has banned travel to one of the popular NBT destinations and sensitive ecosystem "Cheradweep" which is close to St. Martin's island, due to increased negative impacts on the biodiversity (especially marine corals) of the island. Moreover, the government has imposed various restrictions on St. Martin's island to facilitate its ecological restoration process as well as to improve the environmental condition of the island.

Recently, tourists like to explore some new NBT destinations in Bangladesh. Therefore, the popularity of these destinations is increasing day by day. Some of the examples of such new NBT spots are: Sajek valley, Peda Ting, various falls in Rangamati; Konglak para, Tareng, Risang falls in Khagrachari; Bogalek, Nilgiri, Nafakum-Amiakum, Mirinja in Bandarbans; Ratargul swamp forest, Pantumai waterfall, white stones at Bisnakandi, Lovachara, and Lalakhal (Volaganj) in Sylhet; Hijaltala and Satlar beel in Barisal; Char Kukri-Mukri and Char Monpura in Bhola; mango orchards in Chapai Nawabganj; Madhobpur lake, Baikker beel, Hakaluki haor in Moulvibazar; Rupkotha eco-resort in Pabna; Nijhum dweep in Noakhali; among many others.

Coastal and Maritime Area-Based Tourism

Bangladesh is located by the Bay of Bengal, which has made the country important considering her geographic and strategic perspectives. It forms 710 km long coastline that runs parallel to the Bay of Bengal (Ministry of Water Resources, 2005). Such advantageous position and extended coastline have opened multi-dimensional opportunities for coastal and maritime tourism in Bangladesh. Moreover, the defrayal of maritime boundary delimitation dispute with Myanmar and India in 2012 and 2014 respectively provided Bangladesh more sovereignty rights over marine resources and extended better opportunities for maritime-based NBT. Currently, Bangladesh occupies 118,813 km^2 of the territorial sea due to such historical settlements. These coastal and marine zones are especially recognized for their attractive scenic beauty as well as unique biodiversity that attracts a wide range of tourists. Bangladesh's government has taken various initiatives to develop and promote coastal and maritime tourism as one of the means to develop the blue economy of the country (Alam, 2019; Bangladesh Tourism Board, 2018). "Blue Economy Cell (BEC)" - an administrative cell was created in 2017 under the Energy and Mineral Resources Division of the Ministry of Power, Energy and Mineral Resources to take care of various initiatives to develop and promote the blue economy. Additionally, several autonomous NGOs, and private bodies are assisting the government in achieving significant benefits from the blue economy where blue tourism plays a specific role. Coastal and maritime tourism can potentially influence blue economy of Bangladesh which will result to reduce poverty through creating new job opportunities, conserve biodiversity, control environmental pollution, promote sustainable use of coastal and marine natural resources, and so on.

Coastal area-based tourism is mainly concentrated in Cox's Bazar, which is the world's longest unbroken sea beach (120 km), popularly known as the tourism capital city of Bangladesh. There are several spots along with Cox's Bazar-Teknaf beach drives like Inani beach (about 30 km from Cox's Bazar toward Teknaf), which is a magnificent coral reef-based spot, Himchari, Teknaf, and other beach areas like Patengya, Kuakata, Katka-Kochikhali, Dubla, Mandarbaria, and so on are also popular to the national and international tourists. Besides these, several islands ("dweep" in Bengali) like St. Martin's, Maheskhali, Kutubdia, Nijhum dweep, Chera dweep, Char Kukri-Mukri, Char Fashion, Char Monpura, several islands at the Sundarbans (Dublar char, Mandarbaria, Pokkhir char, Bangabondhu char, Putnir char, etc.) are known as popular NBT destinations.

Very recently, the first luxurious cruise ship (Bay One) has been introduced in Bangladesh. Primarily the ship transports passengers following Cox's Bazar- St. Martin's -Cox's Bazar route, which is capable of accommodating more than 2000 people. This ship has added a new milestone in coastal and maritime tourism in Bangladesh. Moreover, there are some other initiatives to introduce cruise ship for Chattogram-Cox's Bazar- St. Martin's route as well as other neighboring islands to develop and promote coastal and maritime tourism in Bangladesh. Additionally, two international luxury cruise ships traveled Bangladesh in 2017 and 2019, which added another milestone in Bangladesh's coastal and maritime tourism history. Such inbound tourism through international cruise ships was expected to happen regularly in Bangladesh. However, the COVID-19 pandemic withheld that journey.

It is evident that private industry is playing significant roles in promoting coastal and maritime tourism by building such ships, various types of accommodations, and recreational facilities as well as establishing partnerships with various stakeholders/ agencies in home and abroad. The government is also trying to support these initiatives of private industry by formulating various policies and coordinating among various concerned stakeholders.

Impacts of COVID-19 on Nature-Based Tourism

According to the Centre for Policy Dialogue, Bangladesh's national poverty rate rose to 35% in FY2019–20 from 20.5% in the previous fiscal year due to the adverse impacts of the coronavirus pandemic where many people lost their jobs or got laid-off (The Daily Star, 2020). Most of these NBT destinations were closed or restricted to travel during the COVID-19 pandemic for nearly 7–8 months, which severely affected the tourism industry. Domestic and international flights were canceled due to the pandemic, which affected around four million people of Bangladesh either directly or indirectly involved in the tourism industry as there was no tourism at these destinations, which incurred a loss of US$ 1.1 bn (The Financial Express, 2020). The majority of the rural people were affected by such pandemic.

Currently, all the NBT destinations are open to tourists. The tourism industry's recovery stage through foreign tourists is still restricted from visiting these destinations due to international travel restrictions. It is forecasted to continue such a situation through 2021 until the COVID-19 vaccines are widely available to the mass people.

Nature got some unique opportunities to become greener and healthier during the COVID-19 pandemic, which is currently attracting more tourists. Moreover, due to on-going strict restrictions to travel abroad, Bangladeshi people are traveling a lot at these NBT destinations (domestic tourism) particularly to overcome their boredom during the pandemic situation. The destination authorities are now under pressure to manage this increased number of tourists, especially in the winter season amid such a pandemic situation. Nowadays, people are not that much caring COVID-19 resulting in their movements to these destinations. Though Bangladesh Tourism Board has prescribed Standard Operating Procedure (SOP) for all the stakeholders related to the tourism industry, many of them are reluctant to maintain such SOP, which is also increasing the vulnerability of COVID-19 contaminations in Bangladesh. However, it is anticipated that domestic demand for NBT will grow throughout the country in the post-COVID-19 pandemic situation.

Investment in Nature-Based Tourism in Bangladesh

Investment in tourism development is a continuous process like other industrial development. Both government and private investors are investing in this industry just after the independence of Bangladesh, which shapes this industry today. However, the investment in the tourism industry was much lower than other industrial investments. This is one reason for the unsatisfactory development of the tourism industry in Bangladesh even after five decades of independence though there is huge potential to develop this industry. Nevertheless, there is no definite account of investment that can exactly inform us particular investment on NBT.

According to the Board of Investment (2015), the investment in the travel and tourism industry was BDT 60.9 bn, which was only 1.4% of the total investment in 2014. In contrast, capital investment was only BDT 42 bn, which was only 0.37% of the GDP at the same time. However, tourism is a growing industry in Bangladesh, which comprised 4.4% of the total GDP in 2018. The contribution of the travel and tourism industry to the GDP of Bangladesh measured 3% in 2019 (6.8% growth over 2018), which generated 2.9% of total employment. This industry has even more potentials, considering the opportunities for tourism entrepreneurship in Bangladesh.

International tourists are expected to increase by 6.1% per year, with a growth in total tourism GDP of 6.2% per year and employment growth of 1.8% per year from 2018 to 2027 (World Travel & Tourism Council, 2020a). However, tourism is considered one of the hardest hit economic industries by the COVID-19 pandemic, which nay faces a drastic decline of 58–78% in international tourist arrivals during 2020 and has 100–120 million direct tourism jobs at risk (United Nations World Tourism Organization, 2020a).

Bangladesh is positioned among the top 25 countries considering tourism growth (World Travel & Tourism Council, 2020b). It is estimated that the overall tourism industry will grow at an annual rate of 6.1% from 2017 to 2027. The Government of Bangladesh allocated BDT 34 bn in FY2019 for the Ministry of Civil Aviation and Tourism, which was more than double the previous FY allocation (Ministry of Finance, 2019), which indicated that the government started to prioritize the importance of the tourism Industry. This consideration has been recognized in the Travel and Tourism Competitiveness Report 2019, where Bangladesh has seen the greatest percentage increase, reaching 120th position from 125th among 140 countries (World Economic Forum, 2019).

Considering the tourism industry's potentials and growth, the Government of Bangladesh took various projects under various ministries and departments/directorates to develop NBT. The Prime Minister herself is directing the concerned ministries for the development of the tourism industry to uplift the socio-economic status as well as the overall image of the country. The Ministry of Civil Aviation and Tourism is the concerned ministry to develop and promote various types of tourism in Bangladesh. Bangladesh Tourism Board and Bangladesh Parjatan Corporation are working under this ministry. However, under the Ministry of Environment, the

Bangladesh Forest Department is responsible for developing and promoting NBT in various forests and forest-based protected areas. Moreover, the Department of Environment under the same ministry implements some NBT projects. Some other ministries, like the Ministry of Shipping, also implements some NBT-based projects. The Ministry of Information is also assisting the concerned ministries in this regard to some extent. Some initiatives have been taken to promote Public-Private Partnership (PPP) to invest in the tourism industry.

As mentioned earlier, the Government of Bangladesh took various initiatives to explore 800 new tourist spots. The government allocated US$ 8.46 m in its Annual Development Programme (ADP) in FY2021 for the tourism industry. The Ministry of Information spent US$ 120,482 on producing pictorial and bilingual handbooks on newly explored 800 tourist spots in six divisions of Bangladesh (Dhaka Tribune, 2020). The proposed international standard tourism center (600 acres) at Meghna island under Chandpur district has planned to invest US$ 721 million through the joint venture of Bangladesh-Japan. This center will accommodate more than 100,000 visitors with 20,000 night-staying accommodation per day. It is expected to generate new job opportunities for around 5000 people (Somoy News, 2020).

Entrepreneurship is the cause of growing an economy that is a significant change agent in social and economic development. Tourism entrepreneurship relates to developing tourism in an area. Generally, the tourism industry is dominated by small and medium-sized enterprises, where entrepreneurship is even more important to this industry's development. Tourism entrepreneurs require the innovations, knowledge, capacity, investment, value-driven leadership, and tools to ensure the sustainability of this entrepreneurship. Proper management of tourism entrepreneurship can serve us significant benefits to facilitate empowered and dignified local community as well to exalt the positive image of Bangladesh in the outer world.

Investment in Academic and Training Facilities on Nature-Based Tourism

Bangladesh is recognized as an emerging and potential tourism destination in the world market (Hassan & Kokkranikal, 2018; Ishtiaque, 2013). Therefore, the demand for trained and competent manpower in the tourism industry has been increased in Bangladesh. To meet-up, the demand several academic and training (both government and private) institutions have invested to include hospitality and tourism within their academic and training curriculums.

Currently, there are eight public universities in Bangladesh where the 4-year undergraduate program in Tourism and Hospitality Management is offered. Two of them offer a postgraduate (Master's) program. Moreover, there are 14 private universities that offer different tourism programs, including both short-course and long-course. Most of these private universities are in the capital city of Bangladesh (i.e. Dhaka treated as more expansive than public universities). Additionally, there are some universities where various tourism (including NBT) and hospitality-related courses are offered to undergraduate and postgraduate students. Moreover, interested students can conduct their research projects at these levels. Additionally, there are 11 private colleges which also offer 4-year professional program like BBA (Honor's) in Tourism and Hospitality Management under Bangladesh National University. Several NBT related courses are taught under these programs offered by these universities and colleges.

Besides these universities and colleges, there are several training institutions that offer various short (Certificate Courses) and long courses (Higher Diploma) on tourism (including NBT), hospitality, and hotel management to generate skilled and competent tourism manpower to meet-up the emerging tourism industry of Bangladesh as well as aboard. There are more than 32 private and public institutes (31 private and 1 public) to produce such skilled manpower.

Tourism and Hospitality Management industry has been recognized as one of the ten priority industries identified by the Skills for Employment Investment Program (SEIP) (under the Ministry of Finance). This program has taken various initiatives for long-term and comprehensive skill development activities. Moreover, Bangladesh Technical Education Board (BTEB) offers some technical and vocational courses on tourism and hospitality. Technical and Vocational Education and Training (TVET) under BTEB has been identified as a tool for enhancing productivity and poverty reduction in Bangladesh, where tourism is considered a potential industry to contribute. Besides these institutional education and training centers, various non-governmental and private organizations in Bangladesh organize various independent training sessions on hospitality and tourism issues, including NBT.

Stakeholders and Institutional Arrangements in Favour of Nature-Based Tourism

Several public, private, non-governmental, and international organizations invest, develop, and promote NBT in Bangladesh. Their initiatives and efforts have been intensified significantly during the last decade. Various private organizations, particularly tour operators and agencies, cooperate with the government by investing their resources to develop and flourish Bangladesh's tourism industry. Several ministries of the government are developing and promoting NBT by formulating various institutional arrangements, infrastructure, and other tourism facilities development. Bangladesh Tourism Board (BTB) and Bangladesh Parjatan Corporation under the Ministry of Civil Aviation and Tourism is responsible for tourism and hospitality development in Bangladesh who also coordinate with other concerned ministries such as Ministry of Home Affairs; Ministry of Foreign Affairs; Ministry of Shipping; Forest Department and Department of Environment under the Ministry of Environment, Forest and Climate Change; Ministry of Cultural Affairs; Ministry of Finance; Ministry of Local Government, Rural Development and

Co-operatives; Ministry of Law, Justice and Parliamentary Affairs; Ministry of Chittagong Hill Tracts Affairs; Ministry of Land; and so on. Besides these government bodies, there are various autonomous (public universities), private (basically various tour operators and agencies and their associations, investors, entrepreneurs, private universities, and so on), non-governmental organisations, donors, and consultants (both national and internationals) who are working for the development and promoting NBT in Bangladesh.

Bangladesh Parjatan Board Act-2010 is considered a landmark institutional development in the field of tourism development in Bangladesh. This important Act has created a new momentum in the country's tourism industry by forming Bangladesh Tourism (Parjatan) Board in 2010, which has initiated diversified tourism development interventions. This body is considered the National Tourism Organisation of Bangladesh responsible for developing, creating, and promoting tourism industry and services, facilitating responsible tourism, including nature-based tourism; and coordination among various national and international stakeholders.

Recently, Bangladesh Tourism Board (BTB) has started preparing "Tourism Master Plan (2021–2041)" for Bangladesh, which will embrace different forms of tourism, including NBT development and promotion. This plan has focused travel and recreation industry as the third-largest economic engine of Bangladesh after the garments and remittance industry. Moreover, the plan has taken various initiatives to create a distinctive brand image of Bangladesh as one of the popular tourist destinations in the world to achieve "Vision 2041" and to emerge as a developed nation. These initiatives have been categorized as short-term (three-year), mid-term (five-year), and long-term (15-year) to implement the master plan (Bangladesh Sangbad Sangstha, 2020) This master plan will also entice the strategies to improve the NBT destination management to enhance tourists' satisfaction.

An extensive literature review on institutional arrangements in favor of NBT in Bangladesh shows a considerable number of legal institutions. These institutions are considered as the platform of NBT development in Bangladesh. Among these institutions, Bangladesh's Constitution (1972) 15(c) is considered the pioneer that recognized the Bangladeshi citizens' right to have reasonable rest, recreation, and leisure as one of the necessities. Subsequently, as per 18A [amended as Constitution (Fifteenth Amendment) Act, 2011], the "State will be responsible for protecting and improving the environment and for preserving and safeguard the natural resources, biodiversity, wetlands, forests, and wildlife for the present and future citizens" which links with 15(c) through the development of responsible NBT (Ministry of Law, Justice and Parliamentary Affairs Government of Bangladesh, 2019).

Following the constitutional arrangements, Bangladesh government formulated various policies where tourism development and promotion have been recognized. Among these institutions, National Tourism Policy, 2010 is the leading one, followed by other policies that directly and indirectly focused NBT development in Bangladesh, such as National Forest Policy, 2016 (Draft); National Environment Policy, 2018; National Water Policy, 1999; National Coastal Zone Policy, 2005; and National Industrial Policy, 2016.

To uphold and implement the above-mentioned national policies, there are several Acts, Orders, and Rules that directly and indirectly assist and guide NBT development and promotion in Bangladesh. Some of the examples of these relevant institutional arrangements are The Bangladesh Parjatan Corporation Order, 1972; Bangladesh Maritime Zones Act, 2019; Bangladesh Tourist Reservation Area and Special Tourism Zone Act, 2010; Bangladesh Economic Zones Act, 2010; Wildlife (Conservation and Security) Act, 2012; The Bangladesh Environment Conservation (Amendment) Act, 2010; The Environment Court Act, 2000; Mobile Court Act (Amendment), 2009; Bangladesh Biodiversity Act, 2017; Protected Area Management Rules, 2017; Environmental Conservation Rules, 1997; and Ecologically Critical Areas Management Rules, 2016.

Moreover, Tourism Vision 2020; Bangladesh Delta Plan 2100; The 7th Five Year Plan; National Biodiversity Strategy and Action Plan (NBSAP); Perspective Plan of Bangladesh 2010–2021 (Vision 2021); Perspective Plan of Bangladesh 2021–2041 (Vision 2041); National Biodiversity Strategy and Action Plan of Bangladesh 2016–2021; Investment and Financing Strategy for Coastal Zone Development in Bangladesh; Coastal Development Strategy, 2006; Teknaf Peninsula Community-Based Ecotourism Strategy; Bangladesh National Conservation Strategy, 2013; Blue economy development of sea resources for Bangladesh; Bangladesh National Programme of Action for Protection of the Coastal and Marine Environment from Land-Based Activities; Voluntary National Review; are some other examples of institutional arrangements which are supporting NBT in Bangladesh.

Prospects of Nature-Based Tourism in Bangladesh

The economy of Bangladesh is growing substantially for few decades. The economic growth was 8.2% in 2019. However, it was declined to 3.8% in 2020 due to the outbreak of the COVID-19 pandemic. Bangladesh is trying hard to recover its damaged economy during the pandemic situation where all industries are putting their combined efforts. According to the UK-based Centre for Economics and Business Research (Cebr) (2020), Bangladesh will become the 25th largest economy in the world by 2035 (currently positioned 41st). The size of GDP will grow to be nearly treble (US$ 855 billion) by the time. The purchasing power parity adjusted GDP per capita of $5139 in 2020 in Bangladesh makes it a lower-middle-income country (Star Business Report, 2020). These statistics are evidence of Bangladesh's growing economy, which will create more disposable income and time for Bangladeshi people, which are very important factors for availing recreation facilities and traveling a tourism destination. The increased per capita income will also facilitate to growth NBT more in the country.

There are the world's largest unbroken sea beach Cox's Bazar, and the world's largest single tract of mangrove forests the Sundarbans. Moreover, there are many other naturals, cultural, religious, and historical sites and attractions which are strongly potential to add values in NBT sector of Bangladesh. Recently, Bangladesh

Parjatan Corporation has explored 800 potential tourist spots in six divisions of Bangladesh, which can attract a great number of tourists from home and abroad. Among these spots, 200 are in Khulna, 111 are in Sylhet, 100 are in Barisal, 200 are in Rangpur, 200 are in Rajshahi, 200 are in Chattogram, and 150 are in Chandpur district (Dhaka Tribune, 2020). Many of them are known as NBT spots. Some of the examples of NBT destinations among these spots are recently explored at Manikpur-Surajpur of Chokoria under Cox's Bazar district, which is claimed to be similar to Darjeeling and Dal lake of Srinagar, Kashmir; Yancha fall at Chokoria; Remakri, Nafakhum at Bandarbans; and so on. Currently, Bangladesh Tourism Board is implementing an initiative by which they are exploring the existing and potential tourism attractions from every district of Bangladesh. After analyzing the collected information, they will be able to plan and develop NBT, which may be a great source of holistic development of the country.

These potential and new NBT destinations are expected to reduce environmental pollution by reducing the tourist pressure on the traditionally popular NBT destinations and improving management actions. Several smaller ethnic groups across the country have been identified to promote cultural tourism in Bangladesh, which warrants to improve their rural economy and uplift their socio-economic status.

Moreover, there are huge potentials for developing non-traditional forms of NBT in Bangladesh, particularly horticulture-based tourism based on various seasons of specific fruits like mango, litchi, jackfruit, pineapple, orange, guava, and so on. Aquaculture-based tourism may also be flourished in Bangladesh, considering her a great portion of water bodies. Similarly, river cruise tourism, blue (coastal and marine) tourism, floriculture tourism, apiculture/honey hunting tourism, agri-tourism, adventure tourism, pro-poor tourism may also be developed here. Moreover, the hospitable people of Bangladesh are the key to the tourism potentials of these destinations. Now, it warrants proper knowledge and skills on tourism entrepreneurship along with the required support from the government to flourish such potentials.

Generally, the NBT destinations of Bangladesh are situated in rural areas. Most of the "Upazillas" (sub-districts) of Bangladesh have some sorts of tourist attractions that have either cultural (heritage) or/and natural/religious/archaeological significance carrying local and/or national importance. Many of these attractions, along with the authentic rural settings of Bangladesh, are not familiar to most of the country's young generation, which makes NBT potential to them along with the international tourists who are also interested in these attractions. Therefore, rural tourism is one of the types of tourism, which means visiting rural (non-urban) areas for leisure and recreation. It refers to all types of tourism activities in a rural area, which depends on the area's overall settings, which may attract tourists from home and abroad. All of these potential NBT forms can reduce rural poverty or improve Bangladesh's socio-economic and environmental situation.

Tourists may have various active and passive opportunities for canoeing, kayaking, swimming, and other water sports, fishing/angling, trekking, hiking, cycling, fruits picking, cooking, having local dresses/food/drinks, and sharing the experience with local culture/traditions/lifestyles/livelihood styles, experiencing the nature including wildlife watching, health activities, and other various authentic activities under NBT. Such tourism activities are capable of facilitating and improve various economic transections via providing various tourism products and services (i.e. accommodation, cuisine, tour guiding, transportation, guiding, selling other local products and services) to tourists. Eventually, rural tourism facilitates to reduce inequalities, inequity, disparity, deprivation, gender disparity (most notably for women and youth), and so on by improving democratic culture and fostering empowerment of an area as various studies mention (e.g. Iorio & Wall, 2012; Komppula, 2014; Nilsson, 2006; Hall & Richards, 2003; Zeng & Ryan, 2012).

NBT based rural tourism encourages small and medium-sized local entrepreneurship that can reduce the unemployment problem of Bangladesh. Moreover, rural tourism provides opportunities for infrastructural development, refurbishment/re-use of abandoned properties, retaining the rural population in their areas, etc. Such tourism can also improve the interactions between the host and guest (tourists), which have immense intangible values (enriched tourism experience) to both by contributing local cultural preservation, fostering the sense of local pride, self-esteem, identity, and so on. Moreover, rural tourism can enhance environmental education awareness to be friendly with our surroundings that can promote biodiversity conservation of forests, water bodies, and other natural areas.

Recently, the Bangladesh government has taken various projects to flourish the potentials of coastal and maritime tourism in Bangladesh. Bangladesh Economic Zones Authority (BEZA), under the Prime Minister's Office has initiated to establish "Sabrang Tourism Park" (proposed area of 1027 acres) at Teknaf sub-district under Cox's Bazar district. This park has been proposed as the first tourism-based "Special Economic Zone (SEZ)" as well as the first exclusive tourism zone in the country. Other initiatives have been taken to establish "Sonadia Ecotourism Park" (proposed area of 8967 acres) and "Naf Tourism Park" (proposed area of 271 acres) in Cox's Bazar district to attract a significant number of domestic and international tourists. Moreover, very recently, another mega project has been initiated at the Meghna island (proposed area of 600 acres) of Chandpur district to establish an international standard tourism center (Somoy News, 2020).

Recently, the Government of Bangladesh has taken initiatives to connect Bangladesh with other South Asian countries, particularly Bhutan, Nepal, and India, through rail transportations to promote inbound NBT in Bangladesh. The government has also taken cross-border initiatives to introduce cruise tourism among Bangladesh-India-Maldives-Sri Lanka.

As discussed earlier, the co-management approach is promoting NBT in various protected areas of Bangladesh. There are even more opportunities to develop and promote NBT in such (both forest and wetland-based) protected areas of Bangladesh through a co-management approach where the grassroots level local people can improve their living and environmental standards. A co-management approach can create a multi-stakeholder platform where stakeholders from different areas, starting from local to international, can have the opportunities to work together. It creates the scopes to leverage resources through collaboration with concerned stakeholders. The existing National Forestry Policy (1994) and new National Forestry Policy (2016) (draft) of Bangladesh also recognized forest-based tourism for its promotion. Though the tourism industry's contributions in the national GDP of Bangladesh are low, various studies have shown that Bangladesh is a prospective country for tourism development (DATABD.CO, 2020; Islam et al., 2011b; Islam, 2010; The Financial Express, 2019).

Challenges in Nature-Based Tourism Development and Investment in Bangladesh

Despite the immense opportunities of NBT in Bangladesh, there are many potential threats that can challenge the successes of NBT. The tourism industry was given comparatively less importance than other industries though there was huge potential due to the negligence of the government as well as political instability and poor political commitments.

Poor governance has been noticed in managing the NBT destinations of Bangladesh. The forest coverage of the country has disappeared over 50% in the last four decades (Nishorgo Support Project, 2008) with an average of 2000 ha or 0.3% per year (2000–2005) (Food and Agriculture Organization, 2006). Moreover, illegal felling, grazing, fishing and poaching cause conflicts (Trakolis, 2001) among the stakeholder resulting deforestation and forest degradation (i.e. damaging the NBT destinations). In addition, the lack of local participation in managing these forests resulted conflicts with the managing authority and deteriorated the potential NBT destinations. However, poverty, illiteracy of the local people, lack of political commitments, and poor enforcement of institutional arrangements were factors responsible for such situations.

Littering and other environmental pollution is known as one of the major challenges of NBT development in Bangladesh. Poor environmental awareness and education of the domestic tourists as well as poor monitoring and regulatory system, are causing such pollution and damaging the landscapes of NBT destinations. Moreover, poor solid waste management strategies at these destinations aggravate the situation.

There is no management plan for each of these NBT destinations, which causes poor visitor, habitat, and traffic management. It also causes seasonal patterns of tourism demand, economic leakages, distortion of local employment structure and local culture, and local price inflation. Due to less use of the concept of sustainability, these destinations could not progress as per the expectations. Moreover, due to overall poor NBT governance safety and security of these destinations are also hampered, which causes the poor reputation of these destinations.

Less skilled manpower in NBT industry is another challenge. Though currently, a considerable number of educational and training institutes are generating skilled and competent manpower, however, they are not rightly placed to manage the NBT destinations. Moreover, systematic studies and research on NBT destinations are also insufficient in Bangladesh. Nevertheless, the existing research findings are seldom considered in sustainably managing these destinations as well as enhancing tourist experiences.

NBT is multidisciplinary and multi-industrial in its nature. It requires good coordination and network among various concerned ministries and stakeholders. However, lack of inter-ministerial coordination and linkages did not progress the industry even after 49 years of the independence of the country. Insufficient private industry patronization is also one of the outcomes of such poor coordination. It causes a poor security system in the country. Moreover, it causes poor tourism marketing in Bangladesh, which restricts international tourists to visit the beautiful country.

Bangladesh positioned 168th in the World Bank's "Ease of Doing Business Ranking 2020" indicating that the regulatory environment is not favorable to a thriving private industry like NBT (The Daily Star, 2020). Moreover, the COVID-19 outbreak is an emerging challenge that Bangladesh is currently facing against NBT development and investment, as discussed earlier. Additionally, the Rohingya influx affects NBT tourism development and investment very seriously in the whole Cox's Bazar region. Such influx is causing forest (including protected areas) destructions and environmental hazards and disturbing the socio-economic and political system of the region and the whole country to some extent.

Conclusion

Tourism has received enhanced recognition from the highest level of the United Nations. One of the examples is illustrated with the recent landmark Policy Brief release on tourism from UN Secretary-General Antonio Guterres where he explained that "for women, rural communities, indigenous peoples and many other historically marginalized populations, tourism has been a vehicle for integration, empowerment and generating income" (United Nations, 2020). Bangladesh has great potentialities to be an attractive NBT destination for both domestic and international tourists. It may harness the economic, socio-cultural, environmental, and institutional manifold benefits of NBT to further develop the country (see Fig. 10.1). Bangladesh government and private sectors have taken various initiatives to adopt

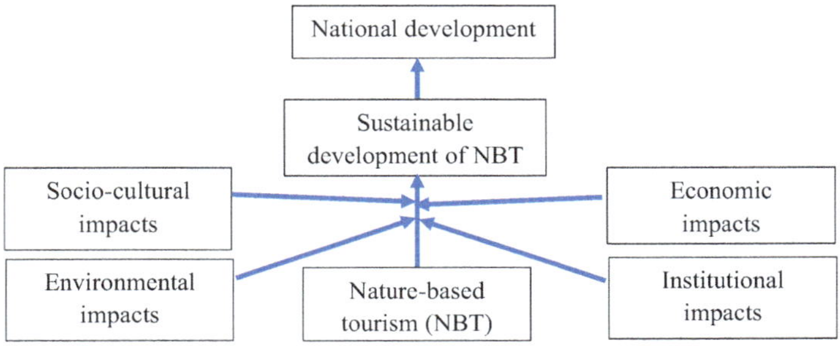

Fig. 10.1 A framework for national development through sustainable NBT. (Source: Author's compilation)

NBT as tool for poverty reduction, growing environment awareness, and education, infrastructure development which will facilitate to achieve sustainable development goals. Additionally, a competent workforce, use of Information and Communications Technology (ICT), research on NBT, and its reflection in tourism planning and implementation are vital to ensure sustainable development through NBT in Bangladesh.

The concern of public involvement in resource management decision-making has been studied by various scientists and researchers (like psychologists, sociologists, geographers, political scientists, and so on. For the sake of the coexistence of protected areas and other NBT destinations with local communities, the holistic relationship is inevitable (Trakolis, 2001). Moreover, being very fragmented in the nature tourism industry requires coordination to make a positive difference by combining various micro and small businesses (United Nations Environment Programme & United Nations World Tourism Organization, 2005). Due to tourism and sustainability complexities, international organizations reccomend to design national governance structures along with strong and functional horizontal and vertical coordination among the public authorities and concerned policies (Dinica, 2009).

It is evident that the active and functional participation of local people as well as getting the shares of benefits from the NBT enterprises are key to the success of NBT enterprises' positive facilitatioon towards local empowerment and environmental conservation.

Most of the tourist destinations of Bangladesh are located in rural areas. A variety of Bangladeshi lifestyles are experienced in these rural areas, where around 63% of the total population (Statista, 2020) live with the treasure of craft, performing arts, and rich traditional norms and values. Therefore, there is a good potential for NBT-based rural tourism. Now, it warrants an effective long-term plan to conserve and preserve all types of resources of these destinations, particularly the natural and cultural resources (safeguarding habitat and endangered species), good understanding for both tourists and local hosts. Moreover, such a development plan should also be integrated with on-going tourism master plan preparation. Bangladesh Tourism Board also needs to work closely with Bangladesh Forest Department, Rural Development and Co-operative Division, along with other relevant departments, divisions, and boards.

NBT destinations of Bangladesh need to follow the concept of "Integrated Conservation and Development Projects (ICDP)" which aims to achieve its biodiversity conservation goals along with other development goals. It is quite difficult job for the development component to promote improved living standards for the local communities. The development process should not necessarily be economically and biologically sustainable but should also conserve the destination's ecosystem. Consequently, the explicit linkages between projects' development components and conservation objectives are required to achieve this challenging demand (Wells et al., 1992). It is one of the concepts which is a harbinger for launching such linkages (Kiss, 2004). NBT can be an incentive for conservation particularly when it generates positive economic change (Stronza & Gordillo, 2008). Unfortunately, though there are many discussions about the potentialities of NBT for the local community's human wellbeing, it lacks much concrete proves on the ground (Langoya & Long, 1998).

The development through NBT can help the rural communities to recover from the COVID-19 crisis after the pandemic is over and tourism restrictions are fully lifted. It is estimated by the United Nations World Tourism Organization (2020b) that by 2050, 68% of the world population will live in urban areas, while 80% of those currently living in "extreme poverty" live outside of towns and cities. NBT might be one of the ways to address the issue.

Moreover, NBT development and investment require more political commitments, improved and diversified livelihoods that are environmentally sustainable and resilient to climate change, more active and responsible participation of private actors, promotion of environmental education and awareness, efficient marketing strategies to attract more international tourists, promoting Public-Private Partnership (PPP), management of these destinations and above all practicing tourism governance to improve NBT development and its investment.

The content of this chapter tried to compile relevant information on investment and development of nature-based tourism with reference to Bangladesh. It is expected that the literature will be useful for the students, researchers, academicians, visitors/tourists, natural resource managers, decision-makers, policymakers of concerned authorities. The content may be useful to provide some research directions on various aspects of investment and development of NBT in Bangladesh. Some of such examples of potential research are effects of an investment in the tourism sector on tourism facility development, destination management, creation and promotion of new NBT destinations, use of ICT in NBT, social-economic wellbeing, environment, governance, the flow of international tourists; the influence of NBT development on the country image, poverty reduction, and community development, addressing climate change effects, local leaderships, protected area development and management, biodiversity conservation, environmental awareness, and education, and so on. Moreover, various research on NBT related to the COVID-19 pandemic may be conducted in the future.

It is worth mentioning some of the limitations of this chapter. This chapter is based on the literature review and practical experiences of the author. The current COVID-19 pandemic situation restricted the author from accessing some of the physical libraries and talking with some key persons. Therefore, the only accessible online database has been considered as the main source of secondary information. More literature review would capture some more and rich information on NBT investment and development of Bangladesh.

References

Ahmed, M. S. (2016). *Sustainable tourism development in Bangladesh: A case study of Cox's Bazar, Bangladesh*. Bachelor's Thesis. Kokkola: Centria University of Applied Sciences.

Alam, M., Furukawa, Y., & Akter, S. (2009). Forest-based tourism in Bangladesh: Status, problems and prospects. *TOURISMOS: An International Multidisciplinary Journal of Tourism, 5*(1), 163–172.
Alam, M. K. (2019). Maritime safety and security in the Bay of Bengal. *Bangladesh Maritime Journal, 3*(1), 25–31.
Alpízar, F. (2006). The pricing of protected areas in nature-based tourism: A local perspective. *Ecological Economics, 56*(2), 294–307.
Ashoka Trust for Research in Ecology and the Environment (ATREE). (2006). *White paper on eco-tourism policy* (Draft) (p. 28). ATREE.
Bangladesh Tourism Board (BTB). (2018). *Master plan for blue economy*. BTB.
Board of Investment (BoI). (2015). *Tourism in Bangladesh*. Retrieved from: https://islamicmarkets.com/publications/tourism-investment-in-bangladesh. Accessed 31 Dec 2020.
Butler, J. R. A., Young, J. C., McMyn, I. A. G., Leyshon, B., Graham, I. M., Walker, I., Baxter, J. M., Dodd, J., & Warburton, C. (2015). Evaluating adaptive co-management as conservation conflict resolution: Learning from seals and salmon. *Journal of Environmental Management, 160*, 212–225.
Centre for Economics and Business Research (CEBR). (2020). *World Economic League Table 2021: A world economic league table with forecasts for 193 countries to 2035*. Retrieved from: https://cebr.com/wp-content/uploads/2020/12/WELT-2021-final-23.12.pdf. Accessed 31 Dec 2020.
Climate-Resilient Ecosystems and Livelihoods (CREL). (2014). *Analysis for the regional development of nature based tourism*. Retrieved from: http://nishorgo.org/wp-content/uploads/2017/02/140106-Final-Nature-based-tourism-Report.pdf. Accessed 27 Nov, 2020.
DATABD.CO. (2020). *Tourism: a possible new driver for the economy of Bangladesh*. Retrieved from: https://databd.co/stories/tourism-a-possible-new-driver-for-the-economy-of-bangladesh-9357. Accessed 19 Oct 2020.
Dhaka Tribune. (2020). *New tourist spots can contribute to rural economy*. Retrieved from: https://www.dhakatribune.com/business/2020/10/31/new-tourist-spots-can-contribute-to-rural-economy. Accessed 31 Oct 2020.
Dinica, V. (2009). Governance for sustainable tourism: A comparison of international and Dutch visions. *Journal of Sustainable Tourism, 17*(5), 583–603.
Elands, B., Islam, M. W., & Van der Duim, R. (2015). Ecotourism in Lawachara National Park in Bangladesh: Perceptions of local people and co-management officials. *South Asian Journal of Tourism and Heritage, 8*(1), 5–20.
Food and Agriculture Organization (FAO). (2006). *Global Forest Resources Assessment 2005: Progress towards sustainable forest management*. FAO.
goHR2. (2011). *What is tourism?* Retrieved from: http://www.go2hr.ca/BCsTourismbrIndustry/WhatisTourism/tabid/71/Default.aspx. Accessed 9 Jan 2013.
Hall, D., & Richards, G. (Eds.). (2003). *Tourism and sustainable community development*. Routledge.
Haque, M., Reza, M., Alam, M., Ahmed, Z. U., & Islam, M. W. (2016). Discovery of a potential site for community-based sustainable ecotourism in the Sundarbans Reserve forests, Bangladesh. *International Journal of Conservation Science, 7*(2), 553–566.
Hassan, A. (2012). Package eco-tour' as special interest tourism product-Bangladesh perspective. *Developing Country Studies, 2*(1), 1–8.
Hassan, A., & Kokkranikal, J. (2018). Tourism policy planning in Bangladesh: Background and some steps forward. *e-Review of Tourism Research, 15*(1), 79–87.
Iorio, M., & Wall, G. (2012). Behind the masks: Tourism and community in Sardinia. *Tourism Management, 33*(6), 1440–1449.
Ishtiaque, A. N. A. (2013). Tourism vision 2020: A case of Bangladesh tourism with special emphasis on international tourist arrivals and tourism receipts. *Journal of Business, 34*(2), 13–36.

Islam, M. W. (2010). Community-based tourism and ecotourism are potential means to facilitate the conservation of natural resources of the Sundarbans – A theoretical perspective. *Khulna University Studies, SESB, 2010*(special issue), 127–140.

Islam, M. W., Ahsan, M. M., & Newaz, A. N. M. Y. (2011a). *Status of tourism at five protected areas of Bangladesh.* Paper presented at the First Bangladesh Forestry Congress 2011. Dhaka: Bangladesh Forest Department, the 19–21 April.

Islam, M. R., Iftekhar, M. S., & Islam, M. W. (2011b). Potential of ecotourism development in Bangladesh coast: An overview. *Tourism Review International, 15*(4), 325–336.

Islam, M. W., Ruhanen, L., & Ritchie, B. W. (2017). Adaptive co-management: A novel approach to tourism destination governance? *Journal of Hospitality and Tourism Management, 37*, 97–106.

Islam, M. W., Ruhanen, L., & Ritchie, B. W. (2018). Tourism governance in protected areas: Investigating the application of the adaptive co-management approach. *Journal of Sustainable Tourism, 26*(11), 1890–1908.

Khandakar, S. (2014). Sustainable tourism development in Bangladesh: Strategies and guidelines. *Journal of World Vision Research, 8*(1), 176–185.

Khanom, S., Shah, M. A. R., & Chaudhary, A. (2011). *Towards ecotourism: Issues of current tourism practices in the Sundarban mangrove forest, Bangladesh.* Paper presented at the Peace, Environment and Tourism Conference. Pokhara: Global Peace Association (GPA), the 20–21 September.

Kirkpatrick, J. B. (2001). Ecotourism, local and indigenous people, and the conservation of the Tasmanian Wilderness World Heritage Area. *Journal of the Royal Society of New Zealand, 31*(4), 819–829.

Kiss, A. (2004). Is community-based ecotourism a good use of biodiversity conservation funds? *Trends in Ecology and Evolution, 19*(5), 232–237.

Komppula, R. (2014). The role of individual entrepreneurs in the development of competitiveness for a rural tourism destination – A case study. *Tourism Management, 40*(0), 361–371.

Kuenzi, C., & McNeely, J. (2008). Nature-based tourism. In O. Renn & K. D. Walker (Eds.), *Global risk governance: Concept and practice using the IRGC framework* (pp. 155–178). Springer.

Langoya, C. D., & Long, C. (1998). *Local communities and ecotourism development in Budongo Forest Reserve, Uganda.* Retrieved from: https://www.odi.org/publications/759-local-communities-and-ecotourism-development-budongo-forest-reserve-uganda. Accessed 27 Nov 2020.

Liu, J., Ouyang, Z., & Miao, H. (2010). Environmental attitudes of stakeholders and their perceptions regarding protected area-community conflicts: A case study in China. *Journal of Environmental Management, 91*(11), 2254–2262.

Ministry of Finance. (2019). *Bangladesh on a pathway to prosperity: Time is ours, time for Bangladesh (budget speech 2019–20).* Retrieved from: http://nbr.gov.bd/uploads/budget/Budget_Speech_2019-20_English_Final_Print.pdf. Accessed 26 Sept 2020.

Ministry of Law, Justice and Parliamentary Affairs. (2019). *The Constitution of the People's Republic of Bangladesh 1972 (Act no. of 1972).* Retrieved from: http://bdlaws.minlaw.gov.bd/act-print-367.html. Accessed 26 Sept 2020.

Ministry of Water Resources (2005). *Coastal zone policy.* Retrieved from: http://lib.pmo.gov.bd/legalms/pdf/Costal-Zone-Policy-2005.pdf. Accessed 27 Nov 2020.

Nilsson, N.-O. (2006). Rural development through ecotourism. In J. Hultman & S. Gössling (Eds.), *Ecotourism in Scandinavia: Lessons in theory and practice* (pp. 134–142). CABI.

Nishorgo Support Project (NSP). (2008). *Nishorgo Support Project.* Retrieved from: http://www.nishorgo.org/nishorgo_support_project.asp. Accessed 1 Jan 2011.

Pacific Asia Travel Association (PATA). (2015). *Nature-based tourism.* Retrieved from: https://sustain.pata.org/sustainable-tourism-online/parks-culture/nature-based-tourism/forms-of-nature-based-tourism/nature-based-tourism/. Accessed 29 Dec 2020.

Scholes, R., & Biggs, R. (Eds.). (2004). *Ecosystem services in southern Africa: A regional assessment.* Council for Scientific and Industrial Research.

Sherwood, D. B. (2009). Community-based wetland Comanagement in Bangladesh. In K. M. Moore (Ed.), *The sciences and art of adaptive management innovating for sustainable agriculture and natural resource management* (pp. 189–208). Soil and Water Conservation Society.

Somoy News. (2020). *6 hajar koti takai parjatan kendra hocche Chandpure* [Tourism centre worth 6 thousand crore Taka is under construction in Chandpur]. Retrieved from: https://www.somoynews.tv/pages/details/255194. Accessed 29 Dec 2020 (In Bengali).

Statista. (2020). *Share of rural population in Bangladesh from 2009 to 2018*. Retrieved from: https://www.statista.com/statistics/760934/bangladesh-share-of-rural-population/. Accessed 28 Dec 2020.

Stronza, A., & Gordillo, J. (2008). Community views of ecotourism. *Annals of Tourism Research, 35*(2), 448–468.

The Daily Star. (2020). *25th largest economy by 2035*. Retrieved from: https://www.thedailystar.net/business/news/25th-largest-economy-2035-2018409. Accessed 28 Dec 2020.

The Financial Express. (2020). *COVID-19 and domestic tourism development in Bangladesh*. Retrieved from: https://www.thefinancialexpress.com.bd/views/views/covid-19-and-domestic-tourism-development-in-bangladesh-1601134828. Accessed 26 Sept 2020.

The Financial Express. (2019). *Tapping tourism potential crucial to BD economy: USAID*. Retrieved from: https://www.thefinancialexpress.com.bd/economy/bangladesh/tapping-tourism-potential-crucial-to-bd-economy-usaid-1574483512. Accessed 27 Nov 2020.

Trakolis, D. (2001). Local people's perceptions of planning and management issues in Prespes Lakes National Park, Greece. *Journal of Environmental Management, 61*(3), 227–241.

United Nations Environment Programme (UNEP) & United Nations World Tourism Organization (UNWTO) (2005). *Making tourism more sustainable – A guide for policy makers*. Retrieved from http://www.unep.fr/shared/publications/pdf/DTIx0592xPA-TourismPolicyEN.pdf. Accessed 27 Nov 2020.

United Nations (UN). (2020). *COVID-19 response*. Retrieved from: https://www.un.org/en/coronavirus/it-imperative-we-rebuild-tourism-sector. Accessed 30 Nov 2020.

United Nations World Tourism Organization (UNWTO). (2020a). *International tourist numbers could fall 60–80% in 2020*. Retrieved from: https://www.unwto.org/news/covid-19-international-tourist-numbers-could-fall-60-80-in-2020. Accessed 17 Oct 2020.

United Nations World Tourism Organization (UNWTO). (2020b). *World tourism day 2020: Global community unites to celebrate "Tourism and rural development"*. Retrieved from: https://www.unwto.org/news/world-tourism-day-2020-global-community-unites-to-celebrate-tourism-and-rural-development. Accessed 17 Oct 2020.

Wells, M., Brandon, K., & Hannah, M. (1992). *People and parks: Linking protected area management with local communities*. IBRD.

World Economic Forum (WEF). (2019). *The travel & tourism competitiveness report 2019: Travel and tourism at a tipping point*. Retrieved from: http://www3.weforum.org/docs/WEF_TTCR_2019.pdf. Accessed 26 Sept 2020.

World Travel & Tourism Council (WTTC). (2020a). *Travel & Tourism: Global economic impact & trends 2020. Bangladesh*. Retrieved from: https://wttc.org/Research/Economic-Impact. Accessed 27 Nov 2020.

World Travel & Tourism Council (WTTC). (2020b). *Travel & Tourism: Global economic impact & trends 2020*. Retrieved from: https://wttc.org/Research/Economic-Impact. Accessed 27 Nov 2020.

World Wide Fund for Nature (WWF). (2014). *Problems with current protected areas*. Retrieved from: http://wwf.panda.org/what_we_do/how_we_work/protected_areas/protected_area_problems/. Accessed 27 Nov 2020.

Zeng, B., & Ryan, C. (2012). Assisting the poor in China through tourism development: A review of research. *Tourism Management, 33*(2), 239–248.

Chapter 11
Gastronomic Tourism in Bangladesh: Opportunities and Challenges

Nikola Naumov and Azizul Hassan

Abstract Food and food traditions are often perceived as important markers of identity, culture, and local heritage. Gastronomy and local culinary traditions provide a direct and more authentic connection to spaces and places while at the same time contributing towards a more memorable and meaningful interaction between "hosts" and "guests". Food is an indispensable part of every community and it is considered as intangible heritage, a valuable tourism resource and, an important tool for destination image and branding. Food and beverage attributes play a special role in Bangladeshi's culture and heritage. Rich culinary traditions, a variety of food, and a unique blend of spices have made Bangladeshi cuisine popular in various parts of the world. Gastronomy provides a window to the country's past and present by giving insights into farming practices, ancient gastronomic traditions, customs, and manners. Largely influenced by ethnic and social diversity and its indigenous communities, Bangladesh is a place with a rather underdeveloped food tourism industry. This chapter seeks to analyze this potential and provide recommendations for how food tourism can be established and further developed as an important pillar of the Bangladeshi tourism industry.

Keywords Gastronomy · Experience · Tourism · Culture · Heritage · Bangladesh

Introduction

Food has always constituted a very important part of our lifestyle as nutrition and personal health are essential to our daily routine and habits. Gastronomy, broadly defined as "the art and science of good eating" (Gillespie, 2002, p. xi) is an area of

N. Naumov (✉)
University of Northampton, Northampton, UK
e-mail: Nick.Naumov@northampton.ac.uk

A. Hassan
Tourism Consultants Network, The Tourism Society, London, UK

A. Hassan (ed.), *Tourism in Bangladesh: Investment and Development Perspectives*, https://doi.org/10.1007/978-981-16-1858-1_11

study that traditionally concerns "discovering, tasting, understanding and writing about food preparation" (Lilholt, 2015, p. 4). The scope of gastronomy, however, has a much broader interdisciplinary ground, including anthropological, sociological, and cross-cultural approaches to food consumption practices and traditions (see, for example, Ellis et al. 2018; Hegarty & O'Mahony, 2001; Kearney, 1970; Medina & Tresserras, 2018).

Culinary traditions, cooking, eating, and drinking have a critical role for our society as they are inextricably linked with traditions, cultures, civilizations, and history in a wider context (Dutta & Naumov, 2020; Pilcher, 2017). Food is linked to culture since the beginning of human civilizations (Chang, 1977) and often serves as an intangible marker to people, places, cultural identity, and religion (Henderson, 2014). Early theorists on food and culture such as Levi-Strauss (1966) and Douglas (1972) argue that food is a distinctive cultural element and one that also shapes patterns of cultural and social behavior.

In the context of tourism development, food and gastronomy play a critical role in both social-cultural and economic dimensions (Hall et al., 2003). Gastronomy has been utilized as an important economic segment for local and regional development (Kim et al., 2010), an influential marketing tool for enhancing destination competitiveness (Hall & Mitchell, 2001; McKercher et al., 2008) while also perceived as a significant marker of cultural identity (Henderson, 2009; Timothy & Ron, 2013). In many destinations worldwide, food tourism has become a priority sector (Liberato et al., 2019), a growing market segment (Horng & Tsai, 2012), and an arena for social innovation, entrepreneurship, and creativity (Naumov & Costandachi, 2020; Prezenza & Chiappa, 2013).

Gastronomy, gastronomic tourism, and food-related experiences play an important part in the "tourist gaze" (Getz et al., 2014; Urry, 1990) in Asia. Largely known as a region with great diversity and variety of food, the Asian continent features some of the most celebrated cuisines and food cultures. Countries such as Taiwan (Chuang, 2009), Japan (Kim & Ellis, 2015), China (Chen & Huang, 2016), Singapore (Dutta & Naumov, 2020; Henderson, 2004), and Thailand (Lunchaprasith & Macleod, 2018) traditionally attract a large number of tourists interested in gastronomy and food traditions, and local food production comprises a major part of the tourism offer from an experiential perspective. However, despite the food traditions and well-known brand image of the region, some countries in that region are yet to promote this ever-growing form of tourism.

This chapter focuses on Bangladesh, a country where food consumption takes a central role in society but also a place with a rather underdeveloped food tourism industry. Food and beverage attributes play a special role in Bangladeshi's culture and heritage. Rich culinary traditions, a variety of food, and a unique blend of spices have made Bangladeshi cuisine popular in various parts of the world. Gastronomy provides a window to the country's past and present by giving insights into farming practices, ancient gastronomic traditions, customs, and manners. This chapter seeks to analyze this potential and provide recommendations for how food tourism can be established and further developed as an important pillar of the Bangladeshi tourism industry.

Background

According to the United Nations World Tourism Organization (UNWTO) and their *Global Report on Food Tourism* (UNWTO, 2012) and *Second Global Report on Gastronomy Tourism* (UNWTO, 2017), food accounts for a significant percentage of tourist expenditure (30–40%). This not only signifies the importance of food from an economic development perspective but also underlines the undisputed importance of food from destination marketing and destination competitiveness perspectives (Ab Karim & Chi, 2010; Williams et al., 2014). The academic literature on food tourism is constantly expanding in both scale and scope (Everett, 2019). Traditional approaches that focus on the role of food tourism as a crucially important pillar of local and national economies tend to dominate the academic scholarship with an explicit focus on tourist motivation and tourist behavior (Kim et al., 2010; Tikkanen, 2007), marketing, and competitiveness (du Rand & Heath 2006; Sidali et al., 2015), innovation, and creativity (Liu et al., 2017), destination image and branding (Folgado-Fernández et al., 2017). However, there is an increased multidisciplinary interest in food tourism as a niche form of tourism that provides a more authentic, cultural, and experiential experience (Everett, 2008; Everett, 2016; Kim & Iwashita, 2016). In line with the shift from a service-based economy to an experience-based economy (Pine & Gilmore, 1999), food tourism has evolved much beyond its traditional focus on consuming food and drinks to consuming culture through food consumption. Gastronomic activities have been developed to address the need to have more memorable and experiential tourism products. Exploring local culinary traditions, taking cooking classes, and learning about local food production are just a few examples of how food tourism has been developed as an experiential tourist activity (Garibaldi & Pozzi, 2018; Mak et al., 2012 Zhang et al., 2019). Therefore, food tourism is not so much about sustenance and fulfilling our basic needs for food and drinks – it is a much more holistic experience that encourages participation, learning about the local tradition and customs, and engagement with the local community.

Food Culture and Culinary Traditions in Bangladesh

Bangladesh is located in South Asia and shares common borders with India (east, west and north), Myanmar (southeast), and the Bay of Bengal (south). With a total area of 147,570 km^2 and a population of around 160 million people, it is one of the most populated countries across the globe (Mazumder, 2016). The country has several environmental concerns related to the global rise in sea level, the monsoon season, and depletion of natural resources due to population pressure (Hall & Page, 2000). Bangladesh is often portrayed as a place with many economic and social challenges such as poverty, hunger, illiteracy, and corruption. However, the country has addressed those issues in the last few decades, achieving lower-middle-income status in 2015 and on track to finally leave the United Nation's Least Developed Countries list in 2024 after noticeable progress in terms of life expectancy, literacy, steady GDP growth and overall economic development (World Bank, 2020).

From a tourism perspective, Bangladesh contains Sundarbans, the world's largest mangrove forest as well as Cox's Bazar, the largest sea beach across the globe (120 km). Despite the variety of tourism facilities and resources, the establishment of the Ministry of Civil Aviation and Tourism (1986), and more recent efforts to categorize tourism sites, the tourism potential of Bangladesh is still under-utilized (Mazumder, 2016). The variety of natural resources and un-spoilt nature create favorable conditions for nature-based tourism activities such as ecotourism, walking tourism, trekking, and mountain sports (Wahab & Majumder, 2011). Rural tourism is also a growing niche market and area of huge potential for development due to a variety of resources, well-preserved traditions, and cultural diversity (Ahmed & Jahan, 2013). Rural tourism as well as ecotourism and other related activities are considered as valuable tools for economic development but also poverty reduction and sustainable development (Islam & Carlsen, 2012). In this chapter, we argue that food tourism can also assist with poverty reduction and also provide a sustainable approach to tourism development and management in Bangladesh.

Although there is a variety of ethnic food available in Bangladesh (e.g. Chinese, Italian, Arabic, Thai, Japanese), it is the traditional Bangla cuisine that provides a window to the ethnic and authentic food culture. Food, beverages, and culinary traditions play an important role in Bangladesh as they showcase the Bangladeshi culture, heritage, traditions, customs, and identity. Moreover, food culture and traditions offer insights into the country's past and present, focusing on the various influences that have shaped the formation of Bangladeshi society. As Mookherjee (2008) argues, Bangladesh is often associated with disasters, poverty, and crisis but at the same time as a place of hospitality, warmth, and good food. The traditional Bangla cuisine is a combination of old traditions, seasonal practices, culinary techniques, and ancestral community customs. It also features various influences arising from trade links with the Arab world and Persia, foreign settlers in the past such as British and Portuguese, as well as distinctive characteristics from neighboring Burma (Myanmar) and India.

The indigenous population of Bangladesh, with its 33 indigenous communities, traditionally consume rice, fish meals, honey, vegetables, and a variety of fruits. The region was an administrative center for the earliest Hindu and Buddhist kingdoms and later for Muslim sultanates. Bangladesh was ruled by Muslim governors from the early thirteenth century to the mid-eighteenth century, and Dhaka was the final destination point for main trade links with Delhi. During this period, Dhaka developed a vibrant and diverse culinary cuisine influenced by the Mughal courts. The Mughlai cuisine developed following the establishment of the Mughal province in 1576 as a part of the Mughal Empire and dominated the food scene with traditional rich sauces and beef (Banerji, 1991).

The British ruled the region between the eighteenth and twentieth centuries, and their influence shaped the development of Bangla cuisine during what we refer to as the "Bangla Renaissance". Unlike other settlers in the region who did not have a great influence, such as the Dutch and French, the British did influence the culinary habits of the region. At the time of the British Raj, a strong culinary influence was brought by Calcutta. West Bengal's Anglo-Indian community brought distinctive flavors. The Baghdadi Jews established the famous Jewish bakery stores and started the baking traditions across the region. Nowadays, this influence can be seen at railway stations where numerous bakery and pastry shops are present. The British also introduced the tea traditions and tea ceremonies (Banerji, 2006; Choudhury, 1993).

Today's Bangla cuisine is a mixture of different flavors, tastes, and influences from various cultures and civilizations. There are notable differences between the regions and various influences that have been inherited from previous generations. The western parts of Bangladesh, including Khulna and Jessore, share food traditions with West Bengal of India, and traditional Bengali recipes dominate the local food scene. The northern parts have a strong influence on Eastern India (mainly Assam and Manipur), and traditional food includes many sweets and fruits such as bananas, mango, and papaya. Eastern parts of Bangladesh are famous for fish and rice as well as the authentic cooking style of the Sylheti community. Southern Bangladesh is inhabited by many indigenous tribes, each one of them with its different cooking and culinary traditions. The Arakan influence that originated in Burma has a strong influence in the most southern parts of the country, and this is well exemplified with the abundance of bamboo shoots, fish, and coconut in many recipes. Central Bangladesh, including the capital city of Dhaka, is famous for fresh fish but also the Nawab Awadhi cuisine. Kebabs, Biriyani, and mutton are particularly distinctive in this part of the country.

Food Tourism: Opportunities and Perspectives

Bangladeshi cuisine is widespread and particularly famous at places outside the country with large expatriate Bangladeshi communities such as United Kingdom, Australia, Canada, and the USA (Pottier, 2014; Sen, 2016). However, Bangladesh is rarely mentioned in the context of food tourism, and this ever-growing type of niche tourism is virtually non-existent on the map of tourist activities. The rich history, culture, and heritage of Bangladesh and particularly the various influences from foreign settlers, the hospitality of the Bangladeshi community as well as the authentic and un-spoilt environment are key factors that underline the potential development of food tourism. Undoubtedly, there is a huge potential to develop good tourism in Bangladesh, but there is a need for a more explicit focus and investment in this sector.

First, food tourism should become a priority area of development and be linked with other growing niche forms of tourism such as ecotourism, cultural tourism, and heritage tourism. Food-related activities should become central to marketing and promotional campaigns, particularly in the western parts of the country where the traditional Bengali recipes dominate but also in the central region and the capital city of Dhaka. Food experiences can provide a platform for further development of cultural activities and ones that tell the story of a destination, region, and community.

Second, a more explicit focus should be put on food tourism as an authentic, memorable, and participatory experience. In this context, food should not be solely understood as a tangible product to be manufactured and sold as a tourism commodity. Instead, the focus should be on the production of food, for example, culinary traditions, authentic recipes, distinctive flavors, etc. Each region in Bangladesh is culturally rich, and promoting food experiences that seek to unfold history, culture, and heritage could greatly enhance the development of tourist activities.

Third, there is a new approach needed to market and develop food-related activities as intangible heritage. Bangladesh only has four elements inscribed on the List of Representative List of the Intangible Cultural Heritage of Humanity, and none of them is linked to a culinary tradition (UNESCO, 2020). Bangladesh is not a part of UNESCO's Creative Cities network, where food is an essential marker for creativity and innovation. The variety and diversity of food traditions across all regions in Bangladesh is a distinctive advantage for cultural and heritage tourism development. Moreover, the various trade links and foreign influences should also be marked separately as distinctive food tourism products.

Challenges and Recommendations for Gastronomic Tourism Development in Bangladesh

For the development of gastronomic tourism in Bangladesh, key challenges are identified, followed by some suggestions in brief.

Authenticity Challenges

Authenticity issues in food culture can be severe in terms of sustainable local food experience offer (Sims, 2009). The efforts and capacities to maintain the authenticity of Bangladeshi cuisine are limited. This is true that the cuisine can find a place in few counties, including the United Kingdom, few regions in the United States of America, and some middle and far east Asian countries. However, the recipe and cuisine are altered either with the locals or with the Indians. As a result, the authenticity of Bangladeshi cuisine is challenged. For example, according to Travel Daily News (2014), the "Adivasi" people are the indigenous people living in specific regions in Bangladesh (i.e. in hilly areas like Chattogram, Sylhet, as well as in Rajshahi, etc.). These people are used to make some exotic and authentic foods that are different from their general neighbors. In specific, the Chittagong Hill Tracts conventionally have a real liking for fish-based flavoring. This resembles the Burmese or Thai cooking slightly shifted towards the South Asian cuisine. This purely Bangladeshi cuisine with the relevant others is missed in most of the countries abroad.

Policy Challenges

The national tourism policy (NTP) of Bangladesh misses the specific directions for the development of tourism niches and gastronomic tourism, in particular. The existing NTP rather is ambiguous and creates confusion (Hassan et al., 2020; Hassan & Kokkranikal, 2018, Hassan & Burns, 2014). The lack of tourism-friendly and timely policy is rather a major challenge for the overall development of this industry, including gastronomy tourism development in the country. Having a tourism-friendly NTP is a major challenge for any country and Bangladesh, in particular.

Awareness Challenges

Awareness of the tourists for local food culture helps to promote gastronomic tourism (Amin & Roy, 2016; Amin 2014). In Bangladesh, the part of her culture is to honor visitors and guests with the creation of an amiable ambiance in the home. They are then served delicious homemade food as the keyway for showing the fervor. In the country, the best food is prepared at home with care, love, and adequate time. However, from the gastronomic tourism context, the richness, diversity, and taste of Bangladeshi food are not very well-known among both the domestic and global tourist base. The Bangladeshi cuisine can face severe challenges for its globalization and for reaching out to the larger gastronomic tourists.

Food Quality, Hygiene, and Pricing Challenges

Maintaining the food quality and hygiene requirements are challenges (Shalini & Duggal, 2015). In Bangladesh, conventionally, food quality and hygiene are challenges and poorly respected in many cases. In terms of attracting foreign gastronomic tourists, this is essential to overcome this challenge. These tourists are very unlikely to compromise with below food standard and hygiene status. As well as, the right food pricing also is a challenge. Gastronomic tourists can be very cautious about food pricing in destinations they travel to, and also, they tend to compare food prices with the destinations they already have visited. According to Huda et al. (2020), innovative skills of chefs and more investment from entrepreneurs can exploit the potentials of culinary tourism in Bangladesh.

Marketing Challenges

Sidali et al. (2015) believe that food tourism is interconnected to niche markets and rural tourism products. Most gastronomic tourists rely on the Internet for getting data and information about the cuisine and food of the destination they are traveling to. Thus, promotion and marketing are important, followed by branding. Bangladesh lacks in all aspects of branding, marketing, and promotion. This is a major challenge for both attracting and retaining both domestic and international gastronomic tourists (Khanam, 2016).

Recommendations

On the basis of the discussions made and arguments of Stewart et al. (2008), some very specific suggestions are offered to effectively minimize the challenges and to promote gastronomic tourism in Bangladesh.

First, the purely Bangladeshi cuisine needs to be made familiar for tourists. This is important for maintaining authenticity. Foreign gastronomic tourists can then be able to identify Bangladeshi food and cuisine. This will also support in ensuring the authenticity of the food culture of Bangladesh before both domestic and international tourists.

Second, the formulation and implementation of a gastronomic tourism-friendly NTP are essential. Without the useful NTP, this is very unlikely that strong gastronomic tourism can be developed in the country.

Third, awareness is important. Advocating for particular information and signage aimed for diverse culinary target market segment sub-sets for dealing with the dissimilarities in consumer preferences and motivations demand adequate attention.

Fourth, maintaining internally acceptable food quality and hygiene is important. Also, for attracting one-time tourists to the country, this is essential for enhancing service with the support of increased service training.

Fifth, timely and effective marketing strategies both on the print and electronic media are essential. The need for both coordination and cooperation within the industry for branding, marketing, and promotion at all levels is important.

Conclusion

This chapter aims to focus on Bangladesh as the country where the consumption of food becomes a key feature of society, as well as a place having an underdeveloped food tourism industry. The chapter makes critical explanations of both gastronomy and food on the global scale that narrows down to the Bangladesh context. This chapter finds rich food culture and culinary traditions in Bangladesh. In support of these findings on the literal notes, Mookherjee (2008) opines that a country is a place of good food, hospitality, and warmth when the conventional Bangla cuisine combines the old traditions with culinary techniques, seasonal practices, and ancestral community customs. The customary food and hospitality tradition is well influenced by the Arabs, Persians, the past foreign settlers like British and Portuguese, even India as the neighbor. The possibilities of gastronomic tourism in Bangladesh are immense, as evidenced by its strong presence in many countries in Europe, America, and far east Asia.

The present Bangla cuisine is thus a combination of different tastes, flavors, and influences of diverse cultures and civilizations. Also, the in-country differences between the regions are notable, as inherited from earlier generations. This research affirms that even with several pertaining challenges, Bangladeshi cuisine is flourishing both domestically and globally. Gastronomy in Bangladesh can represent Bangladeshi present and past gastronomic traditions, customs, and manners to some extent and with limitations. Food tourism in the country has tremendous potentials and can be developed as a useful facilitator of the Bangladeshi tourism industry that future studies can research more elaborately and critically.

Acknowledgments The authors would like to express their gratitude to Md Yusuf Khan and Tanvir Abir from the International University of Business Agriculture and Technology (Dhaka, Bangladesh) for their insights and valuable help with writing the paper. A big thank you to the owner and all staff members at Blue Monsoon restaurant in Winslow, Buckinghamshire (UK), for their invaluable support and expert knowledge on Bangladesh and Bangladeshi traditions and culture. A big thank you also goes to our families and friends for their invaluable support and encouragement offered.

References

Ab Karim, M., & Chi, C. (2010). Culinary tourism as a destination attraction: An empirical examination of destinations' food image. *Journal of Hospitality Marketing & Management, 19*(6), 531–555.

Ahmed, I., & Jahan, N. (2013). Rural tourism-prospects in rustic Bengal. *European Journal of Business and Management, 5*(16), 163–172.

Amin, M. R. (2014). *Domestic tourists' attitude towards responsible tourism management: A case study on Cox's bazar, Bangladesh*. Unpublished Msc thesis. Leeds: Leeds Beckett University.

Amin, M. R., & Roy, B. (2016). Factors influencing tourists' perception towards Bangladeshi foods. *European Journal of Business and Management, 8*(20), 63–74.

Banerji, C. (1991). *Life and food in Bengal*. Weidenfeld and Nicolson.

Banerji, C. (2006). *Feeding the gods: Memories of food and culture in Bengal*. Seagull Books.

Chang, K. C. (1977). *Food in Chinese culture: Anthropological and historical perspectives*. Yale University Press.

Chen, Q., & Huang, R. (2016). Understanding the importance of food tourism to Chongqing, China. *Journal of Vacation Marketing, 22*(1), 42–54.

Choudhury, Y. (1993). *The roots and tales of the Bangladeshi settlers*. Sylheti Social History Group.

Chuang, H. (2009). The rise of culinary tourism and its transformation of food cultures: The National Cuisine of Taiwan. *The Copenhagen Journal of Asian Studies, 27*(2), 84–108.
Douglas, M. (1972). Deciphering a meal. *Daedalus, 101*, 61–82.
Du Rand, G. E., & Heath, E. (2006). Towards a framework for food tourism as an element of destination marketing. *Current Issues in Tourism, 9*(3), 206–234.
Dutta, P., & Naumov, N. (2020). Food culture: Gastronomic experiences in Singapore. *Journal of Gastronomy, Hospitality, and Travel, 3*(1), 20–29.
Ellis, A., Park, E., Kim, S., & Yeoman, I. (2018). What is food tourism? *Tourism Management, 68*, 250–263.
Everett, S. (2008). Beyond the visual gaze? The pursuit of an embodied experience through food tourism. *Tourist Studies, 8*(3), 337–358.
Everett, S. (2016). *Food and drink tourism*. Routledge.
Everett, S. (2019). Theoretical turns through tourism taste-scapes: The evolution of food tourism research. *Research in Hospitality Management, 9*(1), 3–12.
Folgado-Fernández, J. A., Hernández-Mogollón, J. M., & Duarte, P. (2017). Destination image and loyalty development: The impact of tourists' food experiences at gastronomic events. *Scandinavian Journal of Hospitality and Tourism, 17*(1), 92–110.
Garibaldi, R., & Pozzi, A. (2018). Creating tourism experiences combining food and culture: An analysis among Italian producers. *Tourism Review, 73*(2), 230–241.
Getz, D., Robinson, R., Andersson, T., & Vujicic, S. (2014). *Foodies & food tourism*. Goodfellow Publishers.
Gillespie, C. (2002). *European gastronomy into the 21st century*. Butterworth-Heinemann.
Hall, C. M., & Mitchell, R. (2001). Wine and food tourism. In N. Douglas & R. Derrett (Eds.), *Special interest tourism* (pp. 307–329). Wiley.
Hall, C. M., & Page, S. (2000). Developing tourism in South Asia: India, Pakistan and Bangladesh – SAARC and beyond. In C. M. Hall & S. Page (Eds.), *Tourism in South and Southeast Asia: Issues and cases* (pp. 197–225). Butterworth-Heinemann.
Hall, C. M., Sharples, L., Mitchell, E., Macionis, N., & Cambourne, B. (Eds.). (2003). *Food tourism around the world*. Buttertworth-Heinemann.
Hassan, A., & Burns, P. (2014). Tourism policies of Bangladesh – A contextual analysis. *Tourism Planning & Development, 11*(4), 463–466.
Hassan, A., & Kokkranikal, J. (2018). Tourism policy planning in Bangladesh: Background and some steps forward. *e-Review of Tourism Research (eRTR), 15*(1), 79–87.
Hassan, A., Kennell, J., & Chaperon, S. (2020). Rhetoric and reality in Bangladesh: Elite stakeholder perceptions of the implementation of tourism policy. *Tourism Recreation Research, 45*(3), 307–322.
Hegarty, J., & O'Mahony, B. (2001). Gastronomy: A phenomenon of cultural expressionism and aesthetic for living. *International Journal of Hospitality Management, 20*(1), 3–13.
Henderson, J. (2009). Food tourism reviewed. *British Food Journal, 111*(4), 317–326.
Henderson, J. (2014). Food and culture: In search of a Singapore cuisine. *British Food Journal, 116*(6), 904–917.
Henderson, J. C. (2004). Food as a tourism resource: A view from Singapore. *Tourism Recreation Research, 29*(3), 69–74.
Huda, K. N., Uddin, M. J., & Haque, A. E. (2020). Panorama of Bangladeshi culinary tourism: Prospect and challenges. *Hospitality & Tourism Review, 1*(1), 23–34.
Horng, J., & Tsai, C. (2012). Culinary tourism strategic development: An Asia-Pacific perspective. *International Journal of Tourism Research, 14*(1), 40–55.
Islam, F., & Carlsen, J. (2012). Tourism in rural Bangladesh: Unlocking opportunities for poverty alleviation? *Tourism Recreation Research, 37*(1), 37–45.
Kearney, M. (1970). The social meaning of food sharing in Mexico. *Kroeber Anthropological Society Papers, 43*, 32–41.

Khanam, S. (2016). *Marketing of food tourism. Evaluating marketing of food tourism in Finland.* Retrieved from: https://www.theseus.fi/bitstream/handle/10024/114133/Khanam_Shazada.pdf?sequence=1. Accessed 28 Oct 2020.
Kim, S., & Ellis, A. (2015). Noodle production and consumption: From agriculture to food tourism in Japan. *Tourism Geographies, 17*(1), 151–167.
Kim, S., & Iwashita, C. (2016). Cooking identity and food tourism: The case of Japanese udon noodles. *Tourism Recreation Research, 41*(1), 89–100.
Kim, Y. H., Goh, B. K., & Yuan, J. (2010). Development of a multi-dimensional scale for measuring food tourist motivations. *Journal of Quality Assurance in Hospitality & Tourism, 11*(1), 56–71.
Levi-Strauss, C. (1966). Culinary Triangle. In C. Counihan & P. van Esterik (Eds.), *Food and culture – A reader* (pp. 36–41). Routledge.
Liberato, P., Mendes, T., & Liberato, D. (2019). Culinary tourism and food trends. In A. Rocha, A. Abreu, J. de Carvalho, D. Liberato, E. Gonzalez, & P. Liberato (Eds.), *Proceedings of ICOTTS 2019 advances in tourism, technology and smart systems* (pp. 517–526). Springer.
Lilholt, A. (2015). *Entomological gastronomy*. Lulu. com.
Liu, S. Y., Yen, C. Y., Tsai, K. N., & Lo, W. S. (2017). A conceptual framework for agri-food tourism as an eco-innovation strategy in small farms. *Sustainability, 9*(10), 1683.
Lunchaprasith, T., & Macleod, D. (2018). Food tourism and the use of authenticity in Thailand. *Tourism Culture & Communication, 18*(2), 101–116.
Mak, A., Lumbers, M., & Eves, A. (2012). Globalisation and food consumption in tourism. *Annals of Tourism Research, 39*(1), 171–196.
Mazumder, M. (2016). Bangladesh. In J. Jafari & H. Xiao (Eds.), *Encyclopedia of tourism* (pp. 87–88). Springer.
McKercher, B., Okumus, F., & Okumus, B. (2008). Food tourism as a viable market segment: It's all how you cook the numbers. *Journal of Travel & Tourism Marketing, 25*(2), 137–148.
Medina, F., & Tresserras, J. (2018). Introduction. Gastronomy and tourism: Social and cultural perspectives. In F. Medina & J. Tresseras (Eds.), *Food, gastronomy and tourism. Social and cultural perspectives*. Universidad de Guadalajara.
Mookherjee, N. (2008). Culinary boundaries and the making of place in Bangladesh. *South Asia: Journal of South Asian Studies, 31*(1), 56–75.
Naumov, N., & Costandachi, G. (2020). Creativity and entrepreneurship: Gastronomic tourism in Mexico. In J. Rodrigues Soares (Ed.), *Innovation and entrepreneurial opportunities in community tourism* (pp. 90–108). IGI Global Publishing.
Pilcher, J. (2017). *Food in world history*. Routledge Taylor & Francis Group.
Pine, B., & Gilmore, J. (1999). *The experience economy*. Harvard Business School Press.
Pottier, J. (2014). Savoring "the authentic": The emergence of Bangladesh cuisine in East London. *Food, Culture & Society, 17*(1), 7–26.
Prezenza, A., & Chiappa, G. (2013). Entrepreneurial strategies in leveraging food as a tourist resource: A cross-regional analysis in Italy? *Journal of Heritage Tourism, 8*(2), 182–192.
Sen, A. (2016). Food, place, and memory: Bangladeshi fish stores on Devon Avenue, Chicago. *Food and Foodways, 24*(1–2), 67–88.
Shalini, D., & Duggal, S. (2015). A review on food tourism quality and its associated forms around the world. *African Journal of Hospitality, Tourism and Leisure, 4*(2), 1–12.
Sidali, K. L., Kastenholz, E., & Bianchi, R. (2015). Food tourism, niche markets and products in rural tourism: Combining the intimacy model and the experience economy as a rural development strategy. *Journal of Sustainable Tourism, 23*(8–9), 1179–1197.
Sims, R. (2009). Food, place, and authenticity: Local food and the sustainable tourism experience. *Journal of Sustainable Tourism, 17*(3), 321–336.
Stewart, J. W., Bramble, L., & Ziraldo, D. (2008). Key challenges in wine and culinary tourism with practical recommendations. *International Journal of Contemporary Hospitality Management, 20*(3), 303–312.

Tikkanen, I. (2007). Maslow's hierarchy and food tourism in Finland: Five cases. *British Food Journal, 109*(9), 721–734.

Timothy, D., & Ron, A. (2013). Understanding heritage cuisine and tourism: Identity, image, authenticity, and change. *Journal of Heritage Tourism, 8*(2–3), 99–104.

Travel Daily News. (2014). *Bangladesh: Gastronomy.* Retrieved from: https://www.traveldailynews.com/post/8.-bangladesh-%3A-gastronomy%2D%2D51080. Accessed 28 Oct 2020.

UNESCO. (2020). *Bangladesh and the 2003 Convention.* Retrieved from: https://ich.unesco.org/en/state/bangladesh-BD. Accessed 23 Oct 2020.

UNWTO. (2012). *Global report on food tourism.* UNWTO.

UNWTO. (2017). *Second global report on gastronomy tourism.* UNWTO.

Urry, J. (1990). *The tourist gaze.* Sage.

Wahab, M., & Majumder, M. (2011). An appraisal of tourism industry development in Bangladesh. *European Journal of Business and Management, 3*(3). Retrieved from: https://ssrn.com/abstract=3493303. Accessed 23 Oct 2020.

Williams, H., Williams, R., & Omar, M. (2014). Gastro-tourism as destination branding in emerging markets. *International Journal of Leisure and Tourism Marketing, 4*(1), 1–18.

World Bank. (2020). *World Bank in Bangladesh.* Retrieved from: https://www.worldbank.org/en/country/bangladesh/overview. Accessed 23 Oct 2020.

Zhang, T., Chen, J., & Hu, B. (2019). Authenticity, quality, and loyalty: Local food and sustainable tourism experience. *Sustainability, 11*(12), 3437.

Chapter 12
Developing Medical Tourism in Bangladesh: Issues, Challenges and Policies

Md Aslam Mia and Hasanul Banna

Abstract Despite rapid growth in tourism services in recent years, medical tourism remains a lucrative business for many developing countries in the world. In particular, the costlier healthcare facilities in the developed world have pushed many developing countries to invest heavily in medical tourism facilities to attract foreign patients for economic gain. Poor healthcare facilities and recent economic development in Bangladesh have also pushed many patients to get treatment from India, Thailand, Malaysia, Singapore, and other neighboring countries in Asia. As a result, Bangladesh is not only losing economically as medical tourists spend abroad, but also reveals the lack of quality in healthcare facilities. To better understand how Bangladesh could promote its medical tourism domestically and globally, the study evaluates factors affecting medical tourism based on the existing theories and literature. In so doing, the study identified a total of four factors (e.g. healthcare quality, institutional and cultural factors, macroeconomic, and individual) and several sub-factors for medical tourism. To promote medical tourism in Bangladesh, intensive planning, proper coordination among various units (private and public), public awareness, and quality of human resources (e.g. doctors, nurses, medical technologists, etc.) are required. Moreover, appropriate public and private investment in infrastructure is a prerequisite for promoting medical tourism in Bangladesh. Moreover, the study provides some policy implications and guidance to initiate medical tourism for a better and sound healthcare system in Bangladesh, both for domestic and foreign patients.

Keywords Medical tourism · Investment · Healthcare · Pharmaceutical · Bangladesh

---

Md A. Mia (✉)
School of Management (SOM), Universiti Sains Malaysia (USM), Penang, Malaysia
e-mail: aslammia@usm.my

H. Banna
Ungku Aziz Centre for Development Studies, Faculty of Economics and Administration, Universiti Malaya (UM), Kuala Lumpur, Malaysia

A. Hassan (ed.), *Tourism in Bangladesh: Investment and Development Perspectives*, https://doi.org/10.1007/978-981-16-1858-1_12

Introduction

Bangladesh had been in the media forefront in recent years for her remarkable progress in health, education, and socio-economic development (Alam et al., 2020; Barai, 2020; Mahmud et al., 2018). Still, her healthcare sector is yet to perform due to insufficient human resources, overcrowded public healthcare system, corruption, lack of quality of services, and mismanagement (Ali, 2012; Mahmud et al., 2020; Mamun & Andaleeb, 2013). Thus, following the increase in per capita income in recent years, the demand for outbound healthcare services has also increased very significantly in Bangladesh. About 221,751 people traveled to India just in 2017 for medical treatment, and Bangladesh remains the top country for India's inbound medical tourism (Kumar, 2018). Approximately 50% of the inbound medical tourists in India are actually from Bangladesh, based on data from previous years (Kumar, 2018).

Now, the question is: What is medical tourism? Can it be likened to conventional tourism? Medical tourism is often defined as the situation in which people travel overseas to obtain medical, dental, and surgical attention while simultaneously holidaying (Connell, 2006, p. 1094). Moreover, Hopkins et al. (2010) defined health tourism/medical travel or medical tourism as the "cross-border health care motivated by lower cost, avoidance of long wait times, or services not available in one's own country (p. 185)". In other words, Carrera and Bridges (2006) defined medical tourism as "the organized travel outside one's natural healthcare jurisdiction for the enhancement or restoration of the individual's health through medical intervention (p. 447). Based on the types of medical services, medical tourism has been categorized into different dimensions (see Lee & Taggart, 2013 for detailed classification).

With the rapid pace of globalization in recent decades, medical tourism has become a multi-billion-dollar enterprise, with several countries emphasizing its expansion to earn foreign currency and expand the domestic economy (Beladi et al., 2019; Hasan & Hassan, 2013; Sandberg, 2017). According to Business Wire (2020), the worth of medical tourism is projected to be USD 207.9 Billion by 2027, with an average growth rate of 21.2%. Millions of people travel from one country to another for medical treatment, and the number could double if people who travel within the country for such treatment are also considered (Sandberg, 2017). Cheap transportation, more disposable income, longer life expectancy, liberalization of trade in services, and the growing cooperation between the public and private sectors, and unavailability of such services in one's country fuelled the growth of medical tourism globally (Bookman & Bookman, 2007; Connell, 2013).

Due to the global increase in demand for medical tourism (Connell, 2006; Kim et al., 2013), Bangladesh could tap into this market to benefit from its relatively cheap human resources, globally recognized pharmaceutical companies, and recent socio-economic development. There will be at least two significant effects of focusing on medical tourism in Bangladesh. First, attracting foreign tourists for medical treatment will not only enable the country to earn in foreign currency but also help to expand other industries that are highly connected to health tourism. Second, when people come from overseas for treatment in Bangladesh, a positive image and branding among domestic patients are portrayed. This may motivate them to seek domestic treatment as opposed to outbound medical tourism. In this way, the country can save foreign currencies and promote its domestic healthcare facilities to a wider range of customers, as the rate of outbound medical tourism in Bangladesh has become very alarming in recent years. Furthermore, through the export of healthcare services, a country can effectively and efficiently utilize its excess resources, if any (Runnels & Carrera, 2012).

Despite the tremendous benefits of medical tourism in its domestic economy, Bangladesh's medical tourism facilities are still in the nascent stage. Therefore, significant investment and appropriate planning are required for the sector to bring benefits to the country. This chapter aims to identify the factors affecting inbound medical tourism, to give an overview to policymakers and practitioners to enable them to enact effective policies and initiatives by the government and relevant authorities. Moreover, the study also proposes a framework borne from the integration of various ministries and government authorities to promote medical tourism in Bangladesh. By so doing, the chapter will be utilizing secondary types of data, extant literature, and desk research to fulfill the objective of the research.

The organization of the chapter is as follows: section "Factors affecting inbound medical tourism" discusses the main factors affecting inbound medical tourism, section "Sub-dimensions of factors affecting inbound medical tourism" encompasses a thorough discussion on each of the sub-factors, section "Proposed framework to promote inbound medical tourism in Bangladesh" proposes a framework for establishing an efficient and effective medical tourism industry in Bangladesh in a bid to promote inbound tourism, and section "Conclusion" concludes the study with some policy implications and suggestions.

Factors Affecting Inbound Medical Tourism

Since medical tourism is a part of the service industry, several factors such as the quality of healthcare, institutional and cultural factors, macroeconomic factors of the destination country, and individual factors affect such services. It would be impossible to pinpoint the exact factors influencing an individual's preference of seeking medical attention in a country; however, the theory of planned behavior could largely explain this choice to some extent (Seow et al., 2017). Since no quantitative analysis will be dealt with in this chapter, these factors would be briefly discussed in light of existing literature and theories (Fig. 12.1), followed by a thorough discussion in the subsequent sections.

One of the core reasons for seeking healthcare treatment abroad is the quality of healthcare services provided by the country/hospitals (Debata Bikash et al., 2015; Manaf et al., 2015; Musa et al., 2012). Patients will certainly do a review of the quality of medical care provided by the countries or hospitals in question through

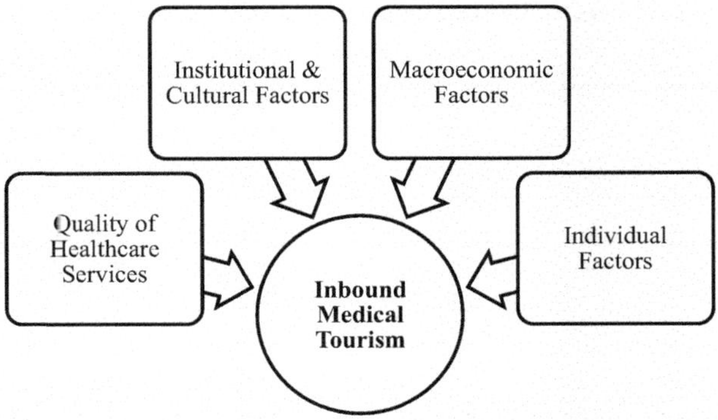

Fig. 12.1 Factors affecting inbound medical tourism. (Source: Authors)

online or third-party consultation before embarking on the decision to travel abroad (Runnels & Carrera, 2012). If the quality of service provided by the hospitals/country abroad is no better than that in the home country, then there is no point seeking healthcare treatment abroad.

Since medical tourism is also linked with recreational activities, it is affected by overall institutional and cultural factors (Esiyok et al., 2017; Liu & Chen, 2013). Institutional factors generally refer to those factors that portray the overall image of a country to foreigners or potential inbound patients. Such image or branding of the destination country will have a profound effect on the inbound medical tourism, as ample researches establishing this fact exists (de la Hoz-Correa & Muñoz-Leiva, 2019). Moreover, a similarity in culture with the destination country gives the patient a better feeling, as it becomes easier for the patient to adapt to the environment during the period of treatment and beyond. As an instance, Malaysia and Indonesia remain the preferred medical tourist destination for patients from many gulf countries, owing to the similarity in culture and religion (Connell, 2013). Similarly, China also attracts medical tourists from Taiwan, Hong Kong, and Macau, due to their geographical connection and similarity in customs (see Esiyok et al., 2017).

The overall development of a country represented by several macroeconomic factors such as the Gross Domestic Product (GDP) per capita, inflation, interest rates, and macroeconomic stability will also have an ensuing effect on a patient's decision to seek medical treatment abroad. There exist several studies, such as Tang & Lau (2017), which have highlighted the significant effect of income and exchange rate on inbound medical tourism in Malaysia. Similarly, Daykhes et al. (2020) highlighted the importance of macroeconomic measures in the enhancement of the quality of healthcare services provided by Russian hospitals.

Apart from these factors, individual-level characteristics and the perceived value of patients also influence the decision to seek medical treatment abroad (Oberoi & Kansra, 2019; Wang, 2012). Usually, a potential medical tourist firstly selects a country based on his or her discretion and subsequently researches for the best medical hospitals, infrastructure, and overall services in the selected country. The economic wellbeing of a patient is a crucial determinant, as the wealthy do not mind traveling far to obtain medical attention. The less wealthy, on the other hand, may prefer to obtain treatment in nearby countries to minimize the cost of travel and other associated expenses. One of the good examples in this case would be the Bangladeshi patients (usually middle income) travelling to India due to close proximity and competitive cost.

Sub-dimensions of Factors Affecting Inbound Medical Tourism

As previously highlighted, a very significant factor affecting the decision of inbound medical tourism is the overall quality of healthcare services rendered by the destination country. In this section, the sub-dimensions of the quality of healthcare are discussed to enable countries to leverage these factors to improve their quality of healthcare and attract more medical tourists, including Bangladesh (Fig. 12.2).

First and foremost is the reputation of the hospitals, as international accreditation agencies around the world provide ratings to hospitals/clinics based on their overall quality (Shaw et al., 2013). Therefore, medical tourists will be convinced if their intended hospitals receive higher ratings from international accreditation agencies or are positively reviewed by existing/past clients. Another quality indicator of

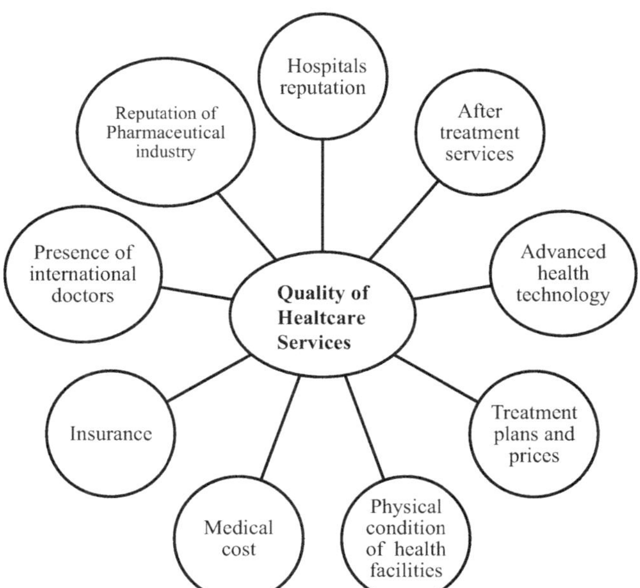

Fig. 12.2 Sub-dimensions of quality of healthcare services. (Source: Authors based on the existing literature)

healthcare services is the presence of international doctors in the hospital/clinic, as this would signal to the client that the hospital/clinic is diversified and have international experts dedicated to the provision of quality services. Nonetheless, the presence of a foreign hospital branch could also enhance the visibility of a country's healthcare facilities to global clients.

Several hospitals provide after-treatment services, follow-up, etc., to medical tourists, and these facilities help to enhance customer confidence and trust. Rapid growth can be observed in the medical technology used in hospitals and clinics (Wu et al., 2019), and one of the reasons people seek medical treatment abroad is due to the absence of such technology in their home country. Therefore, access to the latest and cutting edge medical technology provides a competitive advantage to hospitals and clinics and helps attract foreign medical tourists.

The availability of diverse treatment plans mostly demanded by foreign tourists, such as organ transplant, cosmetic surgery, medical check-ups, etc., will also help to attract inbound medical tourists (Musa et al., 2012). Thus, the establishment of unique and specialized hospitals is a core strength of any country in the attraction of more medical tourists. Also, medical expenses are one of the very significant factors influencing the choice of the destination country by a medical tourist. Potential clients will certainly make a cost and benefits analysis before embarking on the decision to select a country for their treatment. Since the competition for medical tourism is now intense across countries, competitive pricing of medical services should become an ultimate strategy of service providers (Tang & Lau, 2017). To reduce the overall cost of medical treatment, countries such as Singapore, Thailand, India, etc., reduce customs duties and taxes on hospitals attracting foreign tourists, rent out government property at lower than market rates, and provide government incentives to potential tourists and service providers (Bookman & Bookman, 2007; Kim et al., 2013; Singh, 2008).

Nonetheless, the availability of domestic insurance (destination country) and acceptance of international insurance (covered by the patients) in the country would have a significant effect on the reduction of the cost of medical tourism and the provision of flexibility to patients. These factors should be highlighted on the websites/flyers of hospitals to enable easy access of patients to them before arrival.

Institutional and cultural factors are crucial to the attraction of medical tourists (Fig. 12.3). The government must be ready to welcome medical tourists regardless of their race, religion, and country of origin. A friendly government that ensures the safety of foreign tourists and provides quality healthcare through legislation and regulation will certainly be preferred (Tang & Lau, 2017). A reduction in crime rate and safety improvement will enhance the confidence and perception of clients about the destination country. Moreover, some countries are skeptical about receiving foreign tourists, as they find it against their self-interest. Hence, large scale public awareness programs should be enhanced to enlighten people about foreigners. Additionally, several countries have stringent visa requirements for medical tourists, and as such, the ease of visa approval or simplification of visa procedures could also be a stimulating factor driving the demand for medical tourism (Kim et al., 2013). Moreover, the government should liaise with hotel operators, tour agencies, and

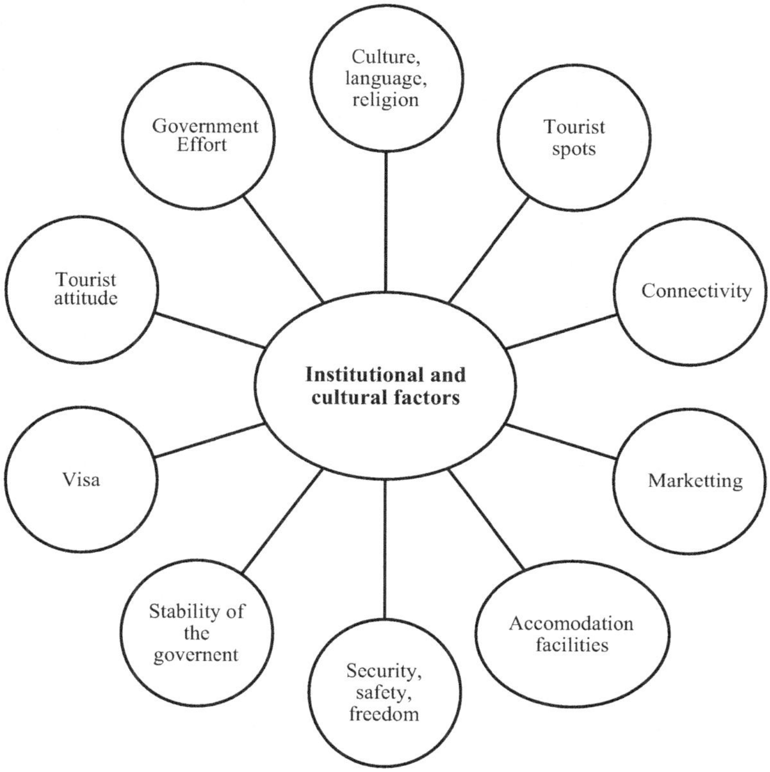

Fig. 12.3 Sub-dimensions of institutional and cultural factors. (Source: Authors based on the existing literature)

local enforcement units to provide a safe, sound, and quality transitions from one part of the country to another for a hassle-free travel experience.

The government should also ensure a better transportation network within the country via air, rail, and road transport, as it is crucial to the promotion of medical tourism. Some studies have found the positive effects of air transport development on overall tourism, such as in Saudi Arabia (Alsumairi & Hong Tsui, 2017). Marketing of medical tourism should be undertaken through the various tour and travel agencies as well as medical tourism facilitators. These marketers should not only highlight the medical services available in the destination countries but also emphasize the ancillary services available to the medical tourists, as observed in the USA (Gan & Frederick, 2011).

Since medical tourists usually spend a long time in the country than normal tourists, a couple of macroeconomic factors may affect their choice of the destination country (Fig. 12.4). As an instance, if the currency of the intended country is relatively expensive, it will ultimately increase the overall cost of their medical treatment. This explains why some Asian countries, particularly India, have become

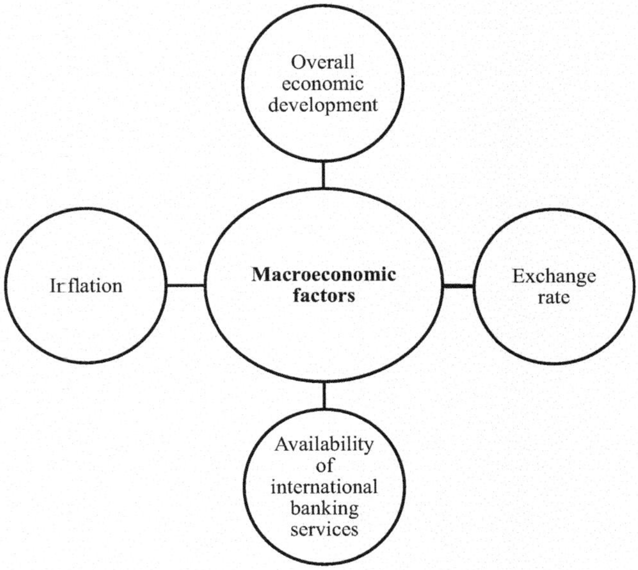

Fig. 12.4 Sub-dimensions of macroeconomic factors. (Source: Authors based on the existing literature)

a choice destination for many westerners due to its relatively low costs, owing to the low value of the Indian currency. Similarly, South Asian countries like India, Bangladesh, Pakistan, etc., will also find medical treatment expensive in some Western and European countries, due to the relatively high value of their currencies. Thus, the depreciation of the local currency against the US$ could boost medical tourism, as it will become relatively cheaper. A similar finding has also been documented in the study by Tang and Lau (2017).

Moreover, rising inflation will not only result in a higher cost of medical treatment but also increase the overall expenses of tourists, as they have to reside in the country for a relatively long period. Thus, policymakers should consider maintaining a stable inflation rate so that the potential medical tourist does not realize the increasing value of tourism expenses by the day. Due to the advancement in technology, tourists prefer to go cashless due to the ease of completing financial transactions through online banking, Automated Teller Machines (ATM), or cash deposit machines. Thus, better, connected, and more reliable banking and financial services should be made available to foreign tourists to facilitate seamless financial transactions (e.g. dollar conversion in their local currencies, availability of ATMs for withdrawals using foreign credit and debit cards, traveler's cheques etc.).

Other factors are influencing the demand for tourism that are beyond the control of the government, and such factors include individual characteristics and preference of the medical tourist (Fig. 12.5). A study conducted by Musa et al. (2012) revealed that the majority of inbound medical tourists in Malaysia are female, middle-aged,

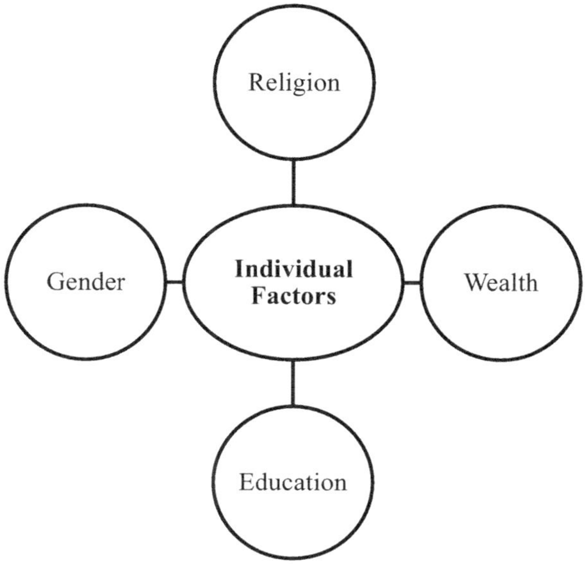

Fig. 12.5 Sub dimensions of individual-level factors. (Source: Authors)

Indonesian, and have an average of two companions. This indicates the importance of the demography of inbound medical tourists, which should be duly considered by the hospitals/clinics when designing medical products and services for clients. This implies that female specialized doctors and relevant entertainment opportunities for the middle age group should be made available in female-friendly hospitals/clinics.

Proposed Framework to Promote Inbound Medical Tourism in Bangladesh

As earlier discussed in the factors and sub-factors of medical tourism, this study proposes a comprehensive and participatory framework to promote inbound medical tourism in Bangladesh (Fig. 12.6). In light of this, we highlight the necessity of establishing a regulatory authority/board to exclusively oversee the inbound medical tourism in Bangladesh, which is currently unavailable. A similar type of government entity is available in neighboring countries like Malaysia (The Malaysian Healthcare Travel Council, 2020), Singapore (Pocock & Phua, 2011), and India (jointly between the Department of Commerce and Services Export Promotion Council-SEPC). Thus, having a similar board in Bangladesh would rapidly boost inbound medical tourism.

In this context, the government of Bangladesh should strategically aim to promote medical tourism and enact a separate entity collaborating with various ministries. In our proposed framework, we recommended that the creation of a separate entity is inevitable to the promotion of inbound medical tourism; hence, the

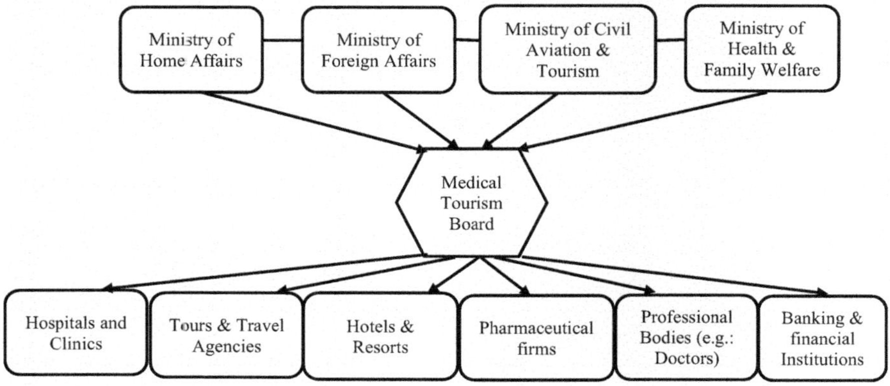

Fig. 12.6 Proposed framework for promoting medical tourism in Bangladesh. (Source: Authors' proposed framework)

ministries of foreign affairs, domestic affairs, civil aviation and tourism, and health and family welfare should jointly establish a 'Medical Tourism Board (MTB)' in Bangladesh (Fig. 12.6). The reasons for this proposal are as follows: since inbound medical tourism requires intensive and aggressive marketing due to the intense global competition, we suggest that the foreign affairs ministry coordinate embassies/missions of Bangladesh abroad to advertise and market the relevant healthcare products/services through various internal and external channels. The domestic affairs ministry should also coordinate with local agencies and enforcement authorities to ensure the safety of tourists from arrival till departure. The civil aviation and tourism ministry will ensure hassle-free trip for inbound medical tourists and provide necessary permissions to airlines that will ultimately enhance the visibility of medical tourism in Bangladesh. The core responsibility of the health and family welfare ministry will be to draft medical services and products to medical tourists based on the availability of human resources, technical expertise, healthcare facilities, and health professionals. Of course, the highlighted ministries are not exhaustive and more ministries (e.g. ministry of trade, commerce, etc.) should step forward to jointly initiate these highly ambitious projects.

The MTB should also interfere in the management of hotels and resorts meant for medical tourists. Incidents were recorded in Bangladesh, whereby hotels/resorts poorly treated foreign tourists (The Financial Express, 2019; British Broadcasting Corporation, 2016). To ensure that inbound medical tourists enjoy the scenic beauty of Bangladesh while receiving medical attention, hotel operators should be unified to ensure the safety of foreign tourists and preserve the image of the country.

Bangladesh has shown a remarkable performance in the development of pharmaceutical products, and a couple of companies now export indigenous and generic medicines on a large scale (Mohiuddin et al., 2015). The integration of the pharmaceutical industry with medical tourism would be a breakthrough, as it will help to expand the domestic pharmaceutical industry and ensure the delivery of quality medicines to inbound medical tourists. Thus, the role of the pharmaceutical industry would be tremendous if quality drugs could be produced at relatively low costs (except for a small number of unavoidably expensive medicines); a quality of Bangladesh, which makes it famous to its international buyers (Azam, 2017; Kasonde et al., 2019).

The availability of quality healthcare services would be impossible without qualified doctors, nurses, technicians, and administrative officers. Thus, it is indeed necessary that the MTB also directly include professional bodies/representatives to represent these segments of the labor market. Once integrated as a stakeholder, they would give their best efforts in serving medical tourists. This has proven to be true in conditions where the leadership of professional directors results in the superior performance of firms (Gaur et al., 2015). Moreover, unionized workers represented by professional bodies will also minimize potential conflicts and issues shortly and any potential interruptions in the servicing of medical tourists.

The availability of developed banking and financial institutions is necessary to provide adequate financial services to medical tourists, including settling of medical fees, international exchange, and other associated services. The existing studies have thoroughly established the complementary relationship between tourism and financial sector development or otherwise (Başarir & Çakir, 2015; Katircioglu et al., 2018; Shahbaz et al., 2017). Nonetheless, insurance service providers in several developed countries also cover costs (total or partial) of international medical services (Snyder et al., 2011). Thus, the MTB should also align with countries' financial institutions to ensure the availability of banking and financial services to medical tourists in areas they mostly visit.

Apart from these tasks, the MTB should also liaise with international accreditation authorities such as the Joint Commission International (based in the US) and Trent Accreditation Scheme (based in UK–Europe) (Gupte & Panjamapirom, 2014), to demonstrate the quality of services of healthcare providers in Bangladesh and safety of the patients. Turner (2010) further argued that medical tourism companies need to first undergo accreditation to obtain state or provincial license before making their services available internationally. By gaining recognition from international authorities, the international visibility of Bangladesh's healthcare service providers will not only be increased, but local patients will also be encouraged to seek treatment within the country, owing to the positive effect of international accreditation and concentration of medical tourism (Woodhead, 2013).

Conclusion

In this chapter, we have comprehensively identified the factors affecting inbound medical tourism based on existing literature, followed by a suggestive policy framework for Bangladesh. To promote inbound medical tourism, intensive planning, proper coordination among various units (private and public), public awareness, and quality of human resources (e.g. doctors, nurses, medical technologists, etc.) are required. Thus, the first step would be a significant investment in the industry through public and private participation to improve the overall healthcare facilities in the country before attracting foreign tourists.

Since the healthcare service in Bangladesh is not at par, competing in the global tourism market will be challenging but not impossible. Hence, a long-term and visionary planning is necessary to ensure the availability of quality healthcare at affordable costs for medical tourists. By so doing, successful countries like India, Singapore, and Malaysia could become a learning ground for Bangladesh to plan, promote and execute inbound medical tourism.

References

Alam, M. S., Islam, M. S., Shahzad, S. J. H., & Bilal, S. (2020). Rapid rise of life expectancy in Bangladesh: Does financial development matter? *International Journal of Finance & Economics (ahead of print)*. https://doi.org/10.1002/ijfe.2046

Ali, M. M. (2012). Outbound medical tourism: The case of Bangladesh. *World Review of Business Research, 2*(4), 50–70.

Alsumairi, M., & Hong Tsui, K. W. (2017). A case study: The impact of low-cost carriers on inbound tourism of Saudi Arabia. *Journal of Air Transport Management, 62*, 129–145.

Azam, M. M. (2017). The impacts of TRIPS on the pharmaceutical regulation and pricing of drugs in Bangladesh: In search of policy direction. *International Journal of Law and Management, 59*(3), 376–393.

Barai, M. K. (2020). Introduction: Construction of a development model for Bangladesh. In M. K. Barai (Ed.), *Bangladesh's economic and social progress* (pp. 3–62). Palgrave Macmillan.

Başarir, Ç., & Çakir, Y. N. (2015). Causal interactions between CO2 emissions, financial development, energy and tourism. *Asian Economic and Financial Review, 5*(11), 1227–1238.

Beladi, H., Chao, C.-C., Ee, M. S., & Hollas, D. (2019). Does medical tourism promote economic growth? A cross-country analysis. *Journal of Travel Research, 58*(1), 121–135.

Bookman, M. Z., & Bookman, K. R. (2007). Introduction to medical tourism. In M. Z. Bookman & K. R. Bookman (Eds.), *Medical tourism in developing countries* (pp. 1–19). Palgrave Macmillan.

British Broadcasting Corporation (BBC). (2016). *Bangladesh siege: Twenty killed at Holey Artisan Bakery in Dhaka*. Retrieved from: https://www.bbc.com/news/world-asia-36692613. Accessed 5 Dec 2020.

Business Wire. (2020). *Global medical tourism market trends report, 2020–2027*. Retrieved from: https://www.businesswire.com/news/home/20200414006133/en/Global-Medical-Tourism-Market-Trends-Report-2020-2027%2D%2D-ResearchAndMarkets.com. Accessed 15 Dec 2020.

Carrera, P. M., & Bridges, J. F. (2006). Globalization and healthcare: Understanding health and medical tourism. *Expert Review of Pharmacoeconomics & Outcomes Research, 6*(4), 447–454.

Connell, J. (2006). Medical tourism: Sea, sun, sand and … surgery. *Tourism Management, 27*(6), 1093–1100.

Connell, J. (2013). Contemporary medical tourism: Conceptualisation, culture and commodification. *Tourism Management, 34*, 1–13.

Daykhes, A. N., Jakovljevic, M., Reshetnikov, V. A., & Kozlov, V. V. (2020). Promises and hurdles of medical tourism development in the Russian Federation. *Frontiers in Psychology, 11*, 1380–1380.

de la Hoz-Correa, A., & Muñoz-Leiva, F. (2019). The role of information sources and image on the intention to visit a medical tourism destination: A cross-cultural analysis. *Journal of Travel & Tourism Marketing, 36*(2), 204–219.

Debata Bikash, R., Patnaik, B., Mahapatra Siba, S., & Sree, K. (2015). Interrelations of service quality and service loyalty dimensions in medical tourism: A structural equation modelling approach. *Benchmarking: An International Journal, 22*(1), 18–55.

Esiyok, B., Çakar, M., & Kurtulmuşoğlu, F. B. (2017). The effect of cultural distance on medical tourism. *Journal of Destination Marketing & Management, 6*(1), 66–75.

Gan, L. L., & Frederick, J. R. (2011). Medical tourism facilitators: Patterns of service differentiation. *Journal of Vacation Marketing, 17*(3), 165–183.

Gaur, S. S., Bathula, H., & Singh, D. (2015). Ownership concentration, board characteristics and firm performance: A contingency framework. *Management Decision, 53*(5), 911–931.

Gupte, G., & Panjamapirom, A. (2014). Understanding medical tourism. In A. J. Culyer (Ed.), *Encyclopedia of health economics* (pp. 404–410). Elsevier.

Hasan, S. R., & Hassan, M. K. (2013). Health tourism: A demographic study on the outbound health tourists from Bangladesh. *Journal of Business, 34*(2), 53–66.

Hopkins, L., Labonté, R., Runnels, V., & Packer, C. (2010). Medical tourism today: What is the state of existing knowledge? *Journal of Public Health Policy, 31*(2), 185–198.

Kasonde, L., Tordrup, D., Naheed, A., Zeng, W., Ahmed, S., & Babar, Z.-U.-D. (2019). Evaluating medicine prices, availability and affordability in Bangladesh using World Health Organisation and Health Action International methodology. *BMC Health Services Research, 19*(1), 383.

Katircioglu, S., Katircioğlu, S., & Altinay, M. (2018). Interactions between tourism and financial sector development: Evidence from Turkey. *The Service Industries Journal, 38*(9–10), 519–542.

Kim, S., Lee, J., & Jung, J. (2013). Assessment of medical tourism development in Korea for the achievement of competitive advantages. *Asia Pacific Journal of Tourism Research, 18*(5), 421–445.

Kumar, N. (2018). *Over 50% medical tourists to India are from Bangladesh.* Retrieved from: https://www.sundayguardianlive.com/news/50-medical-tourists-india-bangladesh. Accessed 14 Dec 2020.

Lee, C. V., & Taggart, L. R. (2013). 36 – Medical tourism. In J. S. Keystone, D. O. Freedman, P. E. Kozarsky, B. A. Connor, & H. D. Nothdurft (Eds.), *Travel medicine* (pp. 343–347). Elsevier.

Liu, I.-C., & Chen, C.-c. (2013). Cultural issues in medical tourism. *American Journal of Tourism Research, 2*(1), 78–83.

Mahmud, M., Otsuka, K., Sawada, Y., & Yamada, E. (2018). Development transformation in Bangladesh: An overview. In Y. Sawada, M. Mahmud, & N. Kitano (Eds.), *Economic and social development of Bangladesh: Miracle and challenges* (pp. 3–26). Springer International Publishing.

Mahmud, M. S., Rahman, M. M., Lima, R. P., & Annie, E. J. (2020). Outbound medical tourism experience, satisfaction and loyalty: Lesson from a developing country. *Journal of Hospitality and Tourism Insights.* https://doi.org/10.1108/JHTI-06-2020-0094

Mamun, M. Z., & Andaleeb, S. S. (2013). Prospects and problems of medical tourism in Bangladesh. *International Journal of Health Service, 43*(1), 123–141.

Manaf, N. H. A., Hussin, H., Kassim, P. N. J., Alavi, R., & Dahari, Z. (2015). Medical tourism service quality: Finally some empirical findings. *Total Quality Management & Business Excellence, 26*(9–10), 1017–1028.

Mohiuddin, M., Rashid, S. F., Shuvro, M. I., Nahar, N., & Ahmed, S. M. (2015). Qualitative insights into promotion of pharmaceutical products in Bangladesh: How ethical are the practices? *BMC Medical Ethics, 16*(1), 80.

Malaysian Healthcare Travel Council. (2020). *Who we are.* Retrieved from: https://www.mhtc.org.my/discover-mhtc/#who. Accessed 14 Dec 2020.

Musa, G., Thirumoorthi, T., & Doshi, D. (2012). Travel behaviour among inbound medical tourists in Kuala Lumpur. *Current Issues in Tourism, 15*(6), 525–543.

Oberoi, S., & Kansra, P. (2019). Factors influencing medical tourism in India: A critical review. *SAMVAD, 17*, 9–16.

Pocock, N. S., & Phua, K. H. (2011). Medical tourism and policy implications for health systems: A conceptual framework from a comparative study of Thailand, Singapore and Malaysia. *Globalization and Health, 7*, 12.

Runnels, V., & Carrera, P. M. (2012). Why do patients engage in medical tourism? *Maturitas, 73*(4), 300–304.

Sandberg, D. S. (2017). Medical tourism: An emerging global healthcare industry. *International Journal of Healthcare Management, 10*(4), 281–288.

Seow, A. N., Choong, Y. O., Moorthy, K., & Chan, L. M. (2017). Intention to visit Malaysia for medical tourism using the antecedents of Theory of Planned Behaviour: A predictive model. *International Journal of Tourism Research, 19*(3), 383–393.

Shahbaz, M., Kumar, R. R., Ivanov, S., & Loganathan, N. (2017). The nexus between tourism demand and output per capita with the relative importance of trade openness and financial development: A study of Malaysia. *Tourism Economics, 23*(1), 168–186.

Snyder, J., Crooks, V. A., Johnston, R., & Kingsbury, P. (2011). What do we know about Canadian involvement in medical tourism?: A scoping review. *Open Medicine, 5*(3), e139–e148.

Shaw, C. D., Braithwaite, J., Moldovan, M., Nicklin, W., Grgic, I., Fortune, T., & Whittaker, S. (2013). Profiling health-care accreditation organizations: An international survey. *International Journal for Quality in Health Care, 25*(3), 222–231.

Singh, P. K. (2008) *Medical tourism: Global outlook and Indian scenario*. Kanishka Publishers.

Tang, C. F., & Lau E. (2017). Modelling the demand for inbound medical tourism: The case of Malaysia. *International Journal of Tourism Research, 19*(5), 584–593.

The Financial Express. (2019). *Bangladesh's tourism: Prospects and lacunae*. Retrieved from: https://www.thefinancialexpress.com.bd/views/reviews/bangladeshs-tourism-prospects-and-lacunae-1577717187. Accessed 5 Dec 2020.

Turner, L. G. (2010). Quality in health care and globalization of health services: Accreditation and regulatory oversight of medical tourism companies. *International Journal for Quality in Health Care, 23*(1), 1–7.

Wang, H. Y. (2012). Value as a medical tourism driver. *Managing Service Quality: An International Journal, 22*(5), 465–491.

Woodhead, A. (2013). Scoping medical tourism and international hospital accreditation growth. *International Journal of Health Care Quality Assurance, 26*(8), 688–702.

Wu, S.-W., Pan, Q. Chen, T., Ge, Y.-Y., Luo, J., & Wang, Q. (2019). Development and evolution of medical technology management in China. *Chinese Medical Journal, 132*(14), 1739-1741.

Chapter 13
Medical Tourism in Bangladesh: Issues, Opportunities and Strategic Marketing Plan Model for Growth and Development

Mohammad Fazle Rabbi and Md. Nekmahmud

Abstract The word "health" refers to a state of complete emotional & physical well-being, and Human health is an uncompromisable substance. Medical tourism is the concept of traveling far and wide for better medical treatment. The globalization of the healthcare sector and the massive demand for cost-effective, higher-quality treatment in recent times have caught the attention of many developing countries that have adequate resources and potential to meet these demands. The main aim of the chapter is to explain pragmatic marketing strategies for healthcare institutions involved in medical tourism in Bangladesh. Moreover, this chapter evaluates the present scenario of the health sector, drivers, and market structure in medical tourism in Bangladesh. Besides, it identifies the major opportunities, investment, growth, and development of healthcare facilities. Using the SWOT analytical model, this paper narrated a pertinent value proportion of Bangladesh's medical tourism. This conceptual chapter analyses the theoretical aspects of medical tourism and marketing. Competitive advantages in the tourism industry such as affordable healthcare costs, related accessible benefits in treatment, the low tuition fee in medical education, low pharmaceutical cost, reasonable accommodation, and competitive labor force opportunities help reduce the operational costs of medical, dental, surgical, paramedical, etc. Adopting alternative medicinal systems like Ayurveda, homeopathic, acupuncture care opens a new door to increase medical tourists' flow.

M. F. Rabbi
Ihrig Károly Doctoral School, Faculty of Economics and Business, University of Debrecen, Debrecen, Hungary

Md. Nekmahmud (✉)
Management and Business Administration Sciences, Doctoral School of Economic and Regional Sciences, Hungarian University of Agriculture and Life Sciences, Gödöllő, Hungary
e-mail: nekmahmud.argon@gmail.com

A. Hassan (ed.), *Tourism in Bangladesh: Investment and Development Perspectives*, https://doi.org/10.1007/978-981-16-1858-1_13

Country-wise diversity of tourist locations is also considered to meet or to exceed the medical tourists' expectations. Government initiatives, develop a strategic marketing plan model, financial investment, new promotional activities, eco-tourism, adaptation of advance technology, the collaboration of the healthcare sector and medical institutions are essential to unlock the full potentiality of this multi-billion industry. Finally, this chapter highlights the strategies that need to be implemented for shaping Bangladesh as one of the key destinations for medical tourism across the globe.

Keywords Medical tourism · Marketing strategic · SWOT analysis · Development · Bangladesh

Introduction

Medical tourism refers to the phenomenon of picking and traveling across the national border to an international one to access modern healthcare services. Medical tourism's main aspects are the availability of highly trained specialized doctors and reputation, modern medical specialization, comparatively, cost-effective treatment, and the vacation opportunity to visit another country (Bookman, 2007). Feasible associations among these variables influence individuals to be a medical tourist.

Formerly, people from developing and least-developed counties travel to developed countries to avail of the complex invasive medical treatment across international boundaries, but after the changes of scenario developed countries, the patient is now traveling to developing countries for low-cost treatment and country with rich cultural heritage. (Horowitz & Rosensweig, 2008) This has now become a rapidly growing global industry that initiated its industry growth since the late 1990s (Han & Hwang, 2018) and has grown past few decades for avoiding the higher cost and long waiting list at the patient's home country (Connell, 2006). In 2018, the estimated value of this industry approximately USD 16,761 million and projected to reach US$ 27,247.6 million by 2024 (More, 2019). However, day by day, medical tourism competition is getting more competitive (Han, 2013). The world's most populated country Bangladesh with over 160 million population is not ready to contribute to this competitive race (Islam & Khan, 2017). Rather, several south Asian developing and developed countries like India, Thailand, Malaysia, and Singapore are in fourth in the medical tourism industry, where Bangladesh is encouraging and promoting to their people to travel to other regions to avail of quality treatment at an affordable cost (Mamun & Andaleeb, 2013).

Most of the Bangladeshi lower and upper-middle-class people prefer to go neighboring country India for better treatment as cosmetic surgeries, dental treatment, cardiovascular, orthopedics, bariatric surgery, fertility treatment, ophthalmic treatment, and wellness evaluations are the main target of medical tourist but also Thailand, Malaysia, Singapore, Turkey preferred as an alternative destination (Noor,

2013). In Bangladesh, there is no obvious guidance in which medical conditions lead to medical instead of accurate treatments and ultimately losing foreign exchange revenue from this potential industry and unable to hold the patient from driving by the transcendent perceived level of treatment abroad (Medhekar & Ali, 2012). But patients are willing to travel to relieve pain, save money, and for extra comfort.

Moreover, Bangladesh is renowned for the ancient cultural heritage and could focus on attracting medical tourists by improving the quality treatment in many specialized treatments. Rather than medical, dental, surgical, paramedical treatment Bangladesh has the full potentiality of traditional treatment like homeopathy, Ayurveda, Unani, and folk medicine (Akber et al., 2011), as well as acupuncture care can add a new dimension in the present treatment system, which is treating as the negligible medical sector. The education system of medicine in Bangladesh is more well-known among Asian countries. Many developing and under-developing country students such as China, Bhutan, Russia, Sri Lanka, Ukraine, Nepal, Kyrgyzstan, and Kazakhstan have come every year for medical study purposes in Bangladesh.

This chapter is to answer the questions emerged: what are opportunities and marketing strategic plan for growth and development in the medical tourism industry in Bangladesh. Therefore, the main objectives of this chapter are to develop pragmatic marketing strategies for healthcare institutions involved in medical tourism in Bangladesh and measure the impact of the government action on the present marketing strategies for foreign patients. Moreover, the chapter focuses on SWOT analysis, investment, growth, and development for the health tourism industry in Bangladesh. At the same time, it identifies the major opportunities and shortcomings of the healthcare facility.

Firstly, this chapter emphasizes the review of the literature on the present scenario of medical tourism in Bangladesh. Secondly, the theoretical aspects of medical tourism techniques are discussed with medical tourism, types of medical tourism, marketing strategy in medical tourism, issues in medical tourism. Finally, SWOT analysis, investment, and development with a conclusion, as well as the recommendation of medical tourism in Bangladesh, have been discussed consecutively.

Present Scenario and Background of the Healthcare Sector in Bangladesh

Bangladesh is unable to allocate adequate quality healthcare services due to a lack of resources (Hena & Hasan, 2019). Furthermore, no treatment facility for complex diseases, higher treatment costs, wrong treatment, shortage of highly qualified doctors, medical staff mistreatment, unskilled and less educated nurse, and other medical staff are considering the prime reasons to divert domestic patients to foreign countries (Hena & Hasan, 2019). Similarly, medical tourists are not getting encouraging to take healthcare services from Bangladesh. Besides, the destination for medical treatment is driven by the nature of a particular country's healthcare service. And the destination attributes should be considered factors patient's decision-making process.

Moreover, the allocation of budget in Bangladesh for the last 10 years (2009–2019) towards the health sector has increased gradually. Last 2018–29 fiscal year, the allocation of funds to healthcare 25,732 core (BDT) and total health expenditure in Bangladesh, as a total percentage of GDP in the year 2019, was 5.8%, and which is comparatively lower in other countries, and it will very less budget to achieve the target goals in the health sector and develop the medical tourism industry. At present, there are a total of 90 recognized medical colleges in Bangladesh, of which 36 are public, and 54 are private. Separately, there are six medical colleges operated by the Bangladesh Armed Forces[1]. On the other hand, in Bangladesh, the total number of hospitals in Bangladesh, including private, public, private, local, was 5816 in 2015 [2].

There are two ways a patient can get healthcare services in Bangladesh: first, "Formal", and second, "Informal". Formal are those who are registered with the government regulatory body and authorized to provide healthcare services. Formally, health service providers through Government and private healthcare institutions including hospital, clinic, dentist, podiatrist, chiropractor, clinical psychologist, optometrist, nurse practitioner, nurse-midwife, or a clinical social worker (Li & Song, 2019).

Informal are those who do not have registration with any government regulatory body. Informal health service providers are Community health workers (CHWs), Unqualified village doctors and drugstore salespeople and drug vendors, "Kabiraj" (Traditional healers), No-secular faith healer who chant religious verses for healing, Trained and non-trained traditional birth attendants, Self-educated homeopathy doctor (Ahmed et al., 2009). Among formal and informal healthcare services, explicitly formal healthcare services are recommended for foreign patients seeking specialized care for healing their diseases.

Medical Tourism and Bangladesh

In general, when people travel to the outside or state of one's residence for taking any kind of medical treatment or service to cure diseases and invest money for treatment and traveling is called medical tourism. Nevertheless, (Kazemi, 2008; Gupta, 2004) stated medical tourism as a cost-effective private medical care system in collaboration with the tourism industry for surgical and other models of specialized treatment. From the problematic framework, there are two broad types of medical

[1] https://en.wikipedia.org/wiki/List_of_medical_colleges_in_Bangladesh

[2] Statistics, B. B. C. (2016). *Statistical yearBook of Bangladesh* (first ed.). Bangladesh

tourists. The first type, those who travel explicitly only for medical treatment. The second type, those who travel for vacation and treatment combinedly (Cohen, 2008).

To expend any tourism industry, the focus should be on tourist type. Their policy should be set to include rules and regulations, guidelines, directives, and strategic marketing plans based on present tourism industry demand (Beeton, 2005). Where till 1992, Bangladesh had no tourism policy apart from a strategic master plan of the United Nations Development Programme (UNDP) and the World Trade Organization (WTO) prepared in 1990. To cope with global demand on 14th December 2009 Bangladesh government adopted the tourism policy framework (Hassan & Burns, 2014).

Medical tourism considers some factors to select a country as his/her tourist destination, such as diagnosis and treatment accuracy of doctors, medical services and performance of doctors and nurses, and adequate modern equipment. Due to the incapability of government hospitals, patients are switching to the private treatment sector, where insurance and increased mobility of patients liberalize treatment facility. Moreover, the booming globalized health services are driven by the upsurge of private healthcare services upsurge due to greater ease of travel and technological advances (Herrick, 2007).

The Medical Tourism Structure in Bangladesh

There are several types of medical tourism. Based on the nature of healthcare service, the medical tourism structure can be categorized into wellness tourism, traditional medicinal systems, cosmetic surgery tourism, advance & life savings healthcare, and rehabilitation, which are exhibited in Figure 13.1.

Wellness Tourism

Wellness tourism evolved to better preserve the body by eliminating complete physical problems and decreasing the disease, including reducing symptoms of mental, psychological, and emotional problems, which are often left untreated. Also,

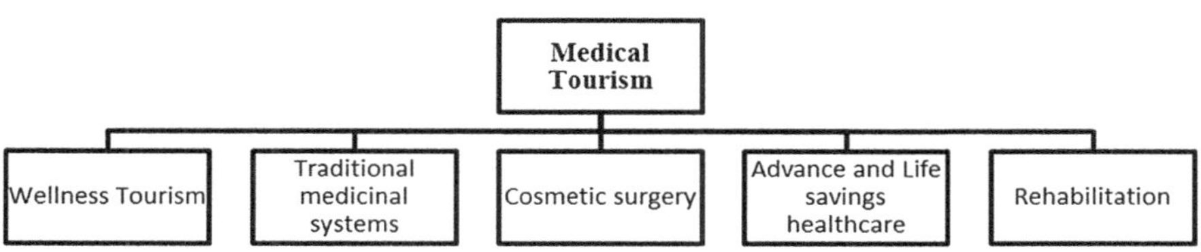

Fig. 13.1 Structure of medical tourism. (Sources: Dawn & Pal, 2011)

traveling for sauna, spa, body massage, and other healthcare activities belongs to wellness tourism (Smith & Kelly, 2006).

Traditional Medicinal Systems

Traditional medicinal systems are pointing to the Ayurveda, which is known as the "Science of life". Almost worldwide, 60% population from developing developed countries are using this treatment system. Most of the practitioners developed their treatment system and formula for healing the patients. But for homeopathy treatment, specific guidelines are set by the concerned authority (Pandey et al., 2013).

Cosmetic Surgery Tourism (CST)

Cosmetic surgery tourism (CST) is a combination of both cosmetic surgery and tourism. It refers to traveling a foreign destination for people, things and practice to gain access to the specific treatment facility for enhancing their physical appearance (Holliday et al., 2015).

Advance and Life Savings Healthcare

Advance and life savings healthcare refer to the complex medical treatment, which is an alternative of medicine, including open transplants, cardiovascular surgery, Replacement of hip and knee, eye surgery, etc. (Dawn and Pal, 2011).

Rehabilitation

Rehabilitation is a cross-border purchase of health services for patient's risky behaviors, including drug addiction, alcohol abuse, smoking, and other addiction programs. Also, counseling and elderly care services are part of rehabilitation healthcare (Hall, 2011).

Table 13.1 Classification of the services spectrum of medical tourism in Bangladesh. (source: Dawn and Pal (2011) and authors' elaboration, 2020)

Wellness Tourism	Traditional medicinal systems	Cosmetic surgery	Advanced and live savings healthcare	Rehabilitation
• Spas • Fitness • Lifestyle/healthy vacations • Nature tourism • Ecotourism • Resorts integrated with health facility	• Homeopathy • Ayurveda • Unani • Folk medicine • Yoga therapy	• Dental care • Plastic surgery • Tummy reduction • Skin treatment	• Open transplants • Cardiovascular surgery • Eye treatment • Hip replacement • Knee replacement	• Addiction programs • Elderly care programs • Counseling services

Potential Medical Tourism Product for Bangladesh Tourist Industry

There are more than a few countries are dominating the medical tourism industry in Asia. And the position of Bangladesh is behind most other Asian countries, which are seeking to enter the market. Proper strategic planning and labeling products and services can open an opportunity to enter this profitable industry. In conventional tourism, the travel industry's overall benefits are a matter of concern, which can be grown on a product basis offer instead of tourist packaging. Product segmentation is required on the nature of healthcare service for the Bangladeshi medical tourism industry that will help to penetrate the Asian medical tourism industry. This table represents the segments of healthcare services like wellness tourism, traditional medicinal systems, cosmetic surgery, advanced and live savings healthcare and rehabilitation, and under each segment of healthcare services that can offer to the medical tourist (Table 13.1). So medical tourists can choose the appropriate service and destination according to their treatment needs.

Issues of Medical Tourism in Bangladesh

Due to overpopulation and lack of treatment service providers' capacity, a huge number of Bangladeshi patients prefer outbound medical tourism to undergo foreign medical treatment. Treatment accuracy another major issue for the patients' experiences in their attempts to receive an inaccurate medical diagnosis. The adverse consequences of incorrect diagnosis range occur with uncontrollable expenses (Floyd, 1997). Medical patients prefer cost-effective but equipped with modern medical equipment and higher quality hospitals, but tourist destinations must be relatively low cost for living, expat-friendly locals and hospital staff, cultural affinity, and accredited by a recognized authority like Joint Commission International (Herrick, 2007).

Among South Asian countries, most the medical tourist prefer Indian as suitable medical tourism to seek treatment for cardiovascular disease, cancer, neurological disorders, and other complex treatments like heart and liver transplants, bone marrow transplants, which are not yet available in Bangladesh except some full-range treatment of triple bypasses, brain tumor removal, complex hip and knee surgery (Mackian, 2003; Mamun & Andaleeb, 2013).

Additionally, there are many obstacles restraint the medical tourism industry, including foreign practitioners entrants, the foreign firm has restriction to incorporate, domestically owned and foreign direct investment in relevant health sectors, foreign ownership restriction, poor infrastructure, and not suitable for investment in medical service firms, people who are not licensed processional restricted to own medical service firm, regulator and capacity constraints (Dee & Dinh, 2009).

Moreover, the government has decreased the fund allocation in medical sector development. In the national budget of 2017–2018, allocated funds were reduced to Taka 13.05 billion from the previous year. Also, the government has declined the health sector's expenditure from 27% to 23%, which is a significant issue in achieving the universal healthcare objective (Ali & Medhekar, 2018).

Marketing Strategy for Medical Tourism

Using traditional or modern marketing tools and strategies in the medical industry are named medical marketing. Medical marketing is attracting medical tourists to satisfy their particular medical service. Das (2017) described Medical marketing as the process of convincing the patients or their guardians in selecting a precise Hospital for their remedy, neglecting the different competitor hospitals in the market. Different marketing tools like marketing mix, communication mix, and marketing strategic model (STP) should apply to the medical tourism industry in Bangladesh to improve the medical sector. These marketing strategies can be very valuable in promoting Medical Tourism in Bangladesh. The proper marketing strategy can help win the patients and increase the revenue over healthcare exports and increase hospitals' reputation. Medical marketing can be effective in attracting patients both nationally and overseas in building a hospital brand, and more importantly, raising awareness among patients around the world for the availability of a specific treatment the life frightening or deadly disease in a specific hospital or clinic in any country (Das, 2017). By applying marketing strategies, including 4Ps', 7Ps', Pricing & promotional strategies of affordable hospitals, and attractive tourist spots, the Bangladesh government can promote its medical industry and increase revenue for the country. Because a now cost of medical service in developed country USA, Canada, Europe is so high. Therefore, people of developed countries are going to less medical cost countries such as India, Malaysia, Thailand, and Singapore. Bangladesh also has good opportunities to develop its medical sector to attract

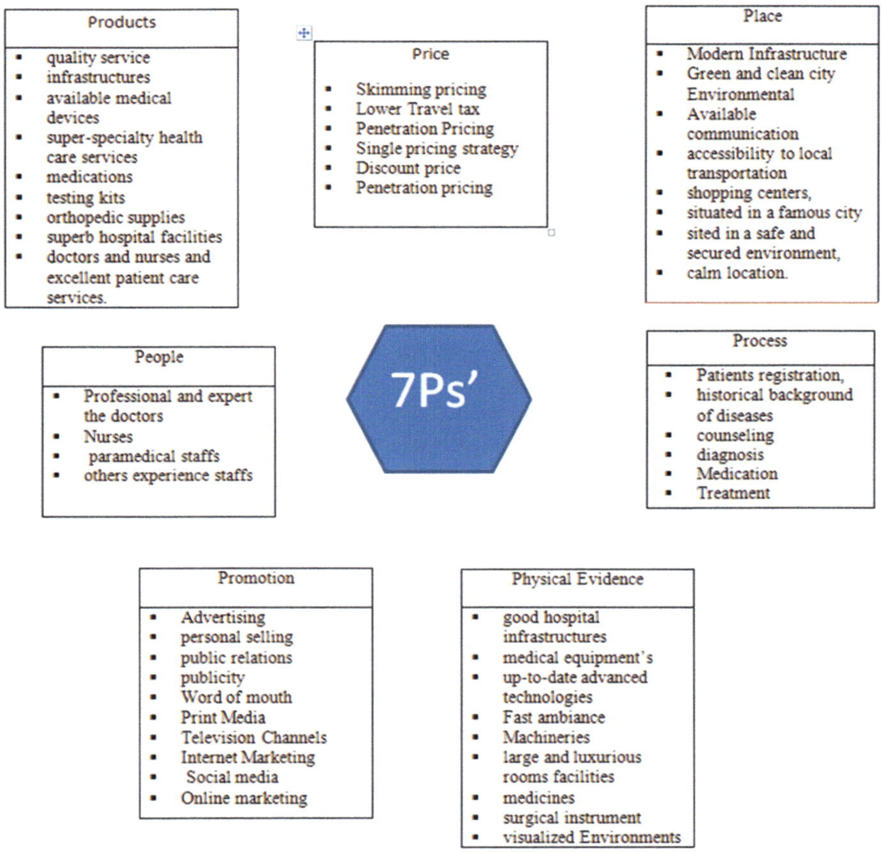

Fig. 13.2 Marketing Mix -7Ps in Medical tourism industry for the development context. (sources: developed by the authors, 2020) ***7Ps of medical tourism in Bangladesh***
4Ps is extended to 3Ps (people, process, and physical evidence).

foreign patients. Figure 13.2 illustrates the marketing Mix -7Ps' in the medical tourism industry.

4 Ps' of Medical Tourism in Bangladesh

Medical marketing is part of service marketing. Similarly, traditional product marketing, Medical marketing, service providers can apply marketing mix such as Marketing mix includes 4Ps product, price, place, and promotion (McCarthy, 1975) and 7Ps along with three (3) additional Ps i.e. People, Process and Physical evidence to attract the patients across the world. If Bangladesh's medical tourism industry wants to develop, it has to focus on marketing strategy 7Ps.

Product

Medical product means the medical treatment and services which the hospitals offer. The product represents goods, services, or ideas provided through a healthcare organization (Spasojeric & Susic, 2011). Moreover, these include nationally and internationally recognized public and private hospitals that provide quick service, excellent hospital facilities, high-tech medical equipment, top physicians and nurses, and excellent patient care services. The healthcare market's products are the quality service, infrastructures, and available medical devices, diversified super-specialty healthcare services, medications, testing kits, orthopedic supplies, and on-call services, etc. The Products or services in medical marketing play a vital role in attracting patients worldwide and are considered the prime factor in choosing the hospitals for treatment in Medical Tourism (Das, 2017).

Price

Price is one of the important factors of the marketing mix in medical marketing. It plays a vital role in fascinating domestic and foreign patients in selecting particular hospitals or clinics to take care of their diseases. The effective pricing strategy will help in deciding to fly the patients from one country to another. On the other hand, price in the medical sector, Patients are out of pocket for their diagnosis and treatment (Das, 2017). Nowadays, the cost of private health services is increasing day by day, and for patients like Bangladesh who don't have any health insurance coverage, price is becoming the greatest concern for patients in Bangladesh. There are two types of pricing strategy: Skimming pricing and Penetration pricing.

In Bangladesh, a penetration pricing strategy is the best way to attract patients in the medical tourism industry.

Place

In the health service sector, the place can describe a location where medical service can be accessed In medical tourism, the place section of hospitals and clinics is very important. The hospital's location should be near to some tourist attractions and shopping centers, access to local transportation, situated in a famous city, and sited in a safe and secure environment, and role in a calm location. But in Bangladesh, hospitals are not located in a good place, and communication is very poor. There have no good opportunities to serve the patients. Therefore, medical policymakers and the owner should consider the place to attract foreign patients also.

Promotion

Promotion is the most important factor in 4Ps in medical marketing. Firstly, promotion can play a role in attending to the patients and guardians. There are several types of promotion tool which can be applied to the medical tourism industry to attract the target country and patients. In general, promotion tools include advertising, personal selling, public relations, publicity is intended to stimulate customer demand and develop sellers' marketing performance (Al-Azzam, 2016; Kumar & Komaraian, 2014). In Bangladesh, the medical service provider can use promotional tools in medical marketing to promote medical tourism, such as 1. Print Media; 2. Radio; 3. Television Channels like BBC, Discovery, CNN, National Geography, etc.; 4. Internet Marketing; 5. Newsletters; 6. Social media; and 7. Online marketing.

People

People are the key variable in the marketing mix of medical tourism. The quality of healthcare depends on the education & experience, language, and communication of the healthcare professionals as doctors, nurses, etc. (Das, 2017). For example, Apollo, Square, Labaid hospitals in Bangladesh have qualified doctors, nurses, and other paramedical staff with an excellent educational background, which can be emphasized to capture and attract international patients' attention.

Process

The process is a part of service marketing and treatment marketing where hospitals are meant to convince patients about the quality of their standard services in a worldwide competitive environment. The healthcare services pass to the patients through several kinds of a process such as patient registration, counseling the patients on the history of their diseases, diagnosis, treatment, etc. (Das, 2017). In the medical tourism industry, quality is considered as most concerning factor among the patients. Before going to select for treatment of their disease, patients always consider the service quality and process of hospitals.

Physical Evidence

Physical evidence is also applicable in medical tourism marketing includes medicines, surgical instruments, technology, infrastructures, etc. Suppose Bangladesh wants to develop the medical tourism industry. In that case, it has to ensure quality full and visualized physical services, for example, fast ambiance service, good hospital infrastructure, medical equipment, advanced technology of modern technology, appliances with large and luxurious rooms, and secure facilities.

Opportunities for Medical Tourism in Bangladesh

Since the glorifying victory in 1971, devastated Bangladesh entered a new era as a sovereign nation. As a holder of ergonomic culture heritage, the Bangladeshi tourism industry is passing through the ups and downs pathway. Despite the fact that the slow growth of the tourism industry is currently poised to be modestly sized. At present, by considering all potentiality of this sector, efforts are ongoing to reflect that in the development policies and programs taken by the Bangladesh government.

Our tourism industry has grown quickly in the last decade. Strategic intervention is required to reveal all the opportunities of this sector. Thus it can prepare itself and enter the global competitive arena, especially with a neighboring country like India.

Medical tourist selects their destination based on some specific comparative advantages over their own country involving specialized treatment facility, the income level of respondents, the aggregate cost of treatment including travel and living expense and social security of destination country.

Bangladeshi medical sector is still a cost-effective sector for domestic and foreign patients. This advantage can be treated as the best opportunity to attract more tourists. Apart from this, the Bangladesh government can attract foreign direct investment to establish a new world-class hospital as well as local branches of foreign hospitals. Due to the overflow of tourists to Indian hospitals, some Indian hospitals have opened its local office in Bangladesh to provide services to the patients who seek higher-quality treatment. Indian flagship hospital, Apollo Hospitals, opened a branch in Dhaka, Bangladesh. So, with the inception of international hospitals, more medical tourists can attract.

SWOT Analysis of Medical Tourism in Bangladesh

SWOT analysis in this paper focused on internal strengths and weaknesses, along with external opportunities and threats of the Bangladeshi medical tourism industry. It represents the weakest medical tourism facility but with lots of potentialities. The typical demonstration is a two-by-two grid with each quadrant indicating one of the Bangladesh medical tourism industry (Fig. 13.3).

	Internal		
Positive	**Strength** • Affordable healthcare cost • Short waiting time • Alternative medical treatment like homeopathy, Ayurveda, Unani, and folk medicine • Low pharmaceutical cost • Low-cost accommodation • Competitive labor force • English speaking medical staff • Land of world heritages sites	**Weakness** • Lack of transparency in government procedures • The slow bureaucratic process associated with procurement practices • Corruption and bribery • Lack of accountability • Lack of collaboration between medical a tourism industry • Low coordination between the various key role player like airlines, hotels, and hospitals • Price disparity across hospitals • Weak infrastructure • Internal level doctor • Expert nurse and medical staff • Unhygienic hospital	Negative
	Opportunities • The higher growth potential of the medical tourist industry • The spread of regional markets • Shortage of supply in the National health system of developed countries (e.g. UK, Canada) • Healthcare facilities demand from underdeveloped countries • Slow but Steady rise in tourism • Foreign direct investment for multinational companies • The positioning of tourist services • Stable foreign exchange	**Threats** • The constraint of internal resources • Strong competition from neighboring countries like India • Foreign relations • Lack of international accreditation • Overseas medical care does not cover by the health insurance provider • Low investment and subsidies for the hospital • Lack of social securities • Risks of travel • Lack of government policies for medical tourists • Political unrest • Economic downturn	
	External		

Fig. 13.3 SWOT analysis of Bangladeshi Medical Tourism. (Source: authors' estimations, 2020)

Strengths

Several factors are affecting the success of medical tourism in Bangladesh. Low-cost medical treatment with a short wait time for an appointment can be considered a significant medical tourism destination. This opportunity easily can attract patients from developed and other south Asian countries. Furthermore, because of the low cost of labor, pharmaceutical, land cost, and management cost, Bangladesh can offer medical service at a lower price than most south Asian and western countries.

Weakness

Though there are several strengths, weakness exists due to incompetency in this sector. Lack of transparency in government procedures, slow bureaucratic process, discriminatory pricing for foreign patients in various procedures, malpractice, corruption, and bribery are the weakest point of this tourism industry. Besides this, they have a lack of accountability in the treatment sector. Doctors' negligence in human life and lack of attention to protecting the patient, weak infrastructure, lack of modern medical equipment, and internal conflict in the hospital made it vulnerable for medical tourists.

Opportunities

Medical tourism is a multi-billion industry whose rapid growth rate contributes to increasing the foreign currency flow and overall GDP of Bangladesh. Bangladesh can stable the flow of foreign from this uncollected revenue sector by creating the opportunity and increasing the attraction for medical tourists. Promotion of foreign direct investment for establishing and operating hospitals in free economic zones, and by launching new tourist spots, new opportunities can be gain.

The alternative treatment system has a great opportunity for the tourism industry of Bangladesh. Improvement of Ayurveda, herbal medicine, homeopathy, Unani can be a new door for the Bangladeshi medical tourism industry.

Threats

Bangladesh is a densely populated country with limited resources and unable to compete internationally to get potential medical tourists. Besides, strong competition from neighboring countries like India is resisting to grow in this sector. Political unrest, lack of social securities and risks of travel, including infection risk with other diseases, have become major threats for growing the medical tourism industry.

International flights may be responsible for the infection of new germs and bacteria for some patients, which may upsurge the travel risk. Some examples are the COVID-19 pandemic worldwide, the SARS outbreak in Asia, and the Swine flu in Mexico (Gan & Song, 2012).

Absence of clear government policies and long-term vision, there is very limited opportunity to grow this industry. Many skilled health professionals are migrating to developed countries, which are the biggest threats to the country's health sector. A prime reason for migration is fewer career opportunities for the health professional in Bangladesh, and even many physicians are providing health services through pharmaceutical dispensaries.

Investment, Growth, and Development

The government can regulate the tourism industry by passing favorable tax laws to encourage entrepreneurs and investors to boost the medical tourism industry. Many new foreign investors can be attracted by providing tax concessions, lowering import duties, and providing subsidies by the Bangladesh government. It will help to grow the industry rapidly. The fund flow of multinationals companies and private funds into destination countries will help to boost the medical tourism industry. There is sharp growing prosperity of the medical tourism industry within south Asian countries, which can contribute to GDP growth and foreign exchange rates.

Malaysia targeted its neighboring countries like Indonesia and Singapore to grow its tourism industry by maintaining favorable currency exchange rates and reasonable healthcare costs. Similarly, Initially, Bangladesh can target Nepal, Bhutan, and other neighboring countries to raise their tourism industry. Later it can develop by expanding the capacity of the health sector and standardize the health services.

Strategic partnership plays a vital role in expanding the medical tourism service in Bangladesh. Initiating partnerships with various key players in the medical tourism industry helps to grow this industry. For example, many popular hospitals that are willing to get medical tourists can make a strategic partnership with international hotel chains for accommodation and with airlines to provide cheap air ticket priority basis to medical tourists.

Without investment and development, we cannot expand medical tourism in Bangladesh. But there are some barriers to development in the medical tourism industry, such as policies and laws, cost, affordability issues, Government assistance, and local community healthcare (Heung et al., 2011). Government initiative and motivation to private investment is important for medical tourism in Bangladesh. Suppose Bangladesh wants to develop the tourism industry. In that case, it needs to take some strategies such as new promotional activities, cooperative efforts by the hospitality sector, and medical institutions to develop medical tourism products. Joint partnerships with medical facilities provide an excellent way for hotels to differentiate their services (Heung et al., 2011; Hume & DeMicco, 2007).

Conclusion and Recommendation

Medical tourism can generate foreign revenue and standardize the whole medical treatment system in Bangladesh, which is already turned into a profitable industry worldwide. Due to a lack of proper vision, planning, and guidelines, Bangladesh is depraving from earning potential revenue from the medical tourism industry.

There are many obstacles, and hindrance must pass to the advancement of the medical sector. Apart from expert doctors, nurses, and medical staff, the hospital should be equipped with the most contemporary medical equipment, and emphasis should be made on the hospital's hygiene and aesthetic appeal. A doctor's quality can be improved by inviting the foreign doctor and facilitate them to work with to extend their capability. The arrangement for special training for existing nurses and medical staff to help to improve service quality.

However, there is a certain level of improvement and reorganization required in all medical sectors, including medical, dental, and paramedical services. Health services include hospitals, medical laboratories, diagnostic centers, and ambulance services, banking, and medical insurance services. Additionally, nationwide recognition and more focus needed to improve homeopathy, Ayurveda, Unani, and folk treatment system to attract more potential medical tourists internationally. For example, we can follow Thailand that expands fame and the number of foreign patients as it separates its hospitals from the presence of high-class hotels, separates indoor public spaces from treatment opportunities. Thailand has also created a comfortable environment, encourages staff to be more responsive to client requests, establishes a constructive relationship between physicians and patients, staff members, a piece of medical equipment, and techniques employed (Cohen, 2008). The government should ensure tourist-friendly policies and regulations for getting medical Visa to visit Bangladesh and reduce tax on medical instruments.

Consequently, we must refurbish our tourist spots to attract more tourists. Because of poor infrastructure and at present, patients get fewer touristic facilities than a medical treatment facility, which is not adequate to attract medical tourists. To solve this issue medical industry must collaborate with the tourism industry to work together. The government can enforce a joint policy to improve collaboration between the medical and tourist industry and can find a way to attract more medical tourists.

References

Ahmed, S. M., Hossain, M. A., & Chowdhury, M. R. (2009). Informal sector providers in Bangladesh: How equipped are they to provide rational health care? *Health Policy and Planning, 24*(6), 467–478.

Akber, M., Seraj, S., Islam, F., Ferdausi, D., Ahmed, R., Nasrin, D., Nahar, N., Ahsan, S., Jamal, F., & Rahmatullah, M. (2011). *A survey of medicinal plants used by the traditional medicinal practitioners of Khulna city, Bangladesh, 5*, 177–195.

Al-Azzam, A. F. M. (2016). A study of the impact of marketing mix for attracting medical tourism in Jordan. *International Journal of Marketing Studies, 8*(1), 139–149.

Ali, M. M., & Medhekar, A. (2018). Healthcare quality of Bangladesh and outbound medical travel to Thailand. *Economy of Region, Centre for Economic Security, Institute of Economics of Ural Branch of Russian Academy of Sciences, 1*(2), 575–588.

Beeton, S. (2005). The competitive destination. *Tourism Management, 26*(2), 204–296.

Bookman, M. (2007). *Medical tourism in developing countries*. Springer.

Cohen, E. (2008). Medical tourism in Thailand. *AU-GSB e-JOURNAL, 1*(1), 24–37.

Connell, J. (2006). Medical tourism: Sea, sun, sand an surgery. *Tourism Management, 27*(6), 1093–1100.

Das, R. (2017). Medical Marketing in Promoting Medical Tourism in India. *ELK Asia Pacific Journal of Marketing and Retail Management, 8*(3), 1–10.

Dawn, S. K., & Pal, S. (2011). Medical tourism in India: Issues, opportunities, and designing strategies for growth and development. *International Journal of Multidisciplinary Research, 1*(3), 7–10.

Dee, P. & Dinh, H. (2009). *Barriers to trade in health and financial services in ASEAN*. Retrieved from: https://www.eria.org/publications/barriers-to-trade-in-health-and-financial-services-in-asean/. Accessed 21 Nov 2020.

Floyd, B. J. (1997). Problems in accurate medical diagnosis of depression in female patients. *Social Science and Medicine, 44*(3), 403–412.

Gan, L. L. and Song, H. (2012). *A SWOT analysis of medical tourism: India and South Korea*. Retrieved from: https://papers.ssrn.com/sol3/papers.cfm?abstract_id=2194856. Accessed 21 Nov 2020.

Gupta, A. (2004). Medical tourism and public health. *Peoples Democracy, 27*(19).

Hall, C. M. (2011). Health and medical tourism: A kill or cure for global public health? *Tourism Review, 66*, 4–15.

Han, H. (2013). The healthcare hotel: Distinctive attributes for international medical travelers. *Tourism Management, 36*, 257–268.

Han, H., & Hwang, J. (2018). Growing competition in the healthcare tourism market and customer retention in medical clinics: New and experienced travellers. *Current Issues in Tourism, 21*(6), 680–702.

Hassan, A., & Burns, P. (2014). Tourism policies of Bangladesh-A contextual analysis. *Tourism Planning and Development, 11*(4), 463–466.

Hena, A., & Hasan, R. (2019). *Cross-border trade in health services between Bangladesh and India : Implications for Global Health systems*. The University of Dhaka.

Herrick, D. M. (2007). *Medical tourism: Global competition in health care*. National Center for Policy Analysis.

Heung, V. C., Kucukusta, D., & Song, H. (2011). Medical tourism development in Hong Kong: An assessment of the barriers. *Tourism Management, 32*(5), 995–1005.

Holliday, R., Bell, D., Cheung, O., Jones, M., & Probyn, E. (2015). Brief encounters: Assembling cosmetic surgery tourism. *Social Science and Medicine, 124*, 298–304.

Horowitz, M. D. & Rosensweig, J. A. (2008). *Medical tourism vs. traditional international medical travel: a tale of two models*. Retrieved from: http://www.imtjonline.com/zine/features/medical-tourism-vs-traditional-international-medical-travel-a-tale-of-two-models. Accessed 21 Nov 2020.

Hume, L. F., & Demicco, F. J. (2007). Bringing hotels to healthcare: A Rx for success. *Journal of Quality Assurance in Hospitality & Tourism, 8*(1), 75–84.

Islam, S., & Khan, M. Z. R. (2017). A review of energy sector of Bangladesh. *Energy Procedia, 110*(December 2016), 611–618.

Kazemi, Z. (2008). *Study of the effective factors for attracting medical tourism in Iran*. Master's Thesis. University of Isfahan.

Kumar, V., & Komaraiah, J. B. (2014). Role of Indian railways in promotion of tourism in India: A case study of foreign tourists. *IOSR Journal of Humanities and Social Science, 19*(1), 22–32.

Li, J., & Song, Y. (2019). Formal and informal care. In D. Gu & M. Dupre (Eds.), *Encyclopedia of gerontology and population aging* (pp. 1–8). Springer.

MacKian, S. (2003). *A review of health seeking behaviour: problems and prospects*. Retrieved from: https://assets.publishing.service.gov.uk/media/57a08d1de5274a27b200163d/05-03_health_seeking_behaviour.pdf. Accessed 21 Nov 2020.

Mamun, M. Z., & Andaleeb, S. S. (2013). Prospects and problems of medical tourism in Bangladesh. *International Journal of Health Services, 43*(1), 123–141.

McCarthy, E. J. (1975). *Basic marketing: A managerial approach*. R.D. Irwin.

Medhekar, A., & Ali, M. M. (2012). A cross-border trade in healthcare services: Bangladesh to India. *The Business & Management Review, 2*(1), 1–13.

More, A. (2019). *Medical tourism market 2019 global industry leading players update, gross margin analysis, size, development history*. MarketWatch.

Noor, R. (2013). *Outbound medical tourism from Bangladesh: Problems and solutions*. The University of Greenwich.

Pandey, M. M., Rastogi, S., & Rawat, A. K. S. (2013). Indian traditional ayurvedic system of medicine and nutritional supplementation. *Evidence-Based Complementary and Alternative Medicine*, 2013.

Smith, M., & Kelly, C. (2006). Wellness tourism. *Tourism Recreation Research, 31*(1), 1–4.

Spasojevic, M., & Susic, V. (2011). Development perspectives of new destinations in medical tourism. *International Journal of Business, Management and Social sciences, 2*(3), 9–13.

Chapter 14
Investment and Development for Agri-tourism in Bangladesh

Md. Alauddin, C. M. Atiqur Rahman, and Azizul Hassan

Abstract Agri-tourism is an important element of the global tourism industry. Bangladesh is a country rich in natural beauty with the potentiality to become a world leader in agri-tourism. To make its agri-tourism industry more profitable, it now requires effective planning and policies. Bangladesh's economic progress can be partly aided by sound, well-thought-out policies of agri-tourism. This chapter gives a quick overview of Bangladesh's agri-tourism business, including its extent, importance, and future prospects. According to our analysis, Bangladesh has a lot of potential for agri-tourism, and if the right steps are taken, the country can make profit from it. Agri-tourism in a developing country like Bangladesh can help to build the economy if careful planning and management are ensured.

Keywords Tourism · Agri-tourism · Scope · Importance · Bangladesh

Introduction

For decades, agri-tourism is recognized as a notion in tourism studies. However, its nomenclature and practice have been evolving for many years. The term agri-tourism in principle denotes the association of visitor amenities and attractions with agriculture that also has connections to farm tourism. In the developed nations, agri-tourism tends to have There are regular accounts, but in Asian countries, this concept has managed relatively less attention both from academics and practitioners.

Md. Alauddin (✉)
Department of Marketing, City University of Hong Kong, Kowloon Tong, Hong Kong

Department of Business Administration, International Islamic University Chittagong (IIUC), Chattogram, Bangladesh
e-mail: malauddin2-c@my.cityu.edu.hk

C. M. A. Rahman
Department of Business Administration, Southern University, Chattogram, Bangladesh

A. Hassan
Tourism Consultants Network, The Tourism Society, London, UK

A. Hassan (ed.), *Tourism in Bangladesh: Investment and Development Perspectives*, https://doi.org/10.1007/978-981-16-1858-1_14

Agri-tourism has links with the social, economic, and environmental sustainability elements and has a strong relatedness to local communities with their attitudes and behavior towards the tourists (Muresan et al., 2016). This consists of diverse leisure and recreational activities for tourists that can be viewed as a unique and innovative diversification strategy for small farms and towns, the farmers' non-economic and economic benefits, communities, and visitors (Tew & Barbieri, 2012). A good number of rural areas are endowed with natural, historical, and social qualities that both create and behold appeal towards tourists. These appeals can be achieved and exploited in line with agricultural farming activities. According to Fleischer and Tchetchik (2005), such phenomenon is referred to by tourism and agriculture specialists of agri-tourism, farm tourism, or rural tourism because of these terms' interchangeable characteristics. Agri-tourism can be conventionally featured as the amalgamation of agricultural production and tourism that aims for encouraging both long and short–term visitors to rural areas or farms having the purposes of education, enjoyment, and/or active involvement in farming life or agricultural production and service activities.

Agri-tourism can be featured as one of the most vital local initiatives that can positively affect the development actions of the rural areas (GEOTEE, 2000; Van der Ploeg et al., 2000; Iakovidou, 1997). Agri-tourism concentrates on the most effective and efficient use of the relative advantages of an area. These advantages generally include both cultural natural environments' protection that is rather important for the agri-tourism products, as well as the promotion in such way, can be sustainable for the rural area's development. In mountainous countries such as Austria, Switzerland, and Sweden, where intensive agricultural production is difficult, a significant proportion of farmers (more than 10%) are involved in various agri-tourism activities (Bazin & Roux, 1997).

Agri-tourism refers to those touristic activities that generally happen in the rural areas and attached to the rural residents' agricultural holdings who may or may not be primarily employed in agriculture (GEOTEE, 2000). In theory, agri-tourism properties are cooperative or family-owned businesses that provide lodging in exchange for participation in agricultural or other outdoor rural activities, as well as using or catering to locally produced agricultural products (Vafiadis et al., 1992). Agri-tourism contributes to economic and social development by providing improved living and earning conditions as a result of many activities (Iakovidou, 1997).

Definition and Scope of Agri-tourism

Agri comes from the Latin term "Ager(agri-gen.-Latin)", which means field, and the Greek word "Agros", which means soil. When tourism is defined as a type of active recreation that takes place outside of one's house and is driven by cognitive, recreational, and athletic demands (Sznajder et al., 2009). Agri-tourism is defined as a commercial activity that generates extra income for the owner of a working farm, ranch, or agricultural industry by entertaining tourists (University of California, n.d.).

Agri-tourism is a multi-disciplinary field of research, and the notion has been referred to as agri-tourism, agricultural tourism, agri-entertainment, farm vacation tourism, and farm tourism in the literature. Agri-tourism is defined as tourism that includes agricultural and rural surroundings, as well as agricultural products. Tourists can buy fresh agricultural commodities and products directly from farmers, bypassing the middleman, as part of the agri-tourism concept (agents) (ATDC, n.d.).

In common, such terms denote community events or small-scale farm enterprises showcasing the products and activities of rural producers with the agricultural production heritage or farming regions to visitors. Agri-tourism offers "rural experiences" towards the tourists to generate revenue for surrounding communities and of course, for the farmers. Agri-tourism is a useful and unique tourism product that presents and exposes visitors to rural areas primarily for recreational and educational purposes. Such rural experience conventionally comes with a range of activities and attractions that feature minimal urban and industrial development, open spaces, and opportunities for visitors to directly experience agricultural, natural, and pastoral environments. From the tourism experience perspective, agri-tourism attaches agricultural or rural settings with it. Agriculture relates human foods' farming with fodder and fiber, beverages, hunting, livestock products, fishing, and forestry. From an agri-tourism viewpoint, agri-tourism scope expands to agro-based small scale industries that are located in the rural areas or countryside. Also, conventional agricultural services, rural culture, values, and belief systems are covered through agri-tourism. A few of the key features of such geographical locations are units of small-scale production, remoteness, low human settlement density, and low degree of specialization (Taware, 2007). Agri-tourism thus refers to tourism activities that attach rural settings with agricultural products, services, or operations on the ground of tourism experiences as paid for by the tourists.

Agri-tourism Is a Tourism Offer

Agri-tourism is a different research area that covers rural environments, followed by its agriculturally produced products and processing and packaging. This tourism type also covers agriculture-dependent services, communities, cultures, and belief systems of the rural area. Agriculture in the latest understanding is divided into two distinct and separate areas: first, urban agriculture, and second, rural agriculture. Thus, agri-tourism denotes the activities that aim at developing the social and economic development of the general countryside and rural regions of a country. Agri-tourism's location widely covers the land areas that are not covered by cities, industrial parks, towns, suburbs, and urban areas. These locations are termed as country, hinterlands, or countryside invariably. Researchers have related other forms of tourism with agri-tourism in few instances. For example, George and Rilla (2011) viewed agri-tourism with nature tourism, or Slocum and Curtis (2018) attached agri-tourism with food tourism. This affirms that agriculture itself can offer tourism in many different forms.

Daugstad and Kirchengast (2013), in their research, explored how summer farmers in Valdres (Norway) and Bregenzerwald (Austria) have been playing their double roles as tourist hosts and farmers. The research, with the support of interviews with farmers, discussed the ways within which such double roles were staging of summer farms and influencing the tourist-host interactions. This research also outlined how the discourses of heritage and authenticity were manifesting themselves therein. The research found that in both Norway and Austria, summer farms were the objects of "idyllisation". The research found that these farms were well suitable for agri-tourism because of their less detrimental to the environment and culture and more intimate capacities than mass tourism. The research suggested that a key characteristic of this agri-tourism settings was a pseudo-backstage arrangement – a frontstage that was seen as an "actual" backstage.

Koutsouris, Gidarakou, Grava, and Michailidis (2014) outlines the (agri-) tourism entrepreneurs' and their businesses' profile followed by the estimation, comparing, and applying a full account approach, the earned cash flows from tourism farming. The research also outlined the owners' and their businesses' socio-economic characteristics' influences in tourism businesses' effectiveness. These identified issues are considered as insufficiently dealt with in the global (agri) tourism literature studies. The findings of this research covered in mountainous Corinth in Greece presented that the relationship between agriculture and agri-tourism is rather weak with non-residents. Tourism development opportunities are mainly exploited, implying the tourism development benefits' leakage from the local economy. When the accommodation business gets differentiated, the owners' relationship to agriculture presents that the combination of farming and tourism supports the farming households. Still, the local farmers were able to capture the opportunities marginally.

The above Table 14.1 elaborately shows agri-tourism as a tourism offer that can have application in Bangladesh's similar context as an agricultural country.

Agri-tourism in Bangladesh

Dhaka Tribune (2020) and the Daily Star (2020) reported that there is some major revenue-generating sources in Bangladesh. These are agriculture, industry, services, and tourism and belong to the service industry. The tourism industry's contribution in the national Gross Domestic Product (GDP) of Bangladesh increased to 9.44% in 2009–10 from only 7% in 2004–05. The tourism industry is believed to have the capacities for ensuring economic development subject to the political will and effective and proper marketing strategy implementation.

Table 14.1 Agri-tourism as a tourism offer

Agri-tourism definition criteria	Definition's components	Advantages to those involved in the development of this activity
Defining agri-tourism on the agricultural activity context	Unites two complex industries' elements. Both agriculture with tourism can be able for playing a crucial role to support agricultural enterprises	Advantages that are oriented for farmers:
		New market niche development
		More interests in locally produced agricultural products
Definition of agri-tourism on the rural community development context	A source of growth and diversification for rural economies that can also help agricultural enterprises stay afloat	Opportunities for using/ maintaining agricultural land
Defining agri-tourism as a stand-alone component	An economic activity in a food industry enterprise or an agricultural farm/holding that provides visitor advantages and satisfaction and also generates additional income for the agriculturalist.	More job creation for the families
		To increase the sustainability of agricultural business enterprises
		Benefit creation for the local communities
		Creating jobs
		Expanding the local markets
		To attract relevant small businesses and industries
		To generate more benefits for the tourism industry
		To diversify the tourist product mix
		To ensure unique positioning within the agri-tourism communities.

Source: Ciolac et al. (2019)

On the ground of activity nature, location type, and duration of the stay, tourism can be classified in diverse ways as cultural tourism, heritage tourism, geo-tourism, historical tourism, adventure tourism, agri-tourism, and eco-tourism. Among these and few others, agri-tourism can be viewed as essential for saving the native fruits, crops, vegetables, fish, flora, and fauna.

Agri-tourism is the vacation type where hospitality is offered both to individuals and firms. This can include the opportunity to view and often assisting with farming during the tours. In these tours, tourists get opportunities for picking vegetables and fruits, riding horses, honey tasting, shopping in farm-stands, or gift-shops for purchasing locally and regionally produced or hand-crafted gifts. Each of such farms commonly offers both memorable and experiences as suitable for almost all tourist types. In some cases, tourists become more interested in learning the ways of food production. Some tourists also tend to meet the farmers, food producers and want to meet and greet them.

Bangladesh is traditionally a country with fertile and alluvial land where more than 70% of the population relies on agriculture. The country can have some selected zones for agri-ecological and sustainable farming. Apart from these, several agriculture enterprises can be established for the well-being of the local goat, poultry, buffalo, and cow farms for eradicating poverty. This can become a good lesson for a tourist to learn the ways of farming in eco-friendly ways.

Students of the schools, colleges, and universities can spend leisure time trying out organic farming in both rural and semi-urban areas. On the other side, both the researchers and students can share views, opinions, and thoughts with the farmers and relevant people. They can teach them and acquire indigenous knowledge about farming from the farmers in return. At present, sustainable cultivation becomes a necessity for ensuring organic foods for the present and future generations. Both the agri-tourists and researchers can often share their experiences with the indigenous farmers about global warming, climate change, crisis for extracting more groundwater, arsenic pollution, bio-fertilizers, the importance of vermicomposting, etc.

Both the agri-tourists and indigenous people can exchange their knowledge and ideas on poultry farming, followed by the importance and benefits of poultry manure in agriculture, etc. Bangladesh is featured as an extreme sufferer from natural calamities and land-based degradation. Thus, both the international and national agri-tourists can share and exchange their views with the local farmers for reducing climatic impacts on water, land, health, forest, and fish resources.

The Government of Bangladesh (GoB) can take up projects to represent the farmers' indigenous capacities for protecting the land, ecology, and environment among the international tourists. This can support attracting them that can enrich the agri-tourism industry in the country. The agricultural products' suitability is different across the country. For example, Barisal and Chandpur are known for fishing, the lands of Kushite are fertile for tobacco, Mymensingh is famous for paddy, Sathkhira for prawn, Dinajpur for its wheat, Jessore for flowers, and so on. Thus, Bangladesh endorses a good number of available options to promote the concept of agri-tourism and to trace out newer means that can have contributions in the national economy.

Bangladesh, as an agricultural country, possesses the capacities for competing with other member countries of the South Asian Association for Regional Cooperation (SAARC). To do this, agricultural development needs to be considered seriously through the formulation of comprehensive strategies and involving the farmers and the local population. Bangladesh has the capacity to increase its agri-tourism potentials with ease with a substantial margin and effective marketing and promotional initiatives. With considerable and effective actions and policy planning and implementation, the tourism industry of Bangladesh thus can be turned into a significant contributor to the national GDP.

For marketing agri-tourism, the application of geographical information systems (GIS) can be highly effective and very important. Through such systems, the relevant tourism authorities can become able to sell agri-tourism products and services, indigenous farming tools, as well as can present suitable localities to incorporate diverse animals and plants, can elucidate geo-demographic features, cultural diversities, and socio-economic conditions of the farmers, indigenous peoples, poultry managers, fish collectors, and others. All these efforts can support knowing the most responsible people for agricultural development.

The gradual intensification of the wider global agricultural business has already started for exterminating the local farming systems around the world. Agri-tourism is thus cannot be viewed as the only way to involve local farmers in the local climatic issue reduction process but this can become an area of concern to consider seriously.

Why Agri-tourism in Bangladesh?

The importance and position of agriculture on the Bangladesh national economy is overwhelming and confirmed to support agri-tourism. Food and Agriculture Organization of the United Nations (FAO) (2020) data show that in Bangladesh GDP, the contributions of agriculture are 17.5%, 70.1% land area used for agriculture, and 48.4% of the total population are employed. Country or rural farmers generate their livelihood and income from agriculture and meet their daily demands of nutrition and food supply. In most cases, these farmers struggle to meet household and relevant expenses or maintain a quality life suffering from lower-income and poor financial conditions. As a reason, the development of agriculture becomes essential in the country to provide the daily needs of these hard-working farmers, increase exports and trading activities, reduce unemployment issues, and offer raw materials for industries that can be achieved with agri-tourism. This type of tourism can also be applied to persuade and inspire the farmer communities to raise their produce in an eco-friendly way and for conserving the farms' biodiversity (Dangol & Ranabhat, 2007). This also allows tourists to come in closer contact with the village, rural dwellers and get engaged with conventional agricultural productions. Tourists also avail the opportunity for knowing the diverse indigenous agricultural practices as crop harvesting, marking and conserving, organic vegetable production, fresh fruit picking and converting into marmalade, and others (Pandey & Pandey, 2011).

In Asia, Nepal is awarded with the agro-biodiversity and exaggerate with the tranquil geographical landscape and location. In Nepal, rice, wheat, and maize are the main crops. "Terai" is the storehouse of Nepal that has tourism potentials. In the high hills, buckwheat, barley, millet is a few of the major crops, and citrus, coffee, tea, cardamom, ginger, etc. are important cash crops. The practice of off-season and on season vegetable production in the entire country can attract agri-tourists. These agri-tourists come closer to the rural and natural activities to be entertained, participate, and feel pleasure for leisure activities.

As an agricultural country, Bangladesh also has immense prospects to develop agri-tourism. Agricultural production techniques differ from region to region. The hard-working Bangladeshi farmers tend to convert the fields into opera house theaters where they are used to sing "Asaregeet", the local songs. These farmers can also exchange feelings, emotions, and experiences with the tourists. The mutual understanding and exchange of the farmers' feelings can fascinate the agri-tourists quickly.

Agri-tourism Potentials in Bangladesh, as Supported by the Ministry of Agriculture and Rural Development

In many developing countries, including Bangladesh, several bottlenecks have blocked the progress of rural and agricultural development. Agri-tourism can become a valid and useful tool to address issues to ensure community participation and sustainable development. The industry should be given top priority. It has the potential to make a significant contribution to Bangladesh's overall economic development. It is important to remember that this is a new and profitable industry. People involved in this industry, both directly and indirectly, need have access to a variety of logistics services. It will provide a new opportunity for the country to establish additional job possibilities. It will also assist in attracting more local and international tourists to these beautiful locations. It's past time to get this firm up and running using cutting-edge technology. Agri-tourism's potential for both rural development and agriculture can have few drawbacks.

Addressing the Issue of Underemployment

Bangladesh has a large number of jobless people. This industry can employ these massive individuals. It will not only serve to raise the level of living, but it will also aid in economic development. As a result, agri-tourism will open up new options for unemployed people to work in the business. They can also serve as tour guides and helpers, and agri-tourism can become a reliable source of revenue and subsistence.

Financial Facilities Arrangements

The financial crisis is quite a common issue for both agriculture and relevant rural industries. In general, industries in the rural areas mostly suffer from liquidity issues. One of the basic reasons for this issue is the uneven distribution of financial arrangements and cash flows between the urban and the rural areas. Financial benefits commonly favor the urban areas in most of the developing countries, including Bangladesh. Capital and cash flow injection by agri-tourism can support in solving liquidity issues to promote agriculture and other rural industries.

Communication System Facilities Development

Bangladesh's communication system has improved, and products can now be easily transferred from one corner to another. People can take advantage of this chance to start an agri-tourism business. Furthermore, better transportation and transportation infrastructure reduce the average cost of producing agricultural products.

Infrastructural Facilities Development

Infrastructure facilities have greatly improved in recent years. It will make producing agro-products easier for them. To grow the agri-tourism industry in Bangladesh, more people need be employed in it. Furthermore, the expansion of infrastructure amenities has made agribusiness very straightforward and convenient.

Expansion of Training Facilities

Because of the training facilities, Bangladesh has a bright future in agri-tourism. In the field of agri-tourism, certain private groups offer various types of training opportunities. Furthermore, several colleges offer a variety of degrees in this field. As a result, suitable training from various public and private training institutions is required to do this company.

Constraints and Issues for Agri-tourism Promotion in Bangladesh

Any country's economic progress can be accelerated by properly utilizing and managing agri-tourism. Many underdeveloped countries were able to rebuild their economies by focusing on agri-tourism. Because Bangladesh is a nation of natural beauty and fertility, agri-tourism is a growing sector. However, the primary impediment to successful utilization is the lack of a coherent strategy and low participation among farmers and other local residents. Bangladesh's agri-tourism development is slowed by the lack of a geographic information system (GIS). The potential for agri-tourism development is further constrained by the local climate problem, which is particularly vulnerable to atmospheric and land-based degradation. Natural climate change has a huge impact on the country's land, water, forest, health, and fish resources. Agri-tourism development may be hampered by a lack of or absence of public utilities (water, canal, and thermal energy) in rural areas, as well as local apathy. One of the key flaws is the infrastructure (energy concerns, low road equipment quality, and a poorly designed utility distribution network). The country's agricultural industry has suffered a shortage of labor force due to job loss, natural disasters, local peoples' migration, and so on. The skilled labor shortage is also the outcome of more attraction to the manufacturing and service industry than agriculture. This increasing workforce and labor shortage for agriculture lead to employment imbalance in the country. Bangladesh can hardly capitalize on the tourism potentials mainly suffering from the lack of local tourism brands and products, lack of skill, and expertise development training, lack of dedicated tourism enterprises and experts' guidance and relevant issues. Climate change, saline intrusion, and frequent droughts are all problems in Bangladesh's south and north. The overwhelming emphasis on urbanization, as well as water constraints in the dry season due to unresolved river-related issues with neighboring nations, have hampered the development of agri-tourism in Bangladesh.

Marketing Facilities that are Available for Agricultural Products

Agri-tourism marketing in Bangladesh is challenging to plan. A number of steps must be followed in order for products to be distributed from manufacturers to customers. In Bangladesh, a lack of suitable storage and warehousing is a key impediment to agri-tourism. Furthermore, manufacturers do not obtain fair pricing for their goods, and as a result, they are sometimes afraid to sell them. Raw material prices change as well, posing a significant challenge for the country's agri-tourism development.

Middlemen's Role

Middlemen are essential in getting items from manufacturers to customers, but they can also be used to manipulate the market. Manufacturers often claim that middlemen do not offer them with reasonable prices for their goods. For their own benefit, they aim to grab the market both directly and indirectly.

Land Scarcity

Bangladesh is a small country with a dense population and one of the world's most densely populated countries. In rural areas, people are also losing their farms owing to floods and other factors. As a result, the average size of cultivable land is shrinking every day, posing a serious threat to Bangladesh's agri-tourism industry. People are losing interest in this industry since the profits are so low, deterring the manufacturers from growing their property.

Education Facilities Development

The literacy rate in Bangladesh is rising every day, yet the new generation is exhibiting little interest in agri-tourism. The majority of educated people are relocating from rural to urban areas in search of better work opportunities.

Lack of Facilities for Credit

Another issue that has arisen in the development of agri-tourism is the availability of loans. Manufacturers are having difficulty obtaining financing from banks and other financial organizations.

Policy Implications for Investment and Development for Agri-tourism in Bangladesh

Bangladesh is an agricultural country with more than 70% of her population depending on agriculture. There are some agro-ecological zones in the country with suitability for sustainable farming. For sustainable farming, Bangladesh needs to formulate comprehensive strategies with the local people and the farmers' involvement. For marketing agri-tourism products and services, Geographic Information Systems (GIS) can be important. GIS can help the authorities and management of the tourism industry sell agri-tourism products and tools of indigenous farming with ease and present suitable localities to incorporate diverse animals and plants. GIS explains geo-demographic features and can simply contain the cultural diversities and socio-economic situations of the rural farmers, poultry managers, fish collectors, and indigenous peoples. This can help for knowing the people having responsibilities to feed the people in Bangladesh. Agri-tourism's success of agri-tourism entirely on proper planning and coordination and execution in a systematic and organized manner. Responsible agencies of the GoB, financial institutions need to play effective and useful roles to develop agri-tourism. The government tries for initiating a project for supporting the famers' indigenous capacities for protecting land. In specific, the Bangladesh Academy for Rural Development (BARD) can help for promoting agri-tourism. The country makes an effort to encourage university and college students to use their imagination to incorporate organic farming into their studies. This incorporation can be an extra-curricular activity dedicated to the community's well-being and service. For strategic promotion, the GoB can publish both creative and attractive advertisements in magazines and travel guides for spreading the word through both the local and overseas travel agents, tour operators, and travel writers.

The development and retention of entrepreneurial spirit within the "natives" can be influenced by agri-tourism. This tourism type can merely be developed under the risk assumption conditions by the relevant rural households. However, agri-tourism relies on the human health and "health" of the environment development. Within such development, agri-tourism can achieve the key objective and that is the rural environment's sustainability. Proper and effective marketing can offer complete and relevant data and information. This can also increase respect for a destination's social, natural, and cultural settings and enhance the customers' satisfaction. Tourism marketing, including direct sales, advertising, and customer service, may benefit from the proper and successful use of technology and the Internet.

Advertisement is portrayed as a non-personal method of mass communication that can provide a product with a greater level of familiarity. Various tourist attractions can be advertised on television, radio, newspapers, and magazines, with persuasive messages to entice potential tourists. Some convincing and appealing advertising, such as "Beautiful Bangladesh/Beautiful Rural Bangladesh," can be shown (The Daily Observer, 2019; the Daily Star, 2017). Through appropriate policy formulation and implementation, proper usage of the fundamental resources—land, water bodies, forests, and human resources—can boost overall production from land, water, and forest, resulting in sustainable development. Agri- tourism's rural to urban migration reduction plan also helped to lower marginal poverty. More attention to rural infrastructure, security measures, safety and security, incentive, and continual training for rural farmers, youth, and women is required for sustainable agri-tourism. The private sector's involvement also ensures agri-long-term tourism's viability. In the context of Bangladesh, effective collaboration between the Internet of Things (IoT) and agri-robotics can help reap the benefits and minimize the negative effects of agri-tourism. This type of application of the most recent technologies needs to be spread across the whole country to promote agri-tourism. Few of the other latest technologies (i.e. Artificial Intelligence (AI), machine learning, big data, etc.) can play a crucial role in promoting sustainable agri-tourism in Bangladesh.

The present GoB stress on food security, nutrition, and the overall development of agriculture as the priority area followed by the implementation of an investment plan to increase food production and secure nutrition supply. The concept of the Country Investment Plan (CIP) is viewed as a strategic tool of the GoB to increase and improve the investment in food security (i.e. agriculture, food access and safety nets, livestock, and fisheries and nutrition activities). This also offers a framework to initiate dialogue and collaboration between relevant partners and interest groups (i.e. the Government, private industries or firms, organization of the farmers, Civil Society Organizations or CSOs and donors) lead by the GoB. the First Country Investment Plan 2011–2015 is followed by the Bangladesh Second Country Investment Plan (CIP2) 2016–2020 (Bangladesh Planning Commission, 2018). CPI2's overreaching goal is for achieving advanced food security and nutrition for everybody and always through turning the food systems as sustainable and nutrition-sensitive. To address climate change effects on agriculture and prioritize investments for improved productivity, mitigation, and resilience in the agricultural industry, the Climate-smart Agriculture investment plan (CSAIP) is jointly launched by the GoB and the World Bank (The Independent, 2019).

Conclusion

The importance of the agri-tourism business to Bangladesh's economic development is enormous. This paper discusses the reasons for agri-tourism, its scale, and the industry's potential to positively impact the country's socioeconomic development. It has the potential to be a watershed moment in Bangladesh's economic development if the right steps are implemented. It will also generate employment opportunities. It is important to remember that this country is rich in natural resources and has enormous potential for growth in this business. Furthermore, the GoB can earn a significant amount of foreign currency from this industry. The study also suggested certain policies that could be implemented to strengthen the industry. Future research studies could concentrate on the public and private sector investment that is needed to grow this business.

References

ATDC (Agri Tourism Development Company). (n.d.). *What is agri tourism all about?* Retrieved from: http://www.agritourism.in/blog-post/what-is-agri-tourism-all-about/. Accessed 3 Oct 2020.

Bangladesh Planning Commission. (2018). *Bangladesh delta plan 2100*. Planning Commission.

Bazin, G., & Roux, B. (1997). *L'Agri-tourisme: un atout pour les zones rurales difficiles mediterraneennes* (pp. 215–236). Ed. Harmmattan. (in French).

Ciolac, R., Adamov, T., Iancu, T., Popescu, G., Lile, R., Rujescu, C., & Marin, D. (2019). Agritourism – A sustainable development factor for improving the 'health' of rural settlements. Case Study Apuseni Mountains Area. *Sustainability, 11*(5), 1467.

Dangol, D. R., & Ranabhat, B. (2007). Developing agro-ecotourism in Nepal. *NTTR*, November, pp. 12–25.

Daugstad, K., & Kirchengast, C. (2013). Authenticity and the pseudo-backstage of agri-tourism. *Annals of Tourism Research, 43*, 170–191.

Dhaka Tribune. (2020). *Why we should invest in agro-tourism*. Retrieved from: https://www.dhakatribune.com/business/2018/02/06/invest-agro-tourism. Accessed 5 Apr 2020.

Fleischer, A., & Tchetchik, A. (2005). Does rural tourism benefit from agriculture? *Tourism Management, 26*(4), 493–501.

Food and Agriculture Organization of the United Nations (FAO). (2020). *Bangladesh*. Retrieved from: http://www.fao.org/asiapacific/perspectives/agricultural-statistics/global-strategy/results-in-the-region/bangladesh/en/. Accessed 5 Apr 2020.

George, H., & Rilla, E. L. (2011). *Agritourism and nature tourism in California*. University of California.

GEOTEE. (2000). *Mountain and less favored areas in Greece: Agro(to)tourisni*. GEOTEE. (in Greek).

Iakovidou, O. (1997). *Agri-tourism in Greece: The case of women agri-tourism co-operatives of Ambelakia*. Retrieved from: https://newmedit.iamb.it/share/img_new_medit_articoli/774_44iakovidou.pdf. Accessed 5 Apr 2020.

Koutsouris, A., Gidarakou, I., Grava, F., & Michailidis, A. (2014). The phantom of (agri)tourism and agriculture symbiosis? A Greek case study. *Tourism Management Perspectives, 12*, 94–103.

Muresan, I., Oroian, C., Harun, R., Arion, F., Porutiu, A., Chiciudean, G., Todea, A., & Lile, R. (2016). Local residents' attitude toward sustainable rural tourism development. *Sustainability, 8*(1), 100.

Pandey, H., & Pandey, P. R. (2011). Socio-economic development through agri-tourism: A case study of Bhaktapur, Nepal. *Journal of Agriculture and Environment, 12*, 59–66.

Slocum, S. L., & Curtis, K. R. (2018). *Food and agricultural tourism: Theory and best practice*. Routledge.

Sznajder, M., Przezbórska, L., & Scrimgeour, F. (2009). *Agritourism*. CABI.

Taware, M. P. T. M. P., & ATDC, P. (2007). *Agri-tourism: Innovative supplementary income generating activity for enterprising farmers*. Retrieved from: http://www.agritourism.in/blog-post/agri-tourism-innovative-supplementary-income-generating-activity-enterprising-farmers/. Accessed 5 Apr 2020.

Tew, C., & Barbieri, C. (2012). The perceived benefits of agritourism: The provider's perspective. *Tourism Management, 33*(1), 215–224.

The Daily Observer. (2019). *Home-grown agro-tourism in BD*. Retrieved from: https://www.observerbd.com/details.php?id=177402. Accessed 5 Apr 2020.

The Daily Star. (2020). *The many benefits of agro-tourism*. Retrieved from: https://www.thedailystar.net/opinion/the-many-benefits-agro-tourism-1454098. Accessed 5 Apr 2020.

The Independent. (2019). *Smart solutions in agriculture: Prospect and challenge*. Retrieved from: http://www.theindependentbd.com/printversion/details/194313. Accessed 5 Apr 2020.

University of California. (n.d.). *What is agritourism?* Retrieved from: http://sfp.ucdavis.edu/agritourism/factsheets/what/. Accessed 3 Oct 2020.

Vafiadis, G., Xristos, K., & Leonidas, P. (1992). *Agri-tourism and equal development*. Agricultural Bank of Greece. (in Greek).

Van der Ploeg, J. D., Renting, H., Brunori, G., Knickel, K., Mannion, J., Marsden, T., De Roest, K., Sevilla-Guzmán, E., & Ventura, F. (2000). Rural development: From practices and policies towards theory. *Sociologiaruralis, 40*(4), 391–408.

Chapter 15
Agri-tourism in Bangladesh: The Investment and Development Perspective

Polin Kumar Saha, Shahida Akhter, Azizul Hassan, Ashraful Alam Anas, and Israt Jahan Shathi

Abstract Emphasis on the tourism industry can drive a nation towards sustainable development, whereas agri-tourism represents an excellent performer in a modern world. But in most countries, sustainable development features lack appropriate strategies to promote investment and development in agri-tourism. This study reviews data and information published by both the government and private sector and aimed to outline the investment and development scenario of agri-tourism in Bangladesh. The study finds that the main source of the rural peoples' income can be agri-tourism. This tourism type can help to reduce poverty and develop the community peoples' lifestyle in a sustainable manner. To ensure socio-cultural and economic advancements, agri-tourism development and investment policy support can help. Findings of this research show that agri-tourism can create job opportunities, can ensure a standard investment environment in rural areas and can support the development of infrastructures to considerably benefit both the rural and urban livelihood generation. The study also finds a number of important aspects such as input, output, outcome, and impact on sustainable tourism development from the application of the sustainability approach.

Keywords Agri-tourism · Development · Tourism · Bangladesh · Sustainability

---

P. K. Saha (✉)
Environment and Sustainability Professional, Dhaka, Bangladesh
e-mail: polin.msls2009@gmail.com; polinksaha@gmail.com

S. Akhter
Department of Development Studies, BRAC Institute of Governance and Development, BRAC University, Dhaka, Bangladesh

A. Hassan
Tourism Consultants Network, The Tourism Society, London, UK

A. A. Anas · I. J. Shathi
Global Research and Marketing (GRM), Dhaka, Bangladesh

A. Hassan (ed.), *Tourism in Bangladesh: Investment and Development Perspectives*, https://doi.org/10.1007/978-981-16-1858-1_15

Introduction

Tourism is a fast-growing industry to obtain higher living standards for the world community through development and economic growth. Developing countries like Indonesia, Kenya, Nepal, Sri Lanka, Thailand, and Zimbabwe have rebuilt their economy with the help of proper utilization and regulation of the tourism industry (The Daily Star, 2017; Dhaka Tribune, 2018; Dhaka Courier, 2019). Bangladesh has the potential space to become a sustainable tourism development in a tourist zone, a land full of natural beauty. As a developing country, Bangladesh is overpopulation. However, a significant part of this population can be successfully involved in the tourism industry. They can cultivate their land as well as can become an element to promote tourism. Agri-tourism can reduce the overpopulation problem when their skills are utilized. Therefore, investment is a sword against the challenges to change the quality engaged in agri-tourism into agricultural, social, cultural, and economic advancement in any country.

In developing countries, lack of cultivable land and population growth can lead to severe food scarcity. Agricultural production followed by transformed agricultural systems has to meet growing markets' demands (Song et al., 2019; Firoozzare et al., 2018; Robinson & Carson, 2015). This growing market demand can rise agri-tourism because Bangladesh has a green countryside, diversified and rich traditions and customs, colorful festivals, handicrafts, historical heritage and folk arts, which can capitalize benefits of the situation (Ahmed & Jahan, 2013). Plan and action on diversification strategies, promoting sustainable management of natural resources, and local community participation is necessary to get a proper outcome from agri-tourism.

From valuable observation based on Veljković and Broćić (2017) data, agri-tourism is associated with a beautiful natural village setting that expresses farming and local craftsmanship. Agri-tourism reveals the use of assets from farm household and expands employment scope for household members, and prioritizes farmers' economic incentives to maintain agricultural land and related natural amenities, such as wildlife, streams, and forests that can be significant for ensuring better quality life style in suburban and exurban areas. As an agricultural country, Bangladesh has enormous opportunities to discover this tourism. The local economy can grow by giving opportunities to visit or stay on the granary and participate in gathering agronomical products and quality information. Veljković and Broćić (2017) informed that there are foreign and domestic visitors or travelers interested in spending weekends or short vacations in the countryside that are consistently increasing (mainly out of town). These foreign visitors have a desire to learn about the culture and tradition of different parts of states and enjoy traditional specialty clothes, foods, dishes, and drinks. A better example can be from the opinion of Goodwin and Chaudhary from 2017 that there are many developing countries have a tourism industry that boost their national economy's performance. Indonesia is one of the countries. Bali, a part of Indonesia with the largest achievement to Indonesian tourism's attainment due to institutional investors in agri-tourism. Moreover, agri-tourism exponential contributions to the process of sustainable rural development. This research explains agri-tourism in Bangladesh in terms of investment and growth.

Agri-tourism Definition and Its Effective Roles

Agri-tourism or agro-tourism has various definitions on different contexts. Among them, the popular explanations are: "rural enterprises which include both a working farm environment and a commercial tourism component" (Weaver & Fennell, 1997: p. 357), and "an alternative farm enterprise which was one of several possible pathways of farm business development" (Srisomyong, 2010: p. 1). Agri-tourism can be distinguished as farmers' businesses inside their working for agricultural holdings to enjoy and educate visitors. Farm stays, picnic areas, farm shops, farm tours, educational demonstrations, pick your farm produce etc. are some typical attractions for tourists at farm destinations. Agri-tourism is considered a form of diversification of a farm that can amplify subsidiary activities to agriculture. The relationships between tourism and agriculture include the reality that tourism can help farm diversification by employment in new and remaining businesses and create fresh markets for agronomical products (Srisomyong, 2010). Agri-tourism is a catalyst that helps to grow the income (supplement) and economy and a booming industry that can increase revenue. Agri-tourism is an activity that is diversified on the natural resources' consumption and the local culture. It also helps to the expansion of personal relationships between the local community and visitors. Still, it can enhance the quality of life by creating employment opportunity, as well as the influence on aspects economic and social; the multifunctional sustainable countryside (rural) development (Sumantra & Yuesti, 2018; Das & Rainey, 2010; Lathiras et al., 2010; Wyporska & Mosiej, 2010; Chesky, 2009).

It is commonly believed that agri-tourism can play a key role in promoting ecosystem functioning and biodiversity to support humans' multiple benefits (Firoozzare et al., 2018). Moreover, agri-tourism or agro-tourism refers to the act that one can visit a working farm or any horticultural, agricultural, or agribusiness operation for the ambition of active involvement in the farm, enjoyment, education, or for operation activities. But the effective agri-tourism or agro-tourism term is difficult which can stand for the development and can be more functional into sustainable agri-tourism in the operational field, as mentioned by Tseng et al, 2019: p.1. For tourism purposes, agri-tourism is an approach that is systematically based on-farm diversification. Quality information is the usual form to assess agri-tourism, where agri-tourism is rural tourism which is a subset that uses farms in a diverse array as tourist destinations. It is an activity that is economic with the potential economic, social, and environmental or triple bottom line. The goal of sustainable agri-tourism is to optimize the benefits in the economy and achieve maximum benefits for the stakeholders, improve quality of life, and promote environmental conservation. In Thailand, agri-tourism is encouraged to support farmers become self-sufficient and is experienced by all professions entire the country. The activities of agri-tourism are based on environmental and ecological preservation to ensure that the natural environment, the ecosystems, and the human population can sustain.

Investment on Agri-tourism and Its Return to the Development

The sustainable development concept was originated in developed first world countries through tourism and agri-tourism. The boosting need for the exercise of natural resources and investment and community participants plan is mandatory to come out economic growth and contribute to agricultural development. Therefore, funds or investment is the first key element to making the excellent business path to rural development. Another study shows that tourism can elicit growth by providing foreign direct investment included many new facilities. The link between capital investment and tourism has been developed based on essential factors like business-friendly environments, accessible human capital, and various amenities through hotels, resorts, parks, transportation, energy, etc. Overall, tourism is extremely needy on power to move towards its diurnal business activities (Khan et al., 2020). The strategy model is forefather in a project that will bring a more profitable impression on any state over a long time (Fig. 15.1).

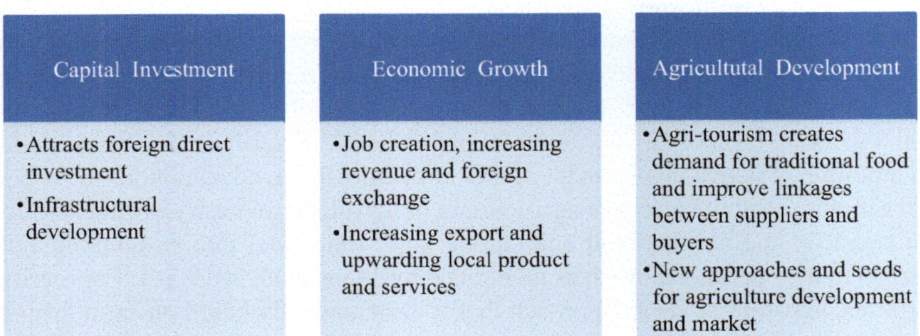

Fig. 15.1 The strategic model on sustainable tourism development and rural development. (Source: Khan et al., 2020)

Tourism and Development for Developing Economies: A Policy Implication Perspective

The majority of developing countries could not be a success in the tourism world. Many possibilities to be a tourism industrial nation need the country's management to think a robust entrepreneurship plan with sufficient funding support to get the lion-size amount back from the tourism industry into agri-tourist.

Notification from the international tourism industry, among the fastest-growing industries tourism is one which contributed more to the world economy uniting the whole world in a village. USD48 billion was increased by International tourism in 2014 which made a record of USD1245 billion, a 3.7% increase in real terms. A 4.4% increase in International tourists' arrival also occurred in 2014, attaining a total of 1135 million, up from 2013 which was 1087 million. Travel and passenger transport of International tourism represents 30% of services from the world's exports and among overall exports, it represents 6% of goods and services (Roy & Hoque, 2015). Concerning the credible note from the article on agri-tourism attracts more visitors to Mekong Delta by Vietnam News/ANN in China Daily (2019), the agri-tourism model is an excellent source of earn and due to an upward trend on income progresses to sustainable agriculture and rural development. And already agri-tourism model garden has been climbed outstanding outcomes, and only fruit farmers earn an annual profit of about USD 17,245 per hectare in several provinces in Vietnam.

According to WTTC (World Travel and Tourism Council) (2015), 8.1 billion USD was the total contribution of tourism in 2014 to Bangladesh GDP compared to neighboring Sri Lanka and India, with contributions of 8.2 billion USD and 125.2 billion USD, respectively. On the word of Ahmed and Jahan's (2013) findings, the particular shape it takes and the way it is organized vary from country to country, such as Indonesia; Sumatra, and Java where rural tourism centers are the plantation areas. Rice planting or rubber tapping etc. activities are main attraction for tourists to visit farms to observe directly. In the form of farm-stay, Korea and Japan offer tourism. The Malaysian government has funded agri-tourism quite massively. Across the country 31 villages has been identified as tourist spots by the ministry of tourism of the Government of India, with the help of state tourism departments and severe attempts has been taken to promote rural tourism. Moreover, the funding sources to support sustainable agri-tourism.

Challenges of Agri-tourism in Bangladesh

Bangladesh is a colorful country to explore tourism development. Yet, Bangladesh could not depict a sustainable tourism development nation, albeit there is a golden garden across the country. Half of the country's inhabitants are farmers who live in poverty because they spend their life with one income source. Their livelihood can turn on better expectancy life to follow Agri-tourism, but obstacles are the culprit. However, the common issues are underdeveloped infrastructure; lack of skilled workforce; political and institutional barriers; administrative complexity of dealing, corruption, and mismanagement of key development institutions; lack of awareness about business opportunities and supporting financial institutions; and poor global image. In rural areas, low priority by the government; resource constraints; lack of communication skills; lack of local involvement/participation; rural representations' absence in decision-making process; lack the relatedness between rural communities and rural government are some of the issues. Bangladesh is a colorful country to explore tourism development. There are some minor challenges to construct the agri-tourism industry in Bangladesh to be addressed. Moreover, several significant challenges can harm agri-tourism, such as:

Poverty

Poverty is an offender to promoted sustainable goals over the world and the spread of the World Bank's work to be successful for poverty reduction. As reported by the World Bank (2020) on poverty is the recent estimate, in 2015, 734 million people or 10% of the world's population spent less than USD1.90 for living in a day. That's down for the people in 1990 from nearly 1.9 billion or 36 percent. Poverty stands as the biggest challenge for development in South Asia, and its root is undeniably strong. Gantait, Mohanty, and Mathew (2018) have quantified many valuable measures in the article that poverty alleviation is the most vital challenge faced by the world and is an indispensable prerequisite to sustainable development. Poverty is still a global phenomenon as 767 million (10.7% of the global population) live below the international poverty line of USD 1.90/day by the year 2013. All these statistics in the background establishes that poverty has become a matter of concern in the South Asian context. Therefore, tourism can act as a catalyst in sustainable development in poverty alleviation (Gantait et al., 2018). The importance of integrating sustainability in poverty eradication requires more likely some modern investment approaches in tourism development where strategic leadership is a push factor in achieving our goal as community money is invested in agri-tourism.

Complex Travel Procedures

Cross-border travel is the most critical problem in the region which includes tour operator regulations, border formalities, airline access, visas permits for internal movement in certain areas, and currency use. It is found that most countries travel procedures around the South Asia are very complicated, complex and restrictive, for both intraregional and international visitors. Although among all travels, land travel is far more cost-effective, in South Asia cross-border land travel is challenging. There are limited facilities at most land border points, such as reservation facilities, transport links, and resting facilities (Rasul & Manandhar, 2009).

Tourism Investment

NTO (National Tourism Organization) is the main actor and serves to coordinate the country's development attempts at home and abroad to govern visitation of diverse tourist spots. Ishtiaque (2014) advocates that investment is still slow in developing countries like Bangladesh. High risk is found for investment in service-based projects, especially in tourism in the majority of developing countries. Thus, though potential natural tourism may have been found in countries, it is still challenging to gain feasible financing for Bangladesh for tourism projects.

Lack of Human Resources in the Tourism Industry

Specifically, in rural areas, the private sector and NGOs expand training programs for tourism. Linking to the government's priorities "Digital Bangladesh", useful aspects of such programs can be to make online tourism training courses more readily available. Study material accessibility of ILO and other main international institutions is also important. Luxury hotels owned by the government, including the Dhaka Sheraton Hotel and the Pan Pacific Sonargaon Hotel, Dhaka, unquestionably should be given more significant positions to employ well-trained Bangladeshis to work as interns and in other suitable positions (Honeck & Akhtar, 2014).

Lack of Safety and Security

The thoughtful matter regarding the security, we have to consider it from the services providers' perspective and the services receivers. In many South Asian countries, the matter is further provoked by terrorist attacks. To control the movement of terrorists, many countries' cross-border activities are brought under the responsible authority's stringent rules and checks. These rules often cause unnecessary hassle at border points. For travelers, a prime concern is to ensure safety and security as a critical factor in determining competitiveness of the tourism and travel sector (World Economic Forum, 2008). In South Asia, significant concerns still remain on the overall security situation, such as terrorist attacks, civil unrest, bomb blasts, insurgencies, war etc. (Rasul & Manandhar, 2009). Ishtiaque (2014) states that tourists' safety is a fundamental factor for any thriving sub-industry related to tourism and should be the primary objective of tourism management and planning. Tourism issues related to safety, whether perceived or real, negatively influence the fame of the host countries. In this respect, political instability and negative perceptions seem to play a disadvantageous role in the possibilities of tourism in Bangladesh.

Run Way of Agri-tourism on Development Process

Many critical issues, plans, and structural investment processes depend on the first step of specific activity areas such as agri- gardening to the outcome for various purposes. The realization came from initiatives from the governmental and private project:

Funding Practice in Agri-tourism into Green Farming

This study presents the information the show realistic scenario and analyze different development plan to evaluate the strategies and to help initiating advancement plans. Villages in Bangladesh have tourism potentials by highlighting agri-tourism and it supplies come into tourist demand who are boring from their daily life. Agri-tourism shows stable expansion of the villages, with the anticipated tendency of even faster expansion in the future (Petrović et al., 2015). This comprises the interests for art, architecture, archaeology, culture, hunting and fishing, hobbies, music, sport, stay in the country, etc. To escalated economic growth, the right target and growing market to the tourist are agri-tourism is the quality spot, and there appear many important things in the Fig. 15.2.

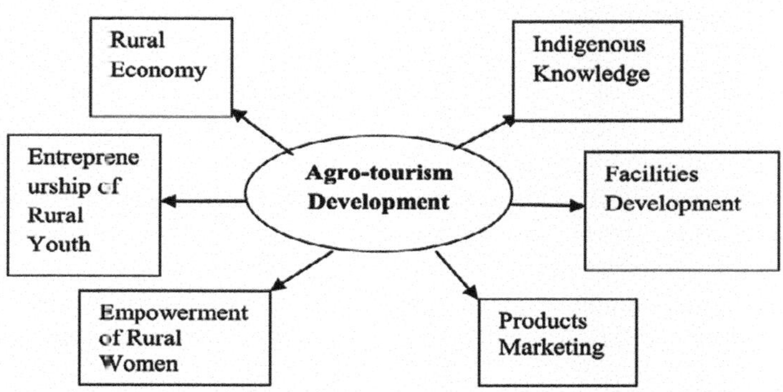

Fig. 15.2 Economics of Home-grown agri-tourism Products in Bangladesh. (Source: Dhaka Courier, 2019)

The Structure of Agri-tourism into Various Indicators

Therefore, the world needs sustainable development through the pillar of economics. However, unfortunately, Bangladesh's tourism industry is still an infant one, and regions authority should think to nurture tourism the industry to achieve a middle-income country tag by SDGs targeting goal. The branches of tourism have been touched skyrocket in the developed countries. However, there are many natural heartlands to explore agri-tourism. Also, developing nations such as Bangladesh did not realize the supplementing benefit from tourism and argi-tourism. Governments of many third-world countries cannot transform their economy in development in the race of right initiative in projection. Still, practical plans and investment in agri-tourism can motivate sustainable tourism development, and green gardening is the best policy to change the nature of tourism. However, the plan and the strategy practice in agri-tourism are the main roles of various aspects. According to Kieselbach and Long (1990) and Okech, Haghiri, and George (2012), some indicators like economic diversification, economic growth, and stabilization are considered the prime income source where the most crucial other secondary income source is important. Also, improvement and maintenance of public services; revitalizing cultural, crafts, and customs identities; protecting and improving both the natural and built environment; enhancing opportunities for the social exchange and contact; increasing acknowledgment of priorities for rural and potential by economic planners and policy-makers have relevancy.

There are many credible data to establish agri-farms to attract visitors in their articles (Okech et al., 2012). Agri-tourism is not only a valid option to generate employment sources but also restores the country's progress' considerably through rural society. Bangladesh has various types of exclusive and expensive places as farm tourism spots. Touristic land areas are connected with various flora-fauna and a good number of facilities which are related to establish the double game in agri-tourism (i.e. production cultivation and enjoying places). Nonetheless, financial support is a rational need here to grow up agri-tourism sources. For example, the Kenya Tourist Board and the Ministry of Tourism from the Government of Kenya are working to develop the circuit of Western Kenyan tourism to turn it into a domestic and international tourist destination. Management and leadership skills can add more resources and skills of human, financial management, setting of planning and goal, and the potentiality to manage individuals effectively. For small entrepreneurs, tourism gives many chances to contribute to the provision of tourism experiences for financial gain.

Moreover, Bangladesh authority can create a model as businesses of Austria. In this country, agri-tourism gets financial aid from their local government. Also, caravans and camping are the major forms of accommodation in rural regions of France that make more colorful and adventurous journeys closer to the nature (Ahmed & Jahan, 2013). A research was conducted in Bandung District of South Kuta, Indonesia that helped to furnish an agriculture model. Patera, Sukarsa, and Wiranatha (2015) found that diverse and intensive activities of tourism from small to large

Table 15.1 The local feature in agri-farm and agri-tourism development

Cultural attribute-farming custom	Natural attribute-rural landscape	Historical attribute-folklore
Local food product	Natural environment	Rural historical
Agricultural products and local handicrafts	Wide open spaces	Rural historical sites
Local lifestyle and festival	Hunting and fishing	Architecture
Local music and dance	Eco-tourism	History
Local educational travel	Rural advance	Heritage
Local sports	Safari drives	
Cooking methods	Natural base activities	

Source: Ayazlar and Ayazlar (2015)

scale firms were energized by beaches, warm air, and attractive natural cliffs in South Kuta. Also, all aspects of tourism development were growing rapidly in such areas. Taxes collected from restaurants and hotels offer the largest contribution of local revenue in South Kuta. These make the area wealthiest in Bandung in Bali. In terms of the new nation, a standard development model can be successful by calculating present condition, formulating a pilot development program, and strategies and implementing the practical plan into the project life cycle to excellent cultivated green garden districts in Bangladesh. Overall, the new farm creator focuses on the figure to be the perfect example for other nations (Table 15.1).

The Daily Star (2017) and Dhaka Tribune (2018) explain the benefits of agri-tourism that to attract international tourists our local government should take on more projects to represent our farmers' indigenous capacities to protect land, environment, and ecology; and by this our agri-tourism industries can be enriched. Rural Development Academy from Bangladesh (BARD) can take the opportunity to promote agri-tourism. One way to innovate agri-tourism can be to integrate global farmers for the reduction of local climatic problems by innovating flood and salinity tolerant agricultural systems. This can also help to share thoughts about marketing on food value and organic manures implementation in land for a productive and collective future (The Daily Sun, 2017).

Farming Progresses in Entrepreneurship

Agri-tourism business or activity invites visitors to visit rural communities or on-farms to enjoy the natural environment of agriculture and its produce, in which it exists. Agri-business is integrated as tourism where agri-tourism generally stays as an addition to the farm (Porcaro, 2010). It is considerably observed that crucially utilizing access of tourists, environmental and agricultural outcome in country sides through agri-tourism entrepreneurship and exploration can consist of business to well income source to develop the capacity of freedom in rural people by reborn of all indicators of human approach, domestically. Agri-tourism helps to gain extensive knowledge in marketing, economics, and management action and it is very important for both urban and rural communities and areas. It can serve several advantages such as employment, raise of income, accommodation use, activities, conservation of natural resources (Zoto et al., 2013). The policy can develop to the front of some critical aims, and these are: first, can be located in countryside; second, can be grounded on rural tourism's attributes; third, can have settlements and buildings on a small scale; fourth, can be associated with local communities and their families; fifth, can be grounded on small towns and villages; finally, can represent intricate environmental, economic, and historical patterns (Ayazlar & Ayazlar, 2015).

For product decision making, design is essential to understand the behavior of consumers. One of key marketing variables is design, which carries considerable competitive significance in overlooked and unattended marketplaces. In Austria, the Alpine tourism is a relevant example. Aesthetic components are found inherently in tourism products (Pikkemaat & Weiermair, 2011). Moreover, Ayazlar and Ayazlar (2015) analyzed important empirical sides about agri-tourism in the article "Rural Tourism: A Conceptual Approach". The findings have determined four market segments in rural tourism. These typologies are divided into four groups based on motivational factors in Spain: visitors seeking serenity, rest, and touch with nature; cultural visitors; proximity-gastronomic and nature visitors; and returning tourists. According to the article, there are three major advantages to rural tourism: first, engagement in rural areas increases the number of options for transactions and businesses. Second, rural tourism can aid economic growth, job creation, outmigration, the development of public services, infrastructure improvements, social interaction, and environmental protection. Finally, rural tourism makes isolated and non-urban locations more accessible.

Establishing Sustainable Model into Development

Agri-tourism is a rigid foundation in sustainable development to bring out many benefits, for example, investor and investment, raise tax revenues, worker opportunities, and contribution of modern technologies in the field. Two elements, such as motivation and monitor & evaluation, are essential to spread up tourism markers. To ensure sustainable development in agri-tourism, the following aspects are very much important. Improving the education quality and training skills, growth and diversification of exportation, implementation and formulation of the national plan for land-use, and for development purposes green growth promoting are also very important.

Tourism spots or farms can be highlighted by the effect of visitors' satisfaction and corporate supply all types of support to visitors and their happiness, creative economic growth through communities' improvement in various ways by subsidy in state GDP. Also, intensive analysis comes from Streimikiene and Bilan (2015) that tourism demand motives are very important for tourism development especially for rural tourism. In the territory, many reasons are found for the development of rural tourism both for success and failure. Several models are apprised to define the leading driving forces on the basis of the supply, demand, and rural communities' motives. Also, an evolutionary approach is applied in most of the theories to identify the leading drivers of development for rural tourism (i.e. different features have a significant impact on the various stages of development for rural tourism). Besides there are factors related to general demand which drive development of rural tourism, tourism services have essential consumer motives. For running a project, the model must be helpful to construct forever for future generations, as reported by Sumantra and Yuesti (2018). The use of a model made it easier for evaluators to know what was happening during policy implementation inputs, implementing activity programs on physical, economic, and social characteristics. However, logic or other models require in-depth data and information about the activities carried out until the results are obtained. As a result, tourism development reveals quality and quantity impact on nations to achieve Sustainable Development Goals with sustainable golden growth in economics.

Ishtiaque (2014) argued that a mechanism can bring durable plans and strategic planning action for compatibility goals. A tourism strategy can be developed by the help of it and this plan can be comprehensive, unified, and integrated. In this regard, the measure of cooperation between the government and private sectors is a related and important issue regarding the co-development of marketing tactics and strategies. As a vehicle for economic growth the preliminary assumption of tourism has focused mainly on its impact on a country's industry and trade, and creating a framework to transform to economic development. Therefore, economic globalization of the world can boost tourism trade by agri-tourism. Surprisingly, this grew a chance for the third world countries to climb the tourism side to fall down the horrific problems and touch the skyrocket in this ever-growing industry. To sustainable tourism into economic growth by work policy has explained in this article (Zoto et al., 2013) that tourism service quality meets all customer demand regarding the cost, the sets undertaken by law, the observance of hygiene and safety, and the natural environment's harmony. Standardization and assertiveness of rural tourism depending on agricultural production is the main instrument of the quality assessment. The purpose of these visits is to review modern and new technologies in rural agricultural farms for production and differentiate these with other countries.

Some of the researchers are influenced by Ishtiaque (2014) to understand the context of evaluating impacts on tourism on the host societies in the form of sociological, environmental, economic, political, anthropological, and holistic perspectives. Also, all these aspects can easily focus on developing the scenario and examining the critical linkages. The Strategic Environmental Assessment (SEA) will be the best-fitted assessment about the growth of agri-tourism in Bangladesh. The linear construction of linkages-effects–externalities is very much significant to understand the SEA system's framework. Moreover, the indicators to achieve it can include responding to social trends and changing economics; responding creatively and strategically to increasing competition; raising tourism dispersal the tourist destination all over the year; and expanding visitation, yield, and satisfaction of the visitors.

The Changes in Nations Development Influenced by Sustainable Agri-tourism

In a developing country as Bangladesh, economic, environmental, and social factors are not adequate for tourism. In such perspective, a technique is studied with its association with countries' development through economic growth for sustainable tourism development. On the similar ground, some indicators as found essential in México are average backwardness of education at home; current per capita income; access to health services, social security; spaces of housing, quality, and food; the degree of social cohesion and road accessibility (García & Michel, 2017). Some of the positive attitudes of sustainable tourism are:

Socio-economic

Tourism promotion brings infrastructure development, foreign currency inflows, regional development, employment generation, and distribution of income through multiplier of economic activities; it also brings social capital as a precondition to develop the tourism industry, and to foster socio-economic development as a whole (Quaranta et al., 2016). The farm income is the best economic door although sadly information in many countries has ignored this stair in development. Above all agri-tourism will open a new normal life in rural people across the country. According to many scholars (Gnanapala & Sandaruwani, 2016), the community is mainly focused by its socio-cultural and it shows long term impacts on the visitors. Therefore, the changes in the value system of collective, societal, and individual, social relationships, behavior and lifestyles, modes of expression, and structure of the community. Through rural tourism, rural community participation and education, and health are tools to solve major problems related to socio-economy in rural and marginalized communities through increasing local activity with active stakeholders' participation to achieve equal distribution of economic and social benefits (Khan et al., 2020). Furthermore, rural tourism motivates acquiescence to an idle pictorial version of community that can result in solidarity of personal ties and community. It will most fruitfully reduce rural people's migration to urban areas (Ahamed & Jahan, 2013). It can be exemplified that a concept has been developed by Agri-tourism Development Corporation that is directing the way towards better livelihoods which will be sustainable for many farmers along with their families (Gantait et al., 2018). It also demonstrates that while the agriculture sector is becoming a

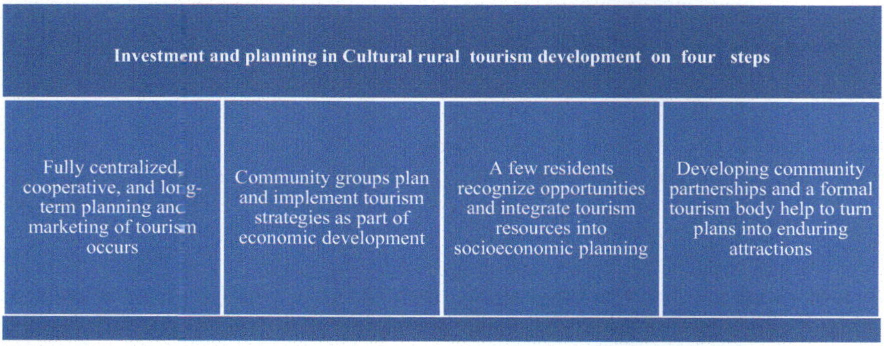

Fig. 15.3 Stages in cultural rural tourism development. (Source: Ayazlar & Ayazlar, 2015)

more challenging and less advantageous industry for the majority population in India, agri-tourism brought out the quality outcome. On the other hand, the practical knowledge of improving socio-economic status can be the best example to Bangladesh by advancing agri-tourism.

Cultural

It is emphasized on the importance of cultural factors that the role of service providers (tourism suppliers) is also important in shaping the rural tourism experience (Ayazlar & Ayazlar, 2015; Kastenholz et al., 2012; Cawley, 2009). They design the physical environment and determine other service features which are more intangible, such as responsiveness and reliability. There is a range of accommodation alternatives for rural tourism, such as guesthouses, private homes, caravan and camping sites, small hotels, and vernacular buildings. Like the French Belief, social norms & values are plenty of valuable indicators in local communities. Those four stages are acceptable when it blooms in investment and planning because it is a right to be enormous beneficial agri-farm (Fig. 15.3).

To create job opportunities, poverty reduction strategies combined with agricultural assets and cultural motivation to the domestic and international tourists in a community can work. Also, both local people and tourists feel comfortable exchanging the cultural nature, and domestic citizens learn exclusive new cultural aspects. Ahmed and Jahan (2013) and other researchers' opinions. Rural people usually earn less from agriculture than the income from tourism which means tourism will be more profitable for them. For that, tourism contributes to many rural areas' education, health, and culture. They will break out of their traditional beliefs and values. With the development of cultural understanding through fairs and festivals will be great attractions for tourists. Moreover, rural tourism will most significantly minimize rural people's migration to urban areas (Ayazlar & Ayazlar, 2015). Village can be tactically situated along the equator which can be very enticing for most of the tourists, both locally and internationally. Moreover, rustic people absorb wondering life, and transformation occurs in cultural elements.

Community Advancement Through Poverty Alleviation

In the twenty-first century, this whole world aims to be a modern society through swift technological advancement from globalization. Still, poverty has not fallen down in South Asia such as Bangladesh, whereas the sun shining in progress will have provided that excellence by tourism into agri-tourism in Bangladesh attain SDGs Goals: zero hunger; no poverty; economic growth and decent work; sustainable cities and communities, innovation, industry and infrastructure; and climate action (Table 15.2).

The maximum level of participation is possible when communities gain the capacity of self-mobilization, which gives community members the opportunity to operate their own form of activities without getting any assistance or help from other ventures, especially foreign business bodies or government (Kamarudin et al., 2016). Job opportunities, especially for the underdeveloped people in terms of economic status of rural communities such as few skilled local residents, can immediately work as retail clerks, food servers, and hospitality workers. Garcia and Michel

Table 15.2 Domestic development through community participants

Self-mobilization	Local residents can make direct contact with customers and produce high-quality products and services on their own. Some initiatives may get funding from non-governmental organizations (NGOs) that are not involved in local decision-making
Empowerment	Empowerment is the ultimate level of community participation, in which local residents have complete authority over all aspects of development without the use of outside force or persuasion. This is good for the community
Partnership	In the participatory process, developers and local residents might come to an agreement. Local organizations choose leaders to represent them and engage with outside developers. In the development process, there is some degree of local influence. The benefits might be delivered in the form of collective benefits, jobs, and money to the community
Interaction	At this stage, people are more involved. At the local level, local people's rights are acknowledged and acknowledged in practice. Projects can be organized by community organizations. It does, however, receive only sporadic backing from government entities
Consultation	People might seek advice in a variety of ways (e.g. involved in a community meeting or even public hearing). Developers may accept local contributions to their projects if they are beneficial to them (e.g. surveying, local transportation, and goods)
Informing	People can provide information to the community's agri-tourism effort. The developers (government agencies/private investors) manage the projects without taking into account the views of the local community

Source: Adopted from Kamarudin (2013)

(2017) opine from different contexts and relate to the tourism enterprise, which in most time is conceiving, is a way to ensure the activity's beneficial results, especially in countries where the economic condition is in an emerging situation. This action is studied in many developed countries where social, economic, and environmental factors are not inadequate for the activities of tourism practice and its connection with the destination's social enrichment. In the destination model's the social prosperity was advanced as a dimension of the competitiveness of the destination tourism that connected with these variables.The tourism industy's traditional economic benefit can be a significant outcome from discovering this study on the agri-tourism conducted by local communities. Agri-tourism is more innovative to increase the number of green jobs, scaling-up salaries; booming local business opportunities, leading to the advent of tourism services (food, culture etc.), and selling to tourists (craftsmen, artists, guides etc.), growing the collective sources of income (through profit, taxes, dividends, donations, etc.) (Marin, 2015).

Therefore, the administration desires that agri-tourism can prosper through a more varied economy by getting support from the diversification of employment in existing and new businesses. It is also hoped that it can check the matter of out-migration, and also help to sustain rural, agricultural and community tourism. To tackle the economic problems of the country, the Thailand government has taken a policy for tourism which is helpful in creating jobs for people as well as raising the country's income. All kinds of tourism have been highly emphasized as a strong root for all geographical areas' economic wealth. As the leading foreign exchange earner, tourism has subsequently surpassed rice of the country and the cornerstone of its economy (Tourism Authority of Thailand, 1997, showed by Srisomyong (2010). Overall, the effect of agri-tourism has changed the mood of the economy by rapidly improving in community peoples' life from the poverty line, and the nature of the development of that county is getting new face by degrees. However, technical thought must have a sustainable development path to be mature in tourism by agri-tourism, and the government will feel proud to achieve a sustainable goal.

The Environmental Impact

tourism profits can be used for the improvements of the environment such as schemes of traffic regulation and village paving, litter disposal and sewage etc. This help will be fruitful in retaining both the existing population with their businesses and attracting new enterprises with their families. Furthermore, rural people will acknowledge how to ensure a healthy environment for both physical and mental health; and along with it they will learn how to preserve and sustain the natural habitats. Kamarudin, Chong and Razak (2016) stated that participation of local community in non-farm development especially in economic sector: the case of two orang asli communities of Royal Belum State Park, donated information in the Table 15.3.

Table 15.3 Community participation into environmental in agri-tourism project

Local issues	Have a direct impact on the tourist experience: local retaliation leads to aggressive conduct toward tourists (Faulkner et al., 1994). As a result, tourist environments should be constructed in accordance with the social climate, so that locals profit from tourism rather than become victims of it (Wahab & Pigram, 1997)
Local assets	Tourism's image is built on the assets of the local community, which include not only the local people but also the natural environment, infrastructure, amenities, and unique events or festivals; thus, the host community's cooperation is required to access and develop these assets correctly (Murphy, 1995)
Local driving force	Public participation serves as a catalyst for protecting the community's natural environment and culture as tourism goods while also encouraging increased tourism-related revenue (Felstead, 2000)
Tourism vulnerability	The tourism industry is sensitive both to internal and external forces. Many tourism development plans are often only partially implemented or not at all (Bovy, 1982). Moreover, even those fully implemented are not always sustainable. Thus, to increase the feasibility and longevity of projects. All plans need to link with the overall socio-economic development of the community.

Source: Adapted from Okazaki (2008)

In the context of environmental issues of agri-tourism, writers have stated that in terms of Malaysia environment in the villages are still protective and not to be a negative issue due to the deterioration of agri-tourism (Kamarudin, 2013). Furthermore, they have also reported that the villagers tried to enhance the beauty of the village to attract the tourist. Establishing and investing in the agri-tourism model project in Bangladesh would be a great initiative to make the environment sustainable and eco-friendly impacts which are positive and potential for tourism can be to secure water quality highly and protect local nature and natural resources better. In a number of targets, tourism supports securing better protection of nature with higher water quality. It can help to develop additional reinforcement to invest in infrastructures and services which are environmentally friendly. Also an active social dimension has been found in agri-tourism. It also seeks to find a way to provide benefits to the local communities by empowering them with control over how to use their lands and resources in more sustainable ways. The more people produce agricultural products, the more they can be able to reduce carbon emissions, which will ensure environmental balance for our planet's sustainability. To ensure non-stop electric supply in the tourist zone's accommodation area, many people install renewable energy initiatives on the farm areas by using the farm land wastages, which helps nature eliminate carbon emissions and balance the total eco-system of the particular locality. There is also scope to build a green building that can be built using non-polluting and energy-efficient construction materials, energy sources, and sewage systems; instead of using concrete that badly affects the environment. Thus, agri-tourism ensures environmental sustainability in its life cycle.

Conclusion and Recommendations

The awesome notice from the UN conference which occurred in Geneva on the 13th–14th March 2013 and the observation is the structural strategy on long-term outcome in a country. Sustainable tourism is related to the policy options where it has a positive impact on sustainable development and economic growth. Particularly, developing countries (low-income) have been benefited by tourism because it has potential for economic growth and poverty reduction. The industry, which is involved with tourism and has the capacity to create employment and income because of its backward and forward connections, makes it very important not only for economic diversification but also for growth. With that it ensures long run opportunities and offers sustainable development.

The aim of this chapter is to analyze the investment and development perspective of agri-tourism in Bangladesh. The tourism industry of Bangladesh is still an infant one. However, in the sector of agri-tourism development, Bangladesh has immense potentiality. Relevant authorities need to think for policies for nursing the industry along with joint venture work among all stakeholders. Tourism can play a vital role in reshaping the economy in many countries along with Bangladesh and complement all of its policies aimed at achieving the tag of a middle-income country by SDGs target goals but governmental interference is compulsory as an investor to achieve success to development perspective. The natural beautify and abundant cultivable land of Bangladesh is awesome and can help the tourism industry to attract more tourists. Still, there are needs for the proper marketing plan and stronger leadership and strategy, which will be a successful plan in sustainable tourism development into agri-tourism, but there is the horrific fact to reach the goals such as public money and aid or grant are a vital part of investing in tourism progress in the development field.

Therefore, the conditional mindful research and a wonderful political policy as an advancement to the greatest number of exclusive issues markedly change the economic, socio-culture aspect to countryside societies and environmental aspects.

Accordingly, small accommodation as a business (i.e. agri-tourism boarding houses) represents the core service, in rural destinations a large share of accommodation is provided by the villagers and it plays a central role in tourism development especially in rural areas rural revitalization and poverty relief. Thereupon, their owner-managers are highly inspired to conserve important social, natural and cultural features that are vital to the sustainability of rural landscapes and destination competitiveness. Moreover, it is crucially finding the causes of this industry's regress, discovering the obstruction of the existing policy measures, and on this policy, suggesting courses of actions which will be effective to undertake by the different industries and other relevant authorities to get the benefits of the opportunities from it. Because of agri-tourism trends to obtain high-quality living standards through communities to domestic development particularly through raising income for rural people who work in the field of agriculture as an example- "Agricultural tourism or Agro-tourism/ Agri-tourism" can have notable outcomes for communities and farmers in the interface of agricultural-urban countries. A visitor study shows that tourism which is related to agriculture has substantial impacts in the context of economy on local economies. Also, agri-tourism delivers economic incentives with diversification for local products, advances social and economic development, and helps guide the public towards the essential inputs of agriculture to the quality of life and country's economy. Sustainable development emerges from many factors, such as the tourism industry's remarkable outcome and markedly notice to developed nations. The majority of advanced states create their earning source but poverty is the expression in developing nations by cause of lacking policy and proper strategy. Therefore, the various recommendation may have come from several researchers (i.e. Khan et al., 2020; Ispas et al., 2019; Stanovčić et al., 2018; Ishtiaque, 2014; Ahmed & Jahan, 2013; Zoto et al., 2013; Rasul & Manandhar, 2009, and also from Lobo et al., 1999).

This research offers some recommendations: first, to inspire conjoint partnerships (such as community participation, public-private partnerships, local economic development strategies, and linking value chain). Second, to use up to date innovative technology (such as smartphones, mobile technologies, user-generated content, GPS, reviews and feedback, use of social media, artificial intelligence and big data). Third, to professionally promote intertwined destinations types of tourism, permitting different attractions for the tourist. Fourth, to ensure orientation of the product by understanding the market better and the market demand. Fifth, to organize the professional training on agri-tourism for farmers in rural areas. Non-governmental organizations can carry out the training. Sixth, foreign investors should be invited to invest by the government in this virgin tourism field in Bangladesh. Seventh, for the tourist movements risk-free and safe security measures should be established. Eighth, to ensure the involvement of women in the process of decision making along with development activities in rural tourism. Ninth, to preserve resources of rural areas with the development of authentic attractions and careful planning are compulsory. Tenth, proper actions to preserve from natural destruction of cultural and archaeological sites, the historical relics, monuments, ancient buildings, and sights of natural importance should be taken to conserve. Eleventh, to encourage intra-regional tourism where a strong political commitment and more efforts for coordination is needed towards poverty reduction among the countries located in South Asia to promote a positive environment for promotion of tourism. Twelfth, to establish strategic alliances where needed to be developed at private sector organizations along with different level agencies of governments and may cover different areas such as marketing, product development, human resource development and infrastructure. Finally, the private sector should come forward to invest along with the government in the tourism industry. Activities of these sectors also need to be harmonious with the programs and plans of poverty reduction in the nation. In order to attract more and more investors to meet the increased demand, tourism industry's investment opportunities need to be liberalized.

References

Ahmed, I., & Jahan, N. (2013). Rural tourism-prospects in rustic Bengal. *European Journal of Business and Management, 5*(16), 163–171.

Ayazlar, G., & Ayazlar, R. A. (2015). *Rural tourism: A conceptual approach. Tourism, environment and sustainability* (pp. 167–184). St. Kliment Ohridski University Press.

Baud-Bovy, M. (1982). New concepts in planning for tourism and recreation. *Tourism Management, 3*(4), 308–313.

Cawley, M. (2009). Rural tourism. In N. J. Thrift & R. Kitchin (Eds.), *The international encyclopedia of human geography* (pp. 313–317). Sage.

Chesky, A. (2009). Can agritourism save the family farm in appalachia? A study of two historic family farms in Valle Crucis, North Carolina. *Journal of Appalachian Studies, 15*(1&2), 87–98.

Daily Sun. (2017). *Role of rural tourism development in Bangladesh economy*. https://www.daily--sun.com/amp/post/204066. Accessed 8 July 2020.

Das, B. R., & Rainey, D. V. (2010). Agritourism in the Arkansas delta byways: Assessing the economic impacts. *International Journal of Tourism Research, 12*(3), 265–280.

Dhaka Courier. (2019). *Economics of home-grown agro-tourism products in Bangladesh*. Retrieved from: https://www.dhakatribune.com/business/2018/02/06/invest-agro-tourism. Accessed 1 June 2020.

Dhaka Tribune. (2018). *Why we should invest in agro-tourism*. Retrieved from: https://www.dhakatribune.com/business/2018/02/06/invest-agro-tourism. Accessed 1 June 2020.

Faulkner, B., Pearce, P., Shaw, R. N., & Betty, W. (1994). Tourism research in Australia: Confronting the challenges of the 1990s and beyond. In B. Faulkner, M. Fagence, M. Davidson, & S. Craig-Smith (Eds.), *CAUTHE 1994: Tourism research and education in Australia: Proceeding of the Australian National Tourism Research and Education conferences* (pp. 3–25). Bureau of Tourism Research.

Felstead, M. L. (2000). *Master plan for community-based eco-tourism in Ulgan Bay, Palawan, Republic of the Philippines*. UNESCO-UNDP-PPC.

Firoozzare, A., Ghorbani, M., Karbasi, A. R., Shahnoushi, N., & Davari, K. (2018). Prioritize agri-environmental measures of water-related ecosystem services: The case of Mashhad. *Journal of Sustainable Development, 11*(4), 240–256.

Gantait, A., Mohanty, P., & Mathew, R. (2018). Tourism-led poverty alleviation in South Asia – An analytical rapport age. *International Journal of Innovative Research & Studies, 8*(1), 67–83.

Garcia, J. I. R., & Michel, J. P. I. (2017). Social prosperity perception in cultural tourism destinations: The case of Pena de Bernal, Huejotzinngo and Yuanhuitlan, Mexico. *Traektoriâ Nauki= Path of Science, 3*(4), 7.1–7.9.

Gnanapala, W. A., & Sandaruwani, J. A. R. C. (2016). Impacts of tourism development in cultural and heritage sites: An empirical investigation. *International Journal of Economics and Business Administration, 2*(6), 68–78.

Goodwin, R. D., & Chaudhary, S. K. (2017). Eco-tourism dimensions and directions in India: An empirical study of Andhra Pradesh. *Journal of Commerce and Management Thought, 8*(3), 436–451.

Honeck, D., & Akhtar, M. S. (2014). *Achieving Bangladesh's tourism potential: Linkages to export diversification, employment generation and the "green economy"* (WTO Staff Working Paper, No. ERSD-2014-15). World Trade Organization (WTO).

Ishtiaque, A. N. A. (2014). A schema of tourism development model: A case of Bangladesh. *Dhaka University Journal of Marketing, 15*, 175–186.

Ispas, A., Untaru, E. N., & Candrea, A. N. (2019). Environmental management practices within agritourism boarding houses in Romania: A qualitative study among managers. *Sustainability, 11*(14), 3887.

Kamarudin, K. H. (2013). Criteria and indicators for sustainable community based rural tourism (CBRT) development: The case of three villages in East Coast (ECER), Malaysia. Unpublished PhD thesis, Oxford Brookes University.

Kamarudin, K. H., Chong, N. O., & Razak, K. A. (2016). Local community participation in non-farm economic development: The case of two orang Asli Communities of Royal Belum State Park, Perak. *Journal of Human Capital Development (JHCD), 9*(2), 79–98.
Kastenholz, E., Carneiro, M. J., Marques, C. P., & Lima, J. (2012). Understanding and managing the rural tourism experience – The case of a historical village in Portugal. *Tourism Management Perspectives, 4*, 207–214.
Khan, A., Bibi, S., Lorenzo, A., Lyu, J., & Babar, Z. U. (2020). Tourism and development in developing economies: A policy implication perspective. *Sustainability, 12*(4), 1618.
Kieselbach, S. R., & Long, P. T. (1990). Tourism and the rural revitalization movement. *Parks & Recreation (Arlington), 25*(3), 62–66.
Lathiras, P., Zopidou, A., Mylonakis, J., Tahinakis, P., Protogeros, N., & Valachis, I. (2010). An evaluation of websites quality factors in agro travel and ecotourism. *Tourism and Hospitality Management, 16*(1), 11–30.
Lobo, R. E., Goldman, G. E., Jolly, D. A., Wallace, D., Schrade, W. L., & Parke, A. (1999). Agritourism benefits agriculture in San Diego County. *California Agriculture, 53*(6), 20–24.
Marin, D. (2015). Study on the economic impact of tourism and agro-tourism on local communities. *Research Journal of Agricultural Science, 47*(4), 160–163.
Murphy, P. E. (1995). Quality management in urban tourism: balancing business and environment. *Tourism Management, 5*(16), 345–346.
Okazaki, E. (2008). A community-based tourism model: Its conception and use. *Journal of Sustainable Tourism, 16*(5), 511–529.
Okech, R., Haghiri, M., & George, B. (2012). Rural tourism as a sustainable development alternative: An analysis with special reference to Luanda, Kenya. *Cultur-Revista de Cultura e Turismo, 6*(3), 36–54.
Patera, I. M., Sukarsa, I. M., & Wiranatha, A. S. (2015). Tourism and poverty in Badung Regency, Bali. *E-Journal of Tourism, 2*(2), 84–95.
Petrović, M. D., & Terzić, A. (2015). An overview of agritourism development in Serbia and European Union Countries. *International Journal of Sustainable Economies Management, 4*(2), 1–14.
Pikkemaat, B., & Weiermair, K. (2011). The aesthetic (design) orientated customer in tourism-implications for product development. In *EIASM 10th international product development management conference*, Brussels, pp. 825–839.
Porcaro, P. (2010). The Italian agritourism model-3 Ps-policy, product and promotion; Recommendations for developing agritourism in Australia. In *CAUTHE 2010: Tourism and hospitality: Challenge the limits* (p. 1149). Hobart.
Quaranta, G., Citro, E., & Salvia, R. (2016). Economic and social sustainable synergies to promote innovations in rural tourism and local development. *Sustainability, 8*(7), 668.
Rasul, G., & Manandhar, P. (2009). Prospects and problems in promoting tourism in South Asia: A regional perspective. *South Asia Economic Journal, 10*(1), 187–207.
Robinson, G. M., & Carson, D. A. (2015). The globalisation of agriculture: Introducing the handbook. In G. M. Robinson & D. A. Carson (Eds.), *Handbook on the globalisation of agriculture* (pp. 1–28). Edward Elgar Publishing.
Roy, B., & Hoque, R. (2015). Building a strong brand image of Cox's bazar as a tourist destination: An empirical analysis on Cox's bazaar. *American Journal of Tourism Management, 4*(2), 27–34.
Song, H., Qiu, R. T., & Park, J. (2019). A review of research on tourism demand forecasting: Launching the Annals of Tourism Research Curated Collection on tourism demand forecasting. *Annals of Tourism Research, 75*, 338–362.
Srisomyong, N. (2010). *Agritourism, rural development and related policy initiatives in Thailand.* Unpublished PhD dissertation. Sheffield Hallam University.
Stanovčić, T., Peković, S., Vukčević, J., & Perović, D. (2018). Going entrepreneurial: Agro-tourism and rural development in Northern Montenegro. *Business Systems Research, 9*(1), 107–117.

Streimikiene, D., & Bilan, Y. (2015). Review of rural tourism development theories. *Transformations in Business & Economics, 14*(2), 21–34.

Sumantra, I. K., & Yuesti, A. (2018). Evaluation of Salak Sibetan agrotourism to support community-based tourism using logic model. *International Journal of Contemporary Research and Review, 9*(1), 20206–202012.

The Daily Star. (2017). *The many benefits of agro-tourism*. Retrieved from: https://www.thedailystar.net/opinion/the-many-benefits-agro-tourism-1454098. Accessed 1 June 2020.

The World Bank. (2020). *Poverty*. Retrieved from: https://www.worldbank.org/en/topic/poverty/overview. Accessed 16 Apr 2020.

Tseng, M.-L., Chang, C.-H., Wu, K.-J., Lin, C.-W. R., Kalnaovkul, B., & Tan, R. R. (2019). Sustainable agritourism in Thailand: Modeling business performance and environmental sustainability under uncertainty. *Sustainability, 11*(15), 4087.

Veljković, B., & Broćić, Z. (2017). Agritourism and rural development. *TISC – Tourism International Scientific Conference Vrnjačka Banja, 2*(2), 480–499.

Wahab, S., & Pigram, J. J. (Eds.). (1997). *Tourism, development and growth, the challenge of sustainability*. Routledge.

Weaver, D., & Fennell, D. (1997). The vacation farm sector in Saskatchewan: A profile of operations. *Tourism Management, 18*(6), 357–365.

World Economic Forum. (2008). *The travel and tourism competitiveness report 2008: Balancing economic development and environmental sustainability*. WEF.

World Travel and Tourism Council (WTTC). (2015). *Travel & tourism in 2015 will grow faster than the global economy – WTTC reports* (p. 30). WTTC.

Wyporska, K., & Mosiej, J. (2010). Technical infrastructure for environmental protection at the level of farms as a factor of sustainable rural development. *Journal Economic and Environmental Studies, 10*(1), 71–84.

Zoto, S., Qirici, E., & Polena, E. (2013). Agrotourism – A sustainable development for rural area of Korca. *European Academic Research, 1*(2), 209–223.

Chapter 16
Developing Garments Tourism in Bangladesh: An Overview

Md Rabiul Alam, Md Aslam Mia, and Hasanul Banna

Abstract Garments tourism, which could be more consonant to fashion tourism, is still in its nascent stage, particularly in the context of developing countries like Bangladesh. Although Bangladesh is one of the largest exporters of apparel suppliers to the global market, most of its export was meant to be for the international fashion companies. By doing so, this has not only brought billions of dollars to the country's economy but also created an avenue for thousands of businessmen/companies to travel to the country. The main purpose of this study is to look at the overall development of the garments industry in Bangladesh and how this industry could be translated into tourism. In so doing, the problems, prospects, and challenges of garments tourism are comprehensively discussed in this study. Overall, we argue that to revive garments tourism, the government should take aggressive measures to promote the branding of this industry by establishing and incentivizing the entrepreneurs, as well as making travel safe and sound for foreigners.

Keywords Garments tourism · Apparel and garments · International firms · Government · Bangladesh

Md R. Alam
Department of Language and Literacy Education, Faculty of Education, Universiti Malaya (UM), Kuala Lumpur, Malaysia

Md A. Mia (✉)
School of Management (SOM), Universiti Sains Malaysia (USM), Penang, Malaysia
e-mail: aslammia@usm.my

H. Banna
Ungku Aziz Centre for Development Studies, Faculty of Economics and Administration, Universiti Malaya (UM), Kuala Lumpur, Malaysia

A. Hassan (ed.), *Tourism in Bangladesh: Investment and Development Perspectives*, https://doi.org/10.1007/978-981-16-1858-1_16

Introduction

The overall development of any country is the result of the combined contributions of many sectors. Similarly, the remarkable economic development of Bangladesh in the last few decades got visibility with the active participation of many sectors where the contribution of ready-made garment (RMG) is remarkable (Islam, 2020; Rahman & Chowdhury, 2020; Rahman & Siddiqui, 2015). Export is a key to the economic development of a country (Peterson Institutes for International Economics, 2015), and many countries across the world are prioritizing export-oriented products to ensure a strong economic structure (International Growth Centre, 2018). Since RMG sector of Bangladesh has become a leading export industry, the Govt. of Bangladesh (henceforth Bangladesh only) is also putting inclusive attention to this sector. More than 80% of the total export of Bangladesh is from RMG sector (Bangladesh Garment Manufacturers and Exporters Association, 2018). The current contribution of this sector to the country's overall economy is not the result of overnight. Instead, it took many breakthroughs.

Just after the liberation in 1971, the main exportable items of Bangladesh were jute and jute-made goods that made less than 6% of the GDP, leather (4.57% of GDP), and tea (2.73% of GDP) (Yunus, 2001). However, when the demand for jute in the global market deteriorated in the 1980s, RMG sector was replaced and gradually became the leading export sector that made 82% of the total export of Bangladesh with 13% of overall GDP in the FY 2015–16 (Export Promotion Bureau of Bangladesh, 2018). Since then, the RMG sector has always had a remarkable contribution to the country's sustainable economic growth. Currently, it constitutes 84% of country's total export with earnings of US$ 34 billion (Bangladesh Garments Accessories and Packaging Manufacturers and Exporters Association, 2019). Besides its continuous and significant contribution to the country's economy, currently, it is thought to be a booming industry for tourism as thousands of buyers and delegates visit Bangladesh as a part of their trade.

The tourism industry has been regarded as one of the most notable revenue-making industries (Kobra et al., 2018). Generally, people visit places during their holidays and other eves to relax and enjoy. However, nowadays, the perception of tourism has changed. So people visit from one country to another is not solely for the purpose of sight-seeing and relaxation, but also they prefer to visit a country for various business deals or opportunities. Thus, tourism can be classified into many dimensions, such as medical tourism, education tourism, religious tourism, business tourism, and so on (Khondker & Ahsan, 2015).

Garments tourism is under the category of business tourism, which means foreign delegates or merchants mainly visit or travel to other countries for their garments trade, and on the same trip, they visit the tourist attractions of those countries for recreational purposes. Hence, with the development of garments trade, Bangladesh can consider RMG sector as a part of tourism that will be a milestone in revenue-generating for the country as every year, 42% of the foreigners visit Bangladesh for their business purpose (Khondker & Ahsan, 2015). Besides generating revenue, tourism enhances a country's branding image to the global context by introducing its tradition, heritage, cultural identity, and historical background. It is an industry that circulates money to every corner of the economy since tourist attractions are scattered throughout the country. Thus, it creates employability for the local people, improves their livelihoods, and subsequently brings an overall structural and economic change for the whole country.

To attract foreign buyers to the tourist attractions of Bangladesh, the country needs to consider a lot of factors that will determine whether garments tourism could be a prospective sector or not in the future. As we know, Bangladesh has many tourist attractions, but lack of proper maintenance, including other reasons, is the obstruction to consider this sector as a mainstream revenue-generating source. Hence, this chapter aims to link garments sector with tourism and discuss various associated factors. The chapter also analyzes factors to be taken into account to boom the industry, and then policy recommendations are provided for government and relevant authorities.

The following sequential order constitutes the organization of the chapter. Section "Conceptual framework: Tourism, garments and investment" highlights the link between RMG sector and tourism in the context of Bangladesh. Section "Investment and development in garments tourism in Bangladesh" notes on the recent development of the garments industry and explains how it could be further developed. Section "Future of garments and apparels tourism" explicates the prospects and constraints of garments tourism, while section "Conclusion with policy recommendations" illustrates the concluding remarks along with policy implications to develop this sector.

Conceptual Framework: Tourism, Garments and Investment

From the conventional perspective, the term tourism refers to traveling to other places only for refreshing minds to enjoy holidays that gives them unlimited fun and amusements (Subadra et al., 2019). During holidays, people go for sightseeing to the attractive and historical places of their own countries. Sometimes they go alone and sometimes with their loving family members. Relatively rich people travel to foreign countries to enjoy holidays. However, in this era of globalization and competition, people lead almost a robotic life where they hardly find any time to spend for recreation. Tourism is the best means of recreation that takes time to enjoy. As people have less time to spend on this, they consider it done during their business trip. Hence, garments tourism could be thought of. The garments industry in Bangladesh is a booming industry for the last few decades. So, Bangladesh can link tourism with the garments industry. This garments tourism can be based upon factors such as economic and infrastructural development, introducing the country's culture, heritage, and history to the foreigner (Fig. 16.1).

In Bangladesh, the tourism industry has been in a rising trend since the mid-nineties that adds revenue to its overall economy. Monetary benefit is a major part

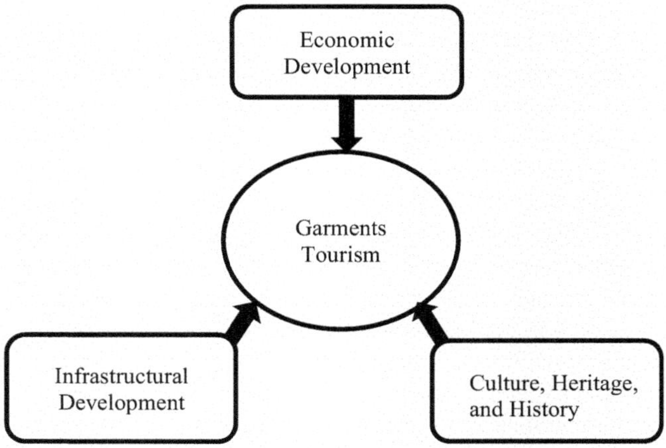

Fig. 16.1 Framework of garments tourism. (Source: Developed by the authors)

of tourism that creates occupations, decreases poverty, and enhances people's living standards (Phillips & Faulkner, 2013). Bangladesh is a developing country and struggling with uprising its economy by maintaining a stable GDP growth over the last few years, mainly focusing on- enhancing foreign reserve, augmenting exports, reducing inflation rate, increasing liquidity, and cutting down interest rate of loans (Kobra et al., 2018). Hence, boosting the garments tourism industry will be a good idea to earn foreign currencies. By developing garments tourism, many types of local entrepreneurship can be created to increase job opportunities and consequently helps to eradicate poverty. Many countries have used this platform, such as Malaysia, Thailand, Switzerland, Singapore, etc. to develop and diversify their economic activities to promote development (Durbarry, 2004; Sinclair, 1998). These countries have consistently emphasized creating a tourism-friendly environment to attract more tourists, so that the domestic economy could be further developed. Other than these, countries like the Philippines, Nepal, Maldives, Vietnam, Sri Lanka, etc. also earn the lion share of their revenue from the tourism industry (Afroz & Hasanuzzaman, 2012).

So, across the world, tourism has been a rapidly growing sector that, till 2019, has contributed US$8.9 trillion to the world's GDP, which is 10.3% of the total GDP. Besides, this sector has created 330 million new jobs, which is 10% of the world's total jobs (World Travel and Tourism Council, 2019). According to the World Bank (2018), Bangladesh is the least developed country (LDC) which national revenue generation mostly depends on the agricultural industry, RMG, and foreign remittance by the workers in different countries. However, its developmental phase is being hindered by many reasons such as fragile rules of law, booming corruption, political unrest, political polarization, unwelcoming regulatory environment, time-wasting bureaucratic procedure, and poor brand image (Fonseca, 2012). To make garments tourism as an index of economic development, many changes need to occur in sectors directly or indirectly pertinent to the tourism industry.

Besides its contribution to the economy of the country, garments tourism can be a milestone towards the infrastructural development of Bangladesh. It is a part of business tourism that has already been regarded as one of the most influential sectors in many countries (Kulendran & Witt, 2003; Nicula & Elena, 2014; Rogerson, 2015; Tichaawa, 2017). To develop this sector in a world-class standard, roads and transportation systems must be developed. As noticed, the recent infrastructural development of Bangladesh is remarkable. The visible structure of Padma Bridge, the expansion of Dhaka-Chittagong highways, the expansion of Dhaka-Mymensingh highways, the metro rail project, various connecting flyovers in big and busy cities are welcoming garments tourism. Through this sector, structural development would be visible in the rural and remote areas with tourist attractions. Small and medium enterprises like hotels, motels, cottages, restaurants, shopping malls, roads and highways, transport systems grow up in the tourist places. The structural visibility of all these sectors creates huge job opportunities, and people become solvent, that consequently helps them upgrade their living standards. The worldwide tourism industry has changed many countries' structural views like Singapore, Malaysia, Nepal, Bhutan, Maldives, Bali of Indonesia, and so forth (Telfer & Sharpley, 2015). Hence, garments tourism in Bangladesh is apprehended to bring a revolutionary infrastructural change that will eventually change its citizen's life style.

Apart from these, through the inclusion of garments tourism, Bangladesh can take the opportunity to introduce its ethnic and cultural identity, historical ups and downs, and long heritage to the foreigners (Lee & Chhabra, 2015; Palmer, 1999). Bangladesh has a long-struggling and tragic history of the mother tongue movement in 1952. With the sacrifice of so many lives that is unique in the world history, it achieved Bangla as its state language. In the memory of those martyrs, the country has the "Central Shahid Minar" to pay solemn homage to those martyrs. Besides, Bangladesh also has a prolonged history of its liberation from West Pakistan in 1971 that was achieved with the sacrifice of a huge number of martyrs. In memory of those martyrs, a national monument is established in Savar, Dhaka. Hence, by visiting these places, foreigners will come to know the struggling history of Bangladesh. Moreover, visiting the Garo Hills of Mymensingh, the Chittagong Hill Tracts, the Sixty Domes Mosque of Bagerhat, and other religious sites across the country will give the tourists a taste of the ethnic and cultural identity of the country too.

Investment and Development in Garments Tourism in Bangladesh

To excel a sector, proper investment is a must. Without timely investment in an appropriate manner, no sector will give expected return. Hence, like other industries, the growth and development of garments tourism depend on proper investment. However, in the tourism industry, the investment level is relatively lower than in other similar sectors that hinder the development trajectory of overall tourism in Bangladesh (Khondker & Ahsan, 2015). Though the capital investment, till the year 2014, was lower in the Bangladesh tourism industry, it steadily goes up from 2015 and is still in a growing trend which is a very good sign for garments tourism. The investment trajectory shows that the Govt. has realized the necessity of developing the tourism industry for the country's economic development.

Investment and Development

According to the World Travel and Tourism Council (WTTC) (2020), the total investment of Bangladesh in its tourism industry is US$ 1.4 billion in 2019. This investment was used for the tourist attractions of Bangladesh. Bangladesh has a good number of naturally beautiful tourist attractions such as the longest sea beach Cox's Bazar, the Sundarbans, the largest mangrove forest, St. Martin's Island, Madhabkund in Sylhet, live coral reef, Patenga sea beach, Kuakata sea beach from where sunset and sunrise is seen clearly, and so forth (Chowdhury, 2020; Majumder & Hossain, 2011; Zaman et al., 2020). If timely investment with proper management is held, foreign buyers who come to Bangladesh for their garments trade purpose will be happy to visit these places for their recreation and refreshment. As there are many naturally beautiful places in Bangladesh, the Govt. can invite direct or indirect foreign investments (Fig. 16.2). Figure 16.3 shows the overall investment in the tourism industry over the period 1995–2019. Besides, it also shows the percentage of the Govt. total expenditure in the tourism industry. It is noticed that the percentage of the total expenditure of the Govt. is also at a rising level. Hence, if the

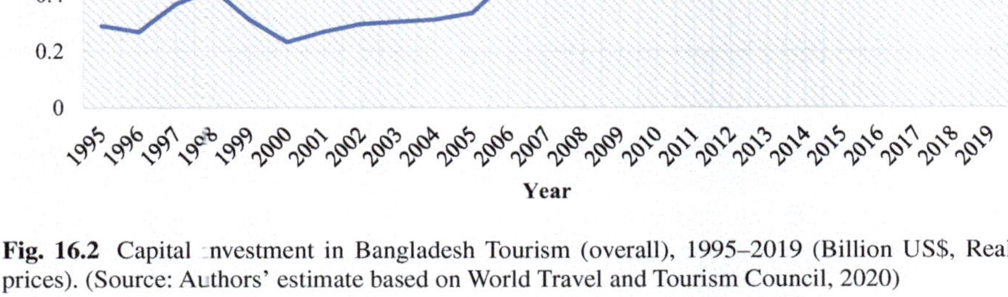

Fig. 16.2 Capital investment in Bangladesh Tourism (overall), 1995–2019 (Billion US$, Real prices). (Source: Authors' estimate based on World Travel and Tourism Council, 2020)

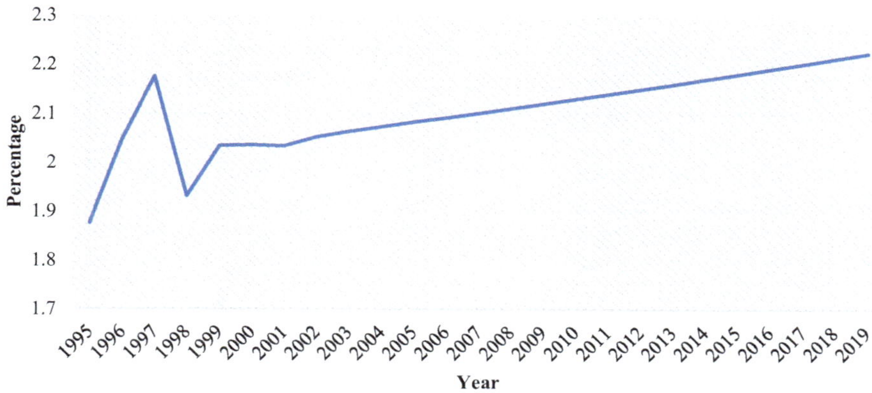

Fig. 16.3 The percentage share of total government individual expenditure, 1995–2019. (Source: Authors' estimate based on World Travel and Tourism Council, 2020)

tourism industry is integrated with the garments industry, the country's overall development is hoped to be enhanced.

The Scenario of Garments and Apparels Industry in Bangladesh

RMG sector in Bangladesh is one the most rapidly growing sectors that bring foreign currencies (Mahmud et al., 2018; World Trade Organization, 2016). For the last two decades, this sector has seen an unprecedented development (Bangladesh Garment Manufacturers and Exporters Association, 2018). Behind this development, so many internal and external factors played and are still playing significant roles. Some internal factors are opening the free market economy, huge supply of unskilled workforce with lower cost, operations of more factories that can produce volumes of orders with timely shipment, developing backward linkage and supporting materials industries, comparative price competitiveness, the supply of natural gas, trade-friendly Govt. policies, and liberalized financial policies. On the other side, some external factors are quota restriction, and the multi-fibre arrangement (MFA) phase-out, and ethnic violence in other export-oriented countries like Sri Lanka that has made the buyers turn to Bangladesh (Ahmed, 2006; Spinanger, 2001), and buyers took it as the safest production hub in South Asia (Feldman, 2009). Another important factor that has stimulated the apparel industry's rapid growth is the tariff and quota-free access of RMG products to the European Union (EU) countries (Bhattacharya & Rahman, 2000). At present, the RMG sector of Bangladesh enjoys duty-free access to most of the developed countries (Bangladesh Garment Manufacturers and Exporters Association, 2018). This duty-free access opportunity has heightened the prospects of RMG sector to a new level.

Basically, RMG sector of Bangladesh is a buyer-driven sector where buyers pay a visit on a regular basis to place orders after investigating the quality of the sample products. RMG industry makes garments items in accordance with the demand and choice of the buyers by ensuring the highest possible qualities. The apparel industry is "the quintessential example of a buyer-driven commodity chain marked by power asymmetries between the suppliers and global buyers of final apparel products" (Fernandez-Stark et al., 2011; p. 7). As it is a buyer-driven sector, garments tourism is, therefore, thought to be one of the prospective revenue-generating sectors for Bangladesh.

Future of Garments and Apparels Tourism

Garments tourism is a novel concept in the context of Bangladesh. It can be a very potential sector if it receives investments. Many opportunities wait for garments tourism industry investments. Bangladesh is such a country that owns three UNESCO World Heritage Sites (WHS): The Historic Mosque City of Bagerhat, Ruins of the Buddhist Vihara at Paharpur, and The Sundarbans (United Nations Educational, Scientific and Cultural Organization, n.d.; Hassan & Kokkranikal, 2018; Hassan & Burns, 2014). Each year, a cohort of foreign tourists pays a visit to these places and enjoy the country's beauty and historical roots. Moreover, the climate of Bangladesh is favorable and moderate. Hence, tourists do not feel the agony of extremely hot, extreme cold, or heavy rain. They can easily move around to enjoy themselves. Bangladesh is also the least carbon emitted country in the world. So, foreign tourists will not feel various health-related problems caused by higher carbon emissions. The beaches (Cox's Bazar, Kuakata, St. Martin's Island) are long with very clean water that undoubtedly can attract foreign tourists. Moreover, Bangladesh has so many exceptionally beautiful mountains. In terms of the political environment, Bangladesh is a liberal democratic country. A country with a liberal democratic system instead of extremism would definitely develop quickly from all respects (Chan, 2002; Leftwich, 2002).

As Bangladesh's rural areas' natural beauty is quite amazing, investors can invest there to establish cottages, resorts, and restaurants with local food flavors. Such an initiative will create job opportunities for the local community, and hence people will be economically solvent. Moreover, there are some long beaches and big hilly areas for parasailing, scuba diving, parachuting, marine diving. Thus, setting up big aquariums like Bali, Malaysia, Singapore, Maldives, and so on could be a stepping stone to attract foreign tourists (Musa, 2000; Wong et al., 2010). These things can easily be set and maintained properly at a lower cost since Bangladesh is the country where labor cost is relatively lower than other countries with the same status. Nonetheless, the country's domestic tourism has been a growing trend that is a perfect sign for garments tourism and attracting foreign investment. The authority has already introduced the public, private, and community partnership (PPCP) that is thought to widen the door of sustainable and garments tourism atmosphere and could attract foreign direct investment (FDI) in the sector.

The infrastructural developments like the completion of Padma Bridge, on-going works of different mega power plants, metro rail etc. are creating a great opportunity for the inclusion of garments tourism. Furthermore, the national industrial policy of 2010 focused on the expansion of the tourism industry as part of its overall development. Hence, through the integration of garments tourism, the tourism industry will undoubtedly be a potential sector for keeping the development flow stable. Moreover, many public and private banks are offering loans for the SME's that could be good support for introducing garments tourism in the country.

Besides the mentioned prospects and opportunities for the inclusion of garments tourism, this industry's development could face various types of constraints or challenges. First, the country's branding image in the global context is very poor due to many reasons. The recent forgery of COVID-19 test reports delivered to the workers working in different countries has intensively fuelled this poor branding image. Hence, it could be quite difficult to attract foreign buyers to visit tourist places by creating a positive outlook. Second, as investment is a major issue for promoting garments tourism, entrepreneurs should be provided loans with lower rates of interest. But unfortunately, the interest rate is very high in Bangladesh, both for long-term and short-term loan schemes that could be a potential threat to garments tourism's blooming. Third, in any country, the development phase is a subject to the political situation of the country. In Bangladesh, political unrest is a threat to garments tourism. Sometimes, in the name of various movements, vehicles are burnt and ransacked, which is alarming and raises fear in the mind of tourists not to be willing to visit tourist places. Fourth, the authority lacks the coordination to maintain the tourist places properly that could hinder the prospect of garments tourism in Bangladesh.

Fifth, sometimes extreme natural catastrophes like earthquakes, cyclones, flood or flash flood, warming etc. can obstruct this tourism industry's glorious journey. Sixth, transportation is a big hindrance in promoting garments tourism. Bangladesh could not grow its transportation system like railways, highways, airways, riverways, etc. that maintain the global standard. However, the situation is improving day by day. Hence, a poor transportation system could demotivate foreign tourists to visit tourist attractions. Moreover, traffic jams in most big cities are bad signs for the blooming of garments tourism. Most of the foreign people are very much time-conscious. But traffic jam will kill most of their time and cannot maintain a time schedule that consequently hinders tourism. Seventh, for the development of the tourism industry, there is no central server system in Bangladesh that exists in most of the tourism friendly countries. Through this central server or information portal, tourists find information regarding their costs, living, food, transportation, etc. in an all-in-one system. So, lack of this might be a constraint for the development of garments tourism.

Eighth, Bangladesh's bureaucratic system has always been questioned by many due to their coordination problem among various ministries and offices. The lengthy process of getting government services for business is also reflected in the 'Ease of

Doing Business' report whereby Bangladesh is not performing well. This type of bureaucracy hinders the sustainable progress of any country. Similarly, this could create an awkward situation to the development of garments tourism. Moreover, though the tourism industry is thought to be a rapidly developing sector, the insufficient budget in this sector could no longer bring any good sign for its overall development.

Ninth, the Internet has made the world easily accessible through which people can get any kind of information they require. Albeit Bangladesh is in the way of making digital Bangladesh, still internet facility is not available in every corner of the country, especially in remote areas. Most of the areas have internet coverage, but it is found to be in a poor state, as argued by many (The Business Standard, 2020). Hence, lack of good internet connection in all the tourist sports could hinder garments tourism. Tenth, most of the local investors are not well aware of the benefit of the overall tourism industry. They are not concerned about the economic impact of garments tourism, and as a result, they do not feel stimulated to invest in the development of this promising sector. Eleventh, one of the remarkable hindrances to promoting garments tourism is the foreign buyers' Visa processing issue. In most cases, it is very alarming that foreign delegates, buyers, and tourists face many sorts of difficulties in getting travel pass that hinders their traveling on time. Such types of difficulties in processing travel pass makes the country the worst victim of great economic loss.

Twelfth, communication problem is a big issue that hinders any kind of development (Banaeianjahromi & Smolander, 2019) and thus creates problems for growing the garments tourism industry. The English language is treated as a lingua franca that is commonly used to communicate (Tweedie & Johnson, 2019). But as a matter of fact, though a good percentage of people living in urban areas have the English communication skill through which they can meet up their daily necessities, it is surprisingly very low for those people who are living in rural or remote areas. Hence, foreign buyers will face problems related to language understanding and communicating with others if deemed necessary. In any distant place, the language barrier is regarded as a primary barrier. Hence, the lack of communication skills of the people, especially those living in adjacent tourist attractions, will obstruct the growing trend of garments tourism. Thirteenth, life security of the foreign people is a prime concern to promoting garments tourism. In Bangladesh, in the recent past years, some incidents took place where many foreign people sacrificed their lives. For instance, the Holy Artisan issue diluted the country's overall image to the global context and resulted in the cancellation of many foreign investments in the country. This had a negative impact on the economic development of Bangladesh. Finally, insufficient numbers of world standard hotels, motels, and restaurants, and best service facilities might hinder this prospective industry's progress.

Conclusion with Policy Recommendations

Though the tourism industry worldwide is a rapidly growing sector, garments tourism is relatively a novel concept. Through this concept of tourism, buyers accomplish two tasks at one trip. Before accelerating garments tourism in a successful manner, the development, safety, and security of the garments industry should be ensured. Because buyers mainly visit for their apparel business. Once they are satisfied with their business mission, they will go for recreation or refreshments to various beautiful tourist places. As mentioned above, the garments tourism industry would be a promising and prospective sector in Bangladesh's context. However, this industry's development may face a number of challenges and constraints that are already mentioned above.

Based on those constraints, the Govt., policymakers, executive bodies, and the concerned authorities might consider the following policies that are hoped to lessen the constraints in various respects. First, since some recent past incidents have ruined the country's brand image in the global context, Bangladesh should continuously strive to regain its positive branding image. In this case, the foreign ministry should work properly by adopting timely policies. They should sit for bilateral discussions where they ventilate the country's positive messages to the global people. Second, the Govt. should relentlessly work to wipe out political unrest. With regard to this, they should sit for unconditional discussions with the oppositions and express solidarity with their legal movements and agendas.

Third, banks and other financial institutions should provide loans to the SME's with a minimum interest rate so that they feel motivated to take loans and use those amounts in the development of garments tourism. Fourth, the Govt. offices should provide fast and one-stop services based on the urgency of the claims/files that will stimulate entrepreneurs to be progressive in their business mentality. Fifth, there should be a central database or service provider where all the necessary and required information of the tourism industry will be available. Make sure the database functions well, is updated regularly, and monitored properly from time to time. Sixth, as the garments industry is a booming sector of Bangladesh, difficulties related to visa/travel passes' approval must come to an end. In this respect, Bangladeshi high commissions in different countries might open another unique section that will merely handle the tourists and buyers' visa issues for ease and comfortability of the potential tourists.

Seventh, sometimes, due to the information gap, many foreigners cannot visit tourist places. Hence, the garments industry's buying houses or offices can print booklets or magazines that will contain all the necessary information of tourist attractions of the country. Besides, they can provide them tourism services by initiating different conducive and inclusive packages. Eighth, roads and highways and transportation facilities should be expanded to minimise traffic jams and attract foreigners to visit tourist places. Ninth, the immigration can open a separate queue both in the inbound and outbound departments that will especially serve the foreign buyers or foreign business delegates that will make them feel valued and attract them to visit the country again. Tenth, as garments tourism could be a prospective revenue-generating sector, the Govt. should allocate a sufficient budget that can be invested to upgrade this sector.

Eleventh, since communication skill is a major obstruction for the development of garments tourism, campaigns and English learning environment with free of charge can be provided to the young generation of those areas where the tourist places are. Twelfth, sometimes taxi drivers, boat drivers, or rickshaw pullers show misconduct with the foreigners. In this case, campaigns and counseling sessions can be taken to make them aware of behaving well with foreign people as they are the country's guests. Thirteenth, adjacent to the tourist places, standard hotels, motels, cottages, shopping malls, and restaurants can be established that will maintain quality and provide the best possible services. Fourteenth, in all the tourist places, strong and high-speed internet facility should be provided. Fifteenth, to ensure the safety and life security of the foreigners, law enforcement agencies should effectively coordinate with local agencies and relevant units for hassle-free movements of foreigners from one place to another.

Finally, above all, an unbiased and proficient monitory body should be formed to monitor and supervise all the activities related to garments tourism and take necessary measures to develop this sector. In doing so, this sector would surely be a prospective sector from where Bangladesh might earn unpredictable foreign currencies.

References

Afroz, N. N., & Hasanuzzaman, M. (2012). Problems and prospects of tourism in Bangladesh: Bandarban district case. *Global Journal of Management and Business Research, 12*(23), 1–9.

Ahmed, N. (2006). *Bangladesh apparel industry and its workers in a changing world economy.* PhD thesis. Wageningen University.

Banaeianjahromi, N., & Smolander, K. (2019). Lack of communication and collaboration in enterprise architecture development. *Information Systems Frontiers, 21*(4), 877–908.

Bangladesh Garment Manufacturers and Exporters Association (BGMEA). (2018). *Trade information*. Retrieved from: http://www.bgmea.com.bd/home/pages/TradeInformation. Accessed 28 Dec 2020.

Bangladesh Garments Accessories and Packaging Manufacturers and Exporters Association (BGAPMEA). (2019). *A glimpse of the growth and status of BGAPMEA*. Retrieved from: http://www.bgapmea.org/index.php/about/statistics. Accessed 25 Dec 2020.

Bhattacharya, D., & Rahman, M. (2000). Bangladesh's apparel sector: Growth trends and the post-MFA challenges. Paper presented at *the National seminar on the Growth of the garment industry in Bangladesh: Economic and social dimensions*. Bangladesh Institute of Development Studies (BIDS) and Oxfam Bangladesh, Dhaka.

Chan, S. (2002). *Liberalism, democracy and development*. Cambridge University Press.

Chowdhury, S. H. (2020). Tourism in Bangladesh: Highlighting the development trend. *International Journal of Science and Business, 4*(5), 23–36.

Durbarry, R. (2004). Tourism and economic growth: The case of Mauritius. *Tourism Economics, 10*(4), 389–401.

Export Promotion Bureau of Bangladesh (EPB). (2018). *Home*. Retrieved from: http://epb.gov.bd/site/view/epb_export_data/. Accessed 25 Dec 2020.

Feldman, S. (2009). Historicizing garment manufacturing in Bangladesh: Gender, generation, and new regulatory regimes. *Journal of International Women's Studies, 11*(1), 268–288.
Fernandez-Stark, K., Bamber, P., & Gereffi, G. (2011). *The fruit and vegetable global value chain: Workforce development and economic upgrading* (p. 7). Duke University.
Fonseca, F. G. (2012). Challenges and opportunities in the world of tourism from the point of view of ecotourism. *Higher Learning Research Communications, 2*(4), 5–22.
Hassan, A., & Burns, P. (2014). Tourism policies of Bangladesh – A contextual analysis. *Tourism Planning & Development, 11*(4), 463–466.
Hassan, A., & Kokkranikal, J. (2018). Tourism policy planning in Bangladesh: Background and some steps forward. *e-Review of Tourism Research (eRTR), 15*(1), 79–87.
International Growth Centre. (2018). *Survey of the literature on successful strategies and practices for export promotion by developing countries*. Retrieved from: https://www.theigc.org/wp-content/uploads/2011/06/Belloc-Di-Maio-2011-Working-Paper.pdf. Accessed 28 Dec 2020.
Islam, M. S. (2020). Ready-made garments exports earning and its contribution to economic growth in Bangladesh. *GeoJournal, 86*, 1301–1309. https://doi.org/10.1007/s10708-019-10131-0
Khondker, B. H., & Ahsan, T. (2015). *Background paper on tourism sector*. Retrieved from: https://d1wqtxts1xzle7.cloudfront.net/52406228/5_Prospect-and-Strategy-for-Tourism-Development.pdf?1490987685=&response-content-disposition=inline%3B+filename%3DBACKGROUND_PAPER_ON_TOURISM_SECTOR.pdf. Accessed 28 Dec 2020.
Kobra, M. K., Bhuiyan, K. H., & Zayed, N. M. (2018). Well and woes of tourism promotion in Bangladesh: Investment perspective. *Academy of Accounting and Financial Studies Journal, 22*(3), 1–8.
Kulendran, N., & Witt, S. F. (2003). Forecasting the demand for international business tourism. *Journal of Travel Research, 41*(3), 265–271.
Lee, W., & Chhabra, D. (2015). Heritage hotels and historic lodging: Perspectives on experiential marketing and sustainable culture. *Journal of Heritage Tourism, 10*(2), 103–110.
Leftwich, A. (2002). Democracy and development. *New Political Economy, 7*(2), 269–281.
Mahmud, W., Ahmed, S., & Mahajan, S. (2018). Economic reforms, growth, and governance: The political economy aspects of Bangladesh's development surprise. In D. Brady & M. Spencer (Eds.), *Leadership and growth* (pp. 227–254). The International Bank for Reconstruction and Development/World Bank.
Majumder, M., & Hossain, T. (2011). An appraisal of tourism industry development in Bangladesh. *European Journal of Business and Management, 3*(3), 287–298.
Musa, G. (2000). Tourism in Malaysia. In C. M. Hall & S. Page (Eds.), *Tourism in South and South East Asia: Issues and cases* (pp. 144–156). Butterworth Heinemann.
Nicula, V., & Elena, P. R. (2014). Business tourism market developments. *Procedia Economics and Finance, 16*, 703–712.
Palmer, C. (1999). Tourism and the symbols of identity. *Tourism Management, 20*(3), 313–321.
Peterson Institutes for International Economics. (2015). *India's rise: A strategy for trade-led growth*. Retrieved from: https://www.piie.com/sites/default/files/publications/briefings/piieb15-4.pdf. Accessed 28 Dec 2020.
Phillips, J., & Faulkner, J. (2013). *Tourism investment and finance accessing sustainable funding and social impact capital*. US Agency for International Development.
Rahman, K. M., & Chowdhury, E. H. (2020). Growth trajectory and developmental impact of ready-made garments industry in Bangladesh. In M. K. Barai (Ed.), *Bangladesh's economic and social progress* (pp. 267–297). Springer.
Rahman, M. H., & Siddiqui, S. A. (2015). RMG: Prospect of contribution in economy of Bangladesh. *International Journal of Scientific and Research Publications, 5*(9), 1–8.
Rogerson, C. M. (2015). The uneven geography of business tourism in South Africa. *South African Geographical Journal, 97*(2), 183–202.
Sinclair, M. T. (1998). Tourism and economic development: A survey. *The Journal of Development Studies, 34*(5), 1–51.

Spinanger, D. (2001). *The WTO, ATC and textiles and clothing in a global perspective: What's in it for Bangladesh?* (Working Paper No. 8). Centre for Policy Dialogue (CPD).

Subadra, I. N., Sutapa, I. K., Artana, I. W. A., Yuni, L. H. K., & Sudiarta, M. (2019). Investigating push and pull factors of tourists visiting Bali as a world tourism destination. *International Journal of Multidisciplinary Educational Research, 8*, 235–269.

Telfer, D. J., & Sharpley, R. (2015). *Tourism and development in the developing world*. Routledge.

The Business Standard. (2020). *When your Internet can't keep pace with Digital Bangladesh*. Retrieved from: https://tbsnews.net/feature/panorama/when-your-internet-cant-keep-pace-digital-bangladesh-123850. Accessed 25 Dec 2020.

Tichaawa, T. M. (2017). Business tourism in Africa: The case of Cameroon. *Tourism Review International, 21*(2), 181–192.

Tweedie, M. G., & Johnson, R. C. (2019). Research directions in medical English as a lingua franca (MELF). *Language and Linguistics Compass, 13*(3), e12312.

United Nations Educational, Scientific and Cultural Organization (UNESCO). (n.d.). *World heritage list*. Retrieved from: https://whc.unesco.org/en/list/&order=country#alphaB. Accessed 12 Jan 2021.

Wong, E. P., Mistilis, N., & Dwyer, L. (2010). Understanding ASEAN tourism collaboration – The preconditions and policy framework formulation. *International Journal of Tourism Research, 12*(3), 291–302.

World Bank (WB). (2018). *World Bank country profile: Bangladesh*. Retrieved from: https://databank.worldbank.org/views/reports/reportwidget.aspx?Report_Name=CountryProfile&Id=b450fd57&tbar=y&dd=y&inf=n&zm=n&country=BGD. Accessed 28 Dec 2020.

World Trade Organization (WTO). (2016). *World trade statistical review*. Retrieved from: https://www.wto.org/english/res_e/statis_e/wts2016_e/wts2016_e.pdf. Accessed 28 Dec 2020.

World Travel and Tourism Council (WTTC). (2019). *Economic impact reports*. Retrieved from: https://wttc.org/Research/Economic-Impact. Accessed 25 Dec 2020.

World Travel and Tourism Council (WTTC). (2020). *Bangladesh at a glance*. Retrieved from: https://tool.wttc.org. Accessed 28 Dec 2020.

Yunus, M. (2001). *Review of Bangladesh's external sector performance*. Retrieved from: https://idlbnc.idrc.ca/dspace/bitstream/10625/30336/1/117790.pdf. Accessed 28 Dec 2020.

Zaman, S., Roy, M., & Chowdhury, M. M. (2020). The involvement of private sector in tourism in Bangladesh. In M. S.-U. Rahman & A. Hassan (Eds.), *Tourism policy and planning in Bangladesh* (pp 123–146). Springer.

Chapter 17
Ecotourism in Bangladesh: Investment and Development Contexts

Md. Alauddin, Md. Aktar Kamal, and Md. Asad Iqbal Chowdhury

Abstract The importance of ecotourism is a rapidly increasing in Bangladesh as the country has natural wonders, panoramic hills, valleys, forests, beaches, lakes, rivers, reserves of glorious creatures etc. These ideally have positioned Bangladesh to become able to compete in the global ecotourism market. Although ecotourism is still relatively new to Bangladesh, several initiatives are taken to expand and promote it in the country. Ecotourism in Bangladesh is gaining popularity as more travelers become aware of the stunning landscapes, mesmerizing wildlife and biodiversity of plants found in this beautiful country. What is more, by offering access offshore islands, rivers, wetlands, "Haors" (water reservoirs), mangrove forests, and diverse bio cultures, Bangladesh possesses potential for development as an ecotourism hub. The aim of this chapter is to define the opportunities for ecotourism services and products in Bangladesh, as well as the current constraints and potential solutions. This chapter will assist policymakers and marketers in putting policies and strategies in place to promote ecotourism products and encourage overall tourism growth in Bangladesh.

Keywords Tourism · Ecotourism · Marketing · Bangladesh

Md. Alauddin (✉)
Department of Marketing, City University of Hong Kong, Kowloon Tong, Hong Kong

Department of Business Administration, International Islamic University Chittagong (IIUC), Chattogram, Bangladesh
e-mail: alauddin4693@gmail.com

Md. A. Kamal
International Tourism and Convention Management, The Hong Kong Polytechnic University, Hung Hom, Hong Kong

Department of Management Studies, Faculty of Business Studies (FBS), Bangladesh University of Professionals, Dhaka, Bangladesh

Md. A. I. Chowdhury
Department of Economics & Banking, International Islamic University Chittagong (IIUC), Chattogram, Bangladesh

A. Hassan (ed.), *Tourism in Bangladesh: Investment and Development Perspectives*, https://doi.org/10.1007/978-981-16-1858-1_17

Introduction

Global tourism offers tremendous potential for export earnings. In 2016, the travel and tourism industry contributed directly to GDP growth of 3.1% and generated USD 7.6 trillion The direct contribution grew to 3.2% in 2018 and is expected to increase to 3.6% by 2029. Moreover, the total contribution to GDP (including direct, indirect, and induced) was 10.4% in 2018 and is projected to grow to 11.5% by 2029 (World Travel and Tourism Council, 2019). Tourism's contribution to employment generation is also noteworthy. In 2016, tourism generated approximately 300 million jobs worldwide (including direct, indirect, and induced contributions). The figure increased to approximately 328 million in 2018 and is projected to grow to around 420 million by 2029 (World Travel and Tourism Council, 2019). Given the ease and accessibility of modern transport, today's tourists increasingly aspire to see and experience new, unique destinations (Yeoman et al., 2006).

In 2014 almost 125,000 tourists visited Bangladesh, and Bangladeshi citizens traveling out of the country showed a similar trend. Overall, travel and tourism investment increased by almost 13.9% in 2017. It contributes to GDP increased by almost 6.2% by direct contribution and 7.2% by total contribution to the GDP in 2017 (Tourismprof.club, 2018). Worldwide, tourist arrivals grew 7.1% in 2017 (approximately 1322 million travelers). The sector is expected to register a growth rate of about 4–5% in 2018 (World Tourism Organization, 2018).

The travel and tourism industry generates employment opportunities and has contributed approximately 10% of all job growth produced across the globe, which is similar to the formation of 109 million employments (World Travel and Tourism Council, 2017). However, this rise accounts for about 8% of overall universal greenhouse gas productions (Lenzen et al., 2018). Public voice to diminish pollution has increased and the industry has recognized the capability to influence social phenomena, such as minimizing the poverty and promoting cultural interactions, as a result 2017 was declared as the "International Year of Sustainable Tourism for Development by United nation" (United Nations Framework Convention on Climate Change, 2017). It is the time that the tourism industry should take initiatives on sustainable development which will ensure the protection of the natural environment from the damaging impact of growing tourism.

History of Ecotourism

According to Oxford Dictionary (2002), "Ecotour was first founded in 1973 and ecotourism in 1982". The roots of the word "ecotourism" are not completely obvious. However, in his study, Hetzer (1965) recognized four pillars of responsible tourism; minimizing the impacts on the environment, respect to host cultures, to maximize the benefit for the local people, and enhancing the satisfaction for the visitors. Here the first pillar is the most critical part of ecotourism.

Other references are found in the early work by (Miller, 1989), originally published in 1978, and the work was on park planning in Latin America ensuring Eco development, and papers produced in Canada to set the roadside "eco tours" which developed during mid of 1979 to early 1980.

Ecotourism was formed "with the womb" of 1970s and 1980s environmental movement. Rapidly increasing environmental concern combined with a growing unhappiness with mass tourism resulted in increased demand for alternative, nature-based experiences. At the same time, developed countries also started to show their concern regarding this issue. By the middle of 1980s, a number of these countries identified ecotourism as a way of achieving both preservation and development objectives (Tourism Notes, 2020). However, the first formal definition was given by (Ceballos-Lascurain, 1987) that defined ecotourism as traveling to relatively undisturbed or uncontaminated natural areas with the specific objective of studying, admiring, and enjoying the scenery and its wild plants and animals. Ecotourism also involves any existing cultural manifestations (both past and present) found in the areas (Wang, 2010).

Meaning of Ecotourism

Ecotourism is a concept that combines two terms: "ecosystem" and "tourism", and it emphasizes on environmentally friendly tourism. The term "ecotourism" comes from the term "nature-based tourism". Ecotourism, according to the International Research Group (2020), is defined as travel for the purpose of enjoying natural attractions. It has supplanted so-called "sight- seeing" as a means of escaping the aggravating crowds, and it has become a significant craze among globetrotters. Visitors who are environmentally concerned have come to embrace environmentally sound tourism, which ensures the preservation of natural quality while avoiding negative environmental impacts. The name "eco" is derived from the Greek word "oikos", which means "habitat" and is an acronym for "ecology". The term "tourism" is derived from the Hebrew word "tora", which means "to study, learn, or research". Ecotourism, according to the International Ecotourism Society (2020), is responsible travel to natural regions that conserves the environment and supports the well-being of local people.

Kabir & Bhuiya (2004) mentioned that ecotourism is to preserve the sustainability of the natural as well as cultural environment. Ahmed & Mollah (2014) suggested that ecotourism includes not only exploring natural wonders but also recreational and archaeological tourism. Some common types of ecotourism are depicted in Table 17.1.

Ecotourism is the most inescapable tourism of all, and it plays a vital role in various countries differently. Surely, it motivates a large number of tourists. The areas that are considered by this tourism include national parks, jungles, and wildlife (Bowker et al., 1999). The tourism industry, especially ecotourism has played an

Table 17.1 Types of ecotourism sites

Types	Sites
Marine	Bays, islands, marine-protected areas, beaches, coral reefs
Terrestrial	Forest areas, mountains, volcanoes, caves, trails
Freshwater	Waterfalls, rivers, lakes, hot and cold springs
Cultural	Reforestation sites, urban parks
Manmade	Festivals, churches, historic sites

Source: Ahmed and Mollah (2014)

important role to make a bridge between humans and community, economic activities, and development (Dehcheshme & Zangibadi, 2008).

The Quebec Declaration of Ecotourism, developed as part of the 2002 United Nations International Year of Tourism, defines ecotourism as tourism that "Contributes actively to the conservation of natural and cultural heritage; includes local and indigenous communities in its planning, developing and operation, and contributes to its well-being; interprets the natural and cultural heritage of the destination to visitors; and lends itself better to independent travelers, as well as to organized tours of small group sizes" (Quebec Declaration on Ecotourism, 2002, p. 12).

According to Tsaur et al., (2006: p. 640), ecotourism is recognized as a form of sustainable tourism, which is expected to contribute to both conservation and development. Ecotourism is defined as nature-based tourism with three added goals: (1) to reduce negative environmental, economic and social impacts that are often associated with mass tourism; (2) to deliver a net positive contribution to environmental conservation; and (3) to develop the livelihoods of the local people (Jamal & Robinson, 2009). According to Narasaiah (2005: p. 81), ecotourism must be responsible, respect the local needs, and improve local welfare. However, to be completely sustainable, ecotourism should complete the ambitions and expectations of all interests of tourism enterprises, organizations, and tourists.

According to Honey (2002) and Cox et al. (2009), there are some elements of authentic ecotourism: first, travel to natural areas; second, minimized impacts; third, building environmental and cultural awareness for hosts and guests; fourth, providing direct financial benefits and empowerment for local communities; fifth, respect of local culture; sixth, sensitivity to the host country's political environment and social climate; and finally, support of human rights and international labor agreements.

Ecotourism is defined in different ways as: Ecotourism is a type of tourism that is primarily inspired by a region's natural history, particularly its indigenous cultures. In the spirit of appreciation, engagement, and sensitivity, the eco tourist explores relatively untouched places. (Ziffer, 1989). Nature-based tourism that emphasizes educational possibilities while also benefiting local and regional communities and exhibiting environmental, social, cultural, and economic sustainability

(Gardner & McArthur, 1995). Ecologically sustainable tourism in natural places that interprets the local environment and cultures, improves tourists' understanding of them, promotes conservation, and improves the local people's well-being (Richardson, 1993). Travel to distant or natural locations with the goal of improving awareness and appreciation of the natural environment and cultural heritage while avoiding damage to the ecosystem and the experience for others (Figgis, 1993). Ecotourism is a type of tourism that emphasizes visiting and observing natural regions while promoting environmental principles (Boyd & Butler 1996). Travel to experience the incredible diversity of natural life and human civilization around the world without harming either (Cater & Lowman, 1994). Ecotourism is a type of tourism and recreation that is both natural and environmentally friendly (Lindberg & McKercher, 1997).

Importance of Ecotourism

The role of ecotourism is immense in our social life. Tourists like this tourism very much especially who like nature. It attracts them to enjoy the scenery. People's increasing interest in the environment and trips focused on fresh air, combined with their growing frustration with mass tourism, has illustrated the need for ecotourism in the tourism industry. Furthermore, the evolution of the word "ecotourism" was aided by an increasing portion of the population's awareness and compliance with the ideals of environmental protection and sustainability. Ecology and cultural values must be ensured in ecotourism. The target of ecotourism is to minimize the environmental impact on any specific tourism area with the combination of cooperation and management, in addition to this traveler may increase the positive impact on the environment and their surroundings while traveling. Ecotourism is described as liable travel to ecological areas that protects the environment while also improving the well-being of locals (Blangy & Wood, 1993) Environmental and growth conflicts, as well as global climate change, necessitate the development of eco-tourism. It is a crucial understanding for tourism-oriented or third-world countries because it effectively contributes to the conservation of natural and cultural resources by including local and indigenous communities in planning and development. It makes an impact to their happiness; it educates visitors about the natural and cultural heritage; and it is well-suited to both individual and small group travel. Ecotourism provides a safe, pollution-free destination, decreases tourism's negative effects, contributes to conservation efforts, collaborates with locals to manage natural areas, and protects local economies and uniqueness. Figure 17.1 showing the importance of ecotourism.

Fig. 17.1 Importance of ecotourism. (Source: Fennell, 2004)

Ecotourism and Sustainable Development

In the developed countries, tourism is considered as a faster growing economic industry. This change can be due to shifting habits in affluent communities, as people pursue more recreation and are more concerned about environmental quality (Veenhoven, 1999). Travelers from developed countries are now also exploring the hidden natural beauty of the developing countries. Exotic travel is motivated by a desire to relax, as well as a desire to experience pristine and serenity natural beauty while learning about diverse cultures. The rapid growth of tourism in developed countries has increased hopes that tourism will help countries develop economically (Tapper, 2001). Mass tourism and alternative tourism are the two broad types of tourism that have been defined (Wood, 2002). Over time, alternative tourism has been categorized into three categories: nature, rural community-based tourism, and cultural events-oriented tourism. As a result, the idea of "ecotourism" arose from natural tourism (Lumsdon & Swift, 1998; Wood, 2002).

Hall (2004) stated that Ecotourism is sometimes considered as a low-cost way into niche tourism industries, based on a seemingly endless supply of natural materials and a desire to promote sustainability and environmental awareness. Hall (2004) also specified that the term's precise definition and meanings are not well

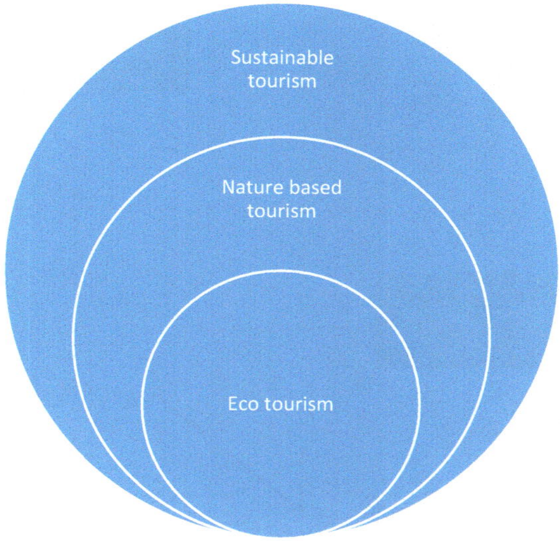

Fig. 17.2 Mechanism of ecotourism ensuring sustainable tourism. (Source: Fennell & Dowling, 2003)

defined. There are at least three different types of ecotourism: Firstly, the chance to engage with nature and people from different cultures; secondly, rather than big (international) tourism firms, the selection of tourism styles that maximize revenue for disadvantaged people; and thirdly, reducing travel's overall ecological influence (Fennell & Dowling, 2003). The relation of ecotourism with sustainable tourism has shown in Fig. 17.2.

Various definitions of ecotourism place different weights on each of these elements, but academics are increasingly emphasizing the combination of these three. According to Björk (2000), ecotourism is a distinct form of tourism that recognizes the close relationship between decent ecosystem and good economy, as opposed to adventure, nature and farm tourism. Wood (2002) and Buckley (1995) explain that travel to natural regions in a responsible manner can protect the environment and ensure the well-being of locals (Fig. 17.3).

Principles of Ecotourism Products

Nature Area Focuses

Focus on providing tourists with the ability to explore nature up close and personal.

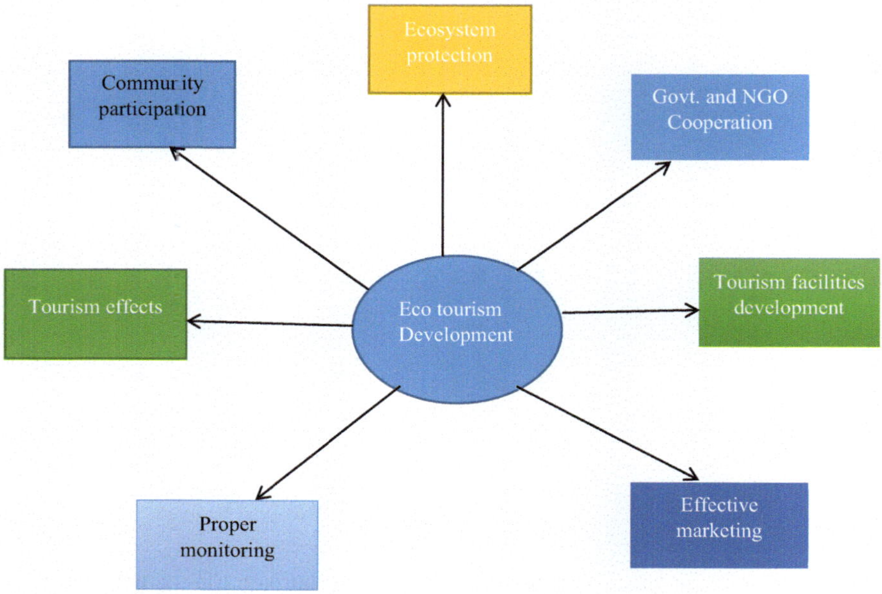

Fig. 17.3 Ecotourism development aspects. (Source: Reza, 2016)

Interpretation

Provide opportunities for people to interact with nature in ways that foster a deeper understanding, appreciation, and enjoyment of it.

Environmental Sustainability Practice

Represent environmentally friendly tourism best practices.

Contribution to Conservation

Make a direct contribution to the preservation of natural areas.

Benefiting Local Communities

Contribute to the local community on a regular basis.

Cultural Respect

Be aware of, interpret, and include the local culture or cultures.

Customer Satisfaction

Meets or exceeds customer needs on a regular basis.

Responsible Marketing

Be truthful and precise in your marketing and promotion so that reasonable standards are established (Eco India, 2020).

Ecotourism in Bangladesh

The idea ecotourism is still new and unexplored in the tourism industry of Bangladesh. Bangladesh is an ecologically rich country with ample opportunity to explore ecotourism as an economic growth engine. However, Bangladesh holds a great possibility for ecotourism growth. The country wholly herself is an ecotourism attraction. There are ample ecotourism destinations across this country. The haors, offshore islands, wetlands, river, mangrove forests, hills, and cultural heritage can be the attractive ecotourism products of Bangladesh. The attractive ecotourism destinations are shown below:

The Sundarbans

The Sundarbans is one of the important and largest areas for ecotourism, covering 10,000 km^2. It is the largest mangrove forest in the world, composed of different islands. Almost 66% of the mangrove forest is in Bangladesh, and the rest 34% of the area is in under India. The mangrove forest's unique environment supports a distinctive anthology of flora and fauna that has made the area a paradise for ecotourism. The Royal Bengal Tigers is the unique attraction of this place. Moreover, spotted deer, crocodiles, wild boar, jungles fowl, lizards, different kinds of monkeys, countless types of lovely birds and ducks have increased the beauty of this forest. Honey and leaf collection, fishing support the employment of local people. Wildlife photography, boating through the forest area, nature study, enjoying the activities of wood-cutters, fishermen, honey-collectors and the mysterious environment of the Sundarbans always invites the tourist, nature lover, researcher and scientists. UNESCO announced the Bangladeshi portion of forest area as a globe heritage site in 1997. Hiron point, Karamjol, Nilkomol, Katka, Dublar Char, Smoronkhola, Kodomtola, Akram Point, Haldibunia, Dobeki, Tiger Point (Kochikhali), Harbaria and Notabeki are the important tourist attraction point in Sundarbans. Ministry of Environment and Forests and Bangladesh Parjatan Corporation (BPC) are jointly working to develop the Sundarbans as an ecotourism destination.

Cox's Bazar

Cox's Bazar, the world lengthiest sea beach, is the most popular and most visited destination in Bangladesh. It is the tourist capital of the country. There is no fear of shark attack, as a result, this beach is good for swimming and sun-bathing. The scenario of the sunset behind the waves of the sea is very charming. Cox's Bazar is the home of many Buddhist pagodas, Rakhaine culture, an ancient monastery, shrimp cultivation, salt production, tribal handicrafts and delicious seafood. The Inani beach, Himchari waterfalls, Maheskhali, Sonadia, Ramu, make this place most attractive for ecotourism. The road in between sea beach and hill on the way to Teknaf through marine drive known to be the home of deer, wild elephants, jungle fowls, and beautiful birds. The river Naf, divided sub-urban town Teknaf and Myanmar, which is situated on the southern tip of Bangladesh. The historical "Mathin Kup" (water reservoir) in Teknaf is an important attraction for tourist. The island in the Bay of Bengal, Maheskhali is in the northwestern site of Cox's Bazar. The century-old Temple Adinath is at the top of the hill on the island. Travel by crossing big wave in the sea, salt production, Rakhaine villages, fishermen villages are some notable attraction of Maheskhali. Sonadia is another beautiful and calm island near Cox's Bazar and very close to Maheskhali famous for dry fish processing. Ramu is the home of the largest rubber gardens of Bangladesh. Buddhist pagodas, Buddhist village, forest area, lake in forest areas increase the scope for ecotourism.

The "Bangabandhu Sheikh Mujib Safari Park" is an attraction of ecotourism, which is the first safari park in Bangladesh. This safari park is situated about 50 km from Cox's Bazar town a sound environment for animal reservation. The green forest is rich with Boilam, Garjan, Chapalish, and Telsur trees. The safari park is full of different wild animals, including elephants, tigers, lions, crocodiles, bears, and different types of monkeys and birds, which has created ample scope for ecotourism. The Saint Martin Island is the only coral island of Bangladesh, which is also known as Narikel Jinjira. This is a small island surrounded by the beach on the northeast side of the Bay of Bengal. Clear green seawater, stunning landscapes, coral colony, sea crabs, coconut groves, the roar of the wave, fresh and unique seaside are notable attractions of visitors. There is another beautiful attraction very closed to St. Martin's named as CheraDwip. The uniqueness of Saint Martin and CheraDwip Island, warmth welcome of the local community, natural resources can be the heaven for ecotourism.

Kuakata Sea Beach

Kuakata is also known as "Sagar Kannya" (daughter of the sea), is the second-largest panoramic sea beach in Bangladesh. It is a sandy sea beach of 3 km^2 wide and 18 km^2 long. This site's unique beauty and main attraction are that it is the only sea beach in Bangladesh where one can enjoy both sunset and sunrise from the same area of Bay of Bengal. Century aged Buddhist community, the lifestyle of tribal Rakhanies are another tourist attraction of this place.

Chattogram Division

Chattogram is the wealthiest division in terms of ecotourism attraction in Bangladesh. The business capital Chattogram city is the second-largest city in Bangladesh. It has the most important and busy seaport. It is famous for its harbor and port, Foy's Lake, Parki beach in Annuara, Patenga beach, Neval Beach, Karnaphuli River, Vatiary Lake, Chattogram Zoo, Batali Hill, the shrine of Bayazid Bostam, Butterfly Park Bangladesh, the beautiful Chandanpura Mosque, Chattogram Commonwealth War Cemetery, Adher Manik Shanti Niketan Buddhist Monastery, Kaibalyadham Ram Thakur Ashram, Chatteshwari Kali Temple, Ethnological Museum, Zia Memorial Museum, etc. The first eco-park of Bangladesh is situated in Sitakunda, which is only 36 km away from Chattogram. Chandranath Hills in Sitakunda is renowned for its botanical garden. The eco-park has been developed to accommodate biodiversity conservation, new plantations, natural rejuvenation, and to encourage nature-based tourism. Hindu pilgrims, a holy place for the Hindu community, is also located here.

There are some wonderful hilly districts under Chattogram divisions, including Rangamati, Bandarban, and Khagrachari, which are home to many tribes with unique traditions and cultures. Rangamati is situated on the south-eastern side of Bangladesh, and its township is on the western bank of Kaptai Lake, 77 km away from Chattogram city. Rangamati is famous for the colorful tribal population, hills, beautiful landscape, Shuvolong Waterfalls, flora, and fauna, hanging bridge, ivory jewelry, homespun textile products, scenic beauty, tribal museum, and Kaptai Lake. Kaptai Lake is the largest man-made lake in the sub-continent with full of natural beauty. Kaptai National Park falls between the Kaptai hill and Karnaphuly. This park is full of different types of old trees planted in 1873, and this place is home to various wildlife like elephant, monkey, jungle cat deer, etc. The lake has been formed due to the establishment of Kaptai Dam on Karnaphuli River's flow under the "Karnaphuli Hydro-electric project". A hanging bridge over this lake has created a tremendous attraction for ecotourism.

Bandarban is considered as one of the attractive tourism destinations in Bangladesh. It is 92 km away from Chattogram city, which offers the beauty of valleys and hills. Nilgiri is a famous tourist attraction in Bandarban. It is the home of Mogh tribe and Morangs. Bohmong, the chief of Mogh tribe, lives in this place. The Moghs are very simple as well as hospitable people. Morangs are famous for their dance and music. The "Buddha Dhatu Jadi Temple" is located here. The topmost three highest mountains of Bangladesh – Keokradong, Mowdok Mual and Tahjindong- are located here. The Raikhiang Lake, known as the highest lake in Bangladesh, Boga Lake, Chimbuk peak and Sangu River, is the highly attractive taste of the country's ecotourism.

Khagrachari is another place of ecotourism, which is also known as Mong Circle or Phalang Htaung and 112 km away from Chattogram city. Most of its land is hilly areas having three rivers, namely Kasalong, Chengi and Maini. It has the Richhang Falls, Alutila Cave and Hanging Bridge, and is mostly famous for Sajek, known as Hill Queen. Sajek is a new attraction for ecotourism which is name after a river that separated Bangladesh and India in this place. Different tribal communities like Chakma, Tripura, Marma, Lushai and Pankua live here.

Sylhet

Sylhet Division has much to present as an ecotourism attraction, which is covered with tropical rainforest, attached tea estates, orange groves, and pineapple plantations. This place is important to the Muslims community as the shrines of "Hazrat Shahjalal (RA)" and "Hazrat Shahporan (RA)" are located there. The climate of Sylhet is the best in the country due to cool air and temperature in the winter and modestly warm in the summer session. There are many recreational resorts and parts with vast green areas boasted as ecotourism sites. There are virgin forests, more than 130 tea estates, national eco-parks, mountain ranges, undiscovered caves, and Khasia and Manipuri tribes, who are well-known for their distinct cultural heritage and folk dances.

Madhabkunda Eco-part has much to offer ecotourism. It's in the Sylhet division's Maulvibazar district. The Madhabkunda waterfall, which is about 200 feet tall and has significant ecological and economic significance, is Bangladesh's largest waterfall. Lawachara is one of Bangladesh's most beautiful tropical forests and natural reserves. This place is located in Maulvibazar district. Bangladesh government declared this place as Lawachara National Park in 1997. The national park is known for its various tree species and wild animals. Bangladesh's largest wetland resource is Hakalukihaor. It is the home of numerous fishes, birds, animals and different flora and fauna. Ratargul Swamp is the only swamp forest of Bangladesh locally known as Amazon of Bangladesh. The green forest is linked with the cannel ChengirKhal and sited by the river Goain. The forest goes around 10 feet deep and rainy season it goes under 20-30 feet water. This place is home of fish, monkey, snakes, birds, insects and lizards. Bisnakandi is a modern ecotourism destination in Bangladesh.

Dhaka

The capital city Dhaka, which is located on the bank of the Buriganga River. Dhaka city is full of heritage, culture and historical place. This place is famous for muslin, Jamdani crafts, silk and pearls. Ahsan Manzil, Lalbager Kella, the Mughal Constructions, the Star Mosque, the Armenian Church, the National Memorial at Savar, Curzon Hall, the Central Shadeed Minar, the National Museum, the National Botanical Garden, the Baldah Gardens, the National Zoo, Sonargaon, Bengal's oldest capital, and other ecotourism destinations can be considered. Many resorts such as Jamuna Resort, Elenga Resort, Padma Resort etc., are built either in the botanical gardens or in the riversides of the Dhaka division.

Rajshahi

Rajshahi is famous for its archaeological heritage and historical places. Paharpur Buddhist Bihar is one of the important tourist attractions. This is a Buddhist monastery located at Paharpur in Naogaon, which was declared as a "UNESCO World Heritage Site" in 1985. Mahasthangarh at Bogra is another Buddhist monastery established in third century BC, which is nevertheless one of Bangladesh's greatest discoveries. The Ramsagor Lake and the Kantaji Temple are two attractive destinations in Dinajpur.

Issues of Ecotourism in Bangladesh

Bangladesh is not well-known in the global ecotourism market. Though Bangladesh has a very good potentiality in ecotourism, it has numerous issues. Afroz and Mahmud (2017), and Khondkar and Anis (2014) have identified the following issues in the ecotourism industry of Bangladesh:

- Frequency of natural catastrophe is one of the important obstacles in ecotourism of Bangladesh.
- A huge number of people are living in a small area. So, population pressure creates a barrier to ecotourism
- Poor physical infrastructural backing makes the destinations difficult for the tourists

- Lack of consciousness, enthusiasm and deficiency of proper market information among the common people do not support the expansion of this industry
- The absence of appropriate guidelines for the tourists
- The absence of calm and isolated areas for tourists
- Poor brand image of tourism destinations of Bangladesh
- The absence of long-term vision both in the private and government sectors
- The absence of sanitation problems in many tourism destination areas
- Poor promotional initiatives as an ecotourism destination in Bangladesh made it as a laggard industry
- Shortage of enough trained and skilled people in this industry
- Lack of adequate professional agencies to coordinate ecotourism operations
- Unplanned construction and development, especially in tourist resorts
- Scarcity of financial allocation in this industry also expanding the problem
- Proper safety and security for the domestic and foreign traveler is the prime concern for the ecotourism growth
- Insufficient accommodation capacity during pic season makes tourism destination more difficult for the traveler
- Political unrest hampers the easy movement of visitors
- Corruption in all stage damaging the potentiality of this industry
- Uncontrolled and mismanagement in transportations sector
- Unplanned urbanization and inadequate institutional capacity
- Absence of stable tourism policies and strategies
- Lack of research institute associated with ecotourism
- The conflict between and among tribal groups especially the Chittagong hill tract area creates a problem for visitors.

Recommendations

Bangladesh is full of natural beauty, and ecotourism can be the dominating force to develop economic growth of the country. If ecotourism can turn as an active professional one, this industry can be the engine of Bangladesh's economic development. A proper corrective measure can help in this regard. Following suggestions can be taken to improve the situation of the ecotourism industry of Bangladesh.

- A long-term vision should be formulated for the growth of this sector. For this, a proper collaboration between the government and other stakeholders is required to promote Bangladesh as an ecotourism destination
- Currently, the entertainment facility is very low in the destination area. In most cases, visitors have no option to enjoy their time after the sunset. A broad tourism plan and policy need to be taken to satisfy the visitors
- Various motivational and awareness creation programs need to be taken to increase the flow of domestic and international tourists

- Public-Private Partnership initiative can help to flourish this industry in Bangladesh
- Different learning and training programs can be arranged for the people who are involved in tourism activities
- Tourist spots should be more secure and safer for domestic and foreign visitors. The appropriate measure for law enforcement authority needs to be taken
- Promotional activities can be broadcasted on global media and social platforms to attract international visitors
- Green and clean environment programs should be taken to make the place more attractive
- Priority should be given to infrastructure development related to ecotourism destinations
- Good connectivity through controlled and proper management in the transportations sector should be incorporated
- Urbanization should be made with proper planning and control
- A sufficient number of suitable restrooms for overnight visitors should be accommodated with proper care. Refreshment zones also need to be developed
- Stable political environment is also required for the growth of this industry
- Economic benefits and training of local people need to be ensured
- Waste management capacity needs to be developed
- Unauthorized resorts, parking, shopping area should be strictly prohibited.

Conclusion

In conclusion, Bangladesh is trying to utilize her full natural beauty to attract many local and foreign tourists. The country should have long-term planning and policy for developing ecotourism. Bangladesh has an enriched culture, strategic geographic location, and magnificent creatures, so it is high time for this country to make it an ideal place for ecotourism development. Moreover, it has the longest sandy beach in the world and the largest mangrove forest in the world which indicates the huge potentiality of the prospects of ecotourism in Bangladesh. For developing this sector, the government should come forward with the private investment sector and prepare a master plan to develop it. The government should also focus to provide a positive look at developing the skills of its human resources required for developing the tourism industry in the country. Furthermore, they can recruit expert professional and technical human resource who can provide better guidelines to contribute positively to satisfy the diversified needs of different tourists. Finally, it is a sector from where the government can earn huge foreign currency from the ecotourism industry.

References

Afroz, N., & Mahmud, M. S. (2017). Analyzing the problem and prospects of ecotourism: A review on Bangladesh. *IOSR Journal of Business and Management, 19*(5), 59–65.
Ahmed, S., & Mollah, M. (2014). A framework for classifying ecotourism initiatives in Bangladesh. *Bangladesh Research Publications Journal, 9*(4), 240-248.
Björk, P. (2000). Ecotourism from a conceptual perspective, an extended definition of a unique tourism form. *International Journal of Tourism Research, 2*(3), 189–202.
Blangy, S., & Wood, M. E. (1993). *Developing and implementing ecotourism guidelines for wildlands and neighboring communities* (pp. 32-54). The Ecotourism Society.
Bowker, J. M., English, D. B., & Cordell, H. K. (1999). Projections of outdoor recreation participation to 2050. In C. H. Ken, B. Carter, J. M. Bowker, D. B. English, S. H. Mou, J. C. Bergstrom, R. J. Teasley, M. A. Tarrant, & J. Loomis (Eds.), *Outdoor recreation in American life: A national assessment of demand and supply trends* (pp. 323–351). Sagamore Publishing.
Boyd, S. W., & Butler, R. W. (1996). Managing ecotourism: An opportunity spectrum approach. *Tourism Management, 17*(8), 557–566.
Buckley, R. (1995). Where tourism and ecology meet. In *Conference proceedings ecotourism: Concept, design and strategy*, Sri Nakharinwirot University, 6–8th February, p. 114.
Cater, E., & Lowman, G. (1994). *Ecotourism, a sustainable option?* Wiley.
Ceballos-Lascurain, H. (1987). The future of ecotourism. *Mexico Journal, 1*, 13–14.
Cox, L. J., Saucier, M., Cusick, J., Richins, H., & McClure, B. (2009). *Achieving sustainable tourism in Hawai'i using a sustainability evaluation system*. University of Hawaii.
Dehcheshme, M. M., & Zangibadi, A. (2008). The feasibility of ecotourism potential of CharMahal and Baktiari Province, by using the SWOT method. *Journal of Environmental Studies, 34*(47), 1–10.
Eco India. (2020). *Home*. Retrieved from: http://www.ecoindia.com/. Accessed 18 Sep 2020.
Fennell, D. A. (2004). *Ecotourism: An introduction*. Routledge.
Fennell, D. A., & Dowling, R. K. (2003). *Ecotourism policy and planning*. CABI Publishing.
Figgis, P. J. (1993). Ecotourism: Special interest or major direction? *Habitat Australia*, February, pp. 8–11.
Gardner, T., & McArthur, S. (1995). Guided nature – Based tourism in Tasmania's forests: Trends, constraints and implications. *Tourism Recreation Research, 20*(1), 53–56.
Hall, D. (2004). Rural tourism development in South-Eastern Europe: Transition and the search for sustainability. *International Journal of Tourism Research, 6*, 166–176.
Hetzer, W. (1965). Environment, tourism and culture. *Links*, July, pp. 1–3.
Honey, M. (Ed.). (2002). *Ecotourism and certification: Setting standards in practice*. Island Press.
Jamal, T., & Robinson, M. (2009). *The SAGE handbook of tourism studies*. Sage.
Kabir, S. M. H., & Bhuiya, M. M. M. (2004). Ecotourism development in Bangladesh: An overview. *Pakistan Journal of Social Sciences, 2*(1), 45–52.
Khondkar, M., & Anis, A. (2014). Bangladesh as an ecotourism destination. *Dhaka University Journal of Marketing, 17*, 77–90.
Lenzen, M., Sun, Y. Y., Faturay, F., Ting, Y. P., Geschke, A., & Malik, A. (2018). The carbon footprint of global tourism. *Nature Climate Change, 8*(6), 522–528.
Lindberg, K., & McKercher, B. (1997). Ecotourism: A critical overview. *Pacific Tourism Review, 1*(1), 65–79.
Lumsdon, L. M., & Swift, J. S. (1998). Ecotourism at a crossroads: The case of Costa Rica. *Journal of Sustainable Tourism, 6*(2), 155-172.
Miller, K. R. (1989). *Planning national parks for ecodevelopment. Methods and cases from Latin America*. Peace Corps.
Narasaiah, M. L. (2005). *Water and sustainable tourism*. Discovery Publishing House.
Oxford English Dictionary. (2002). *Ecotour*. Oxford University Press.
Quebec Declaration on Ecotourism. (2002). World ecotourism summit. *Quebec City*. Retrieved from: http://www.gdrc.org/uem/ecotour/quebec-declaration.pdf. Accessed 18 Sept 2019.

Reza, S. (2016). Integrated marketing strategy and development of tourism industry: A case of Bangladesh. *Thoughts on Banking and Finance, 5*(2), 114–129.

Richardson, J. (1993). *Ecotourism and nature-based holidays*. Simon and Schuster.

Tapper, R. (2001). Tourism and socio-economic development: UK tour operators' business approaches in the context of the new international agenda. *International Journal of Tourism Research, 3*(5), 351–366.

The International Ecotourism Society. (2020). *What is ecotourism?* Retrieved from: www.ecotourism.org. Accessed 1 Mar 2020.

Tourism Notes. (2020). *Eco tourism*. Retrieved from: https://tourismnotes.com/eco-tourism/. Accessed 1 Mar 2020.

Tourismprof.club. (2018). *Bangladesh tourism statistics*. Retrieved from: https://www.tourismprof.club/bangladesh-tourism-statistics/. Accessed 2 Feb 2020.

Tsaur, S., Lin, Y., & Lin, J. (2006). Evaluating ecotourism sustainability from the integrated perspective of resource, community and tourism. *Tourism Management, 27*, 640–653.

United Nations Framework Convention on Climate Change. (2017). *Is international year of sustainable tourism for development?* Retrieved from: https://unfccc.int/news/2017-isinternational-year-of-sustainable-tourism-for-development. Accessed 1 Mar 2020.

United Nations World Tourism Organisation. (2018). *International tourism results: The highest in seven years*. Retrieved from: http://media.unwto.org/press-release/2018-01-15/2017-international-tourism-results-highest-seven-years. Accessed 1 Mar 2020.

Veenhoven, R. (1999). Quality-of-life in individualistic society. *Social Indicators Research, 48*(2), 159–188.

Wang, X. (2010). The development research of ecotourism. In *2010 international conference on management and service science* (pp. 1–4). IEEE.

Wood, E. M. (2002). *Ecotourism: Principles, practices & policies for sustainability*. UNEP Publication.

World Travel and Tourism Council (WTTC). (2017). *Travel & tourism global economic impact & issues 2017*. Retrieved from: https://www.wttc.org/-/media/files/reports/economic-impact-research/2017-documents/global-economic-impact-and-issues-2017.pdf. Accessed 5 Mar 2020.

World Travel and Tourism Council (WTTC). (2019). *Travel & tourism global economic impact & issues 2019*. Retrieved from: https://www.wttc.org/-/media/files/reports/economic-impact-research/regions-2019/world2019.pdf. Accessed 14 Feb 2020.

Yeoman, I., Munro, C., & McMahon-Beattie, U. (2006). Tomorrow's: World, consumer and tourist. *Journal of Vacation Marketing, 12*(2), 174–190.

Ziffer, K. (1989). *Ecotourism: The uneasy alliance, conservation international*. Ernst and Young.

Part V
Facility Offers

Chapter 18
Tourism Industry in Bangladesh: An Assessment of Advanced SWOT Model and TOWS Matrix

Mallika Roy

Abstract Bangladesh's historical, cultural, and natural possibilities are diverse. The topic of this paper is the marketing strategies prepared with the aim of creating and developing tourism in Bangladesh. The research conducts SWOT analysis and TOWS matrix on tourism development strategies and communication of Bangladesh as a tourist destination. With the progression of time, the tourism industry has moved toward becoming a valuable piece of our typical life. Bangladesh is blessed with excellent scenic beauty the rich culture and is known for its various tourist spots, which help the economy to develop. Because of certain threats and challenges, Bangladesh has failed to utilize its strengths and opportunities and to present itself as a tourist destination country. The purpose of this chapter is to overview the concept, impacts, types, and tourism industries of Bangladesh generally. The major concern of the writer is to highlight the current situation of tourism according to the national budget, income-expenditure, and contribution. This study also highlights strengths, weaknesses, opportunities, threats by using SWOT analysis. The main purpose of this paper is to analyze the strengths, weaknesses, opportunities, and threats of tourism in Bangladesh. The primary data was collected from 360 respondents, and the secondary data was taken from various national and international reports, journals, books, magazines, and newspapers. A few investigations have been done on the qualitative analysis of Tourism in Bangladesh. But, very few studies have been undertaken to analyze tourism in Bangladesh rigorously and to assess both positive and negative economic, social, cultural, and environmental impacts. Additionally, no single initiative was undertaken to find out the optimal solution by using SO, WO, ST and WT strategies. This study suggests SO strategy is best for the tourism industry in Bangladesh.

M. Roy (✉)
Department of Economics and Finance, City University of Hong Kong, Kowloon Tong, Hong Kong

Department of Economics, University of Chittagong, Chattogram, Bangladesh
e-mail: mallikroy2-c@my.cityu.edu.hk

A. Hassan (ed.), *Tourism in Bangladesh: Investment and Development Perspectives*, https://doi.org/10.1007/978-981-16-1858-1_18

Keywords Likert method · SWOT · TOWS matrix · Tourists' satisfaction · Bangladesh

Introduction

Over time, tourism has evolved into the most vibrant service industry trend and a multibillion-dollar industry. Bangladesh has a subtropical monsoon climate, which includes rainy winds, relatively warm temperatures, and high humidity. Bangladesh has some globally significant natural resources, including the world's largest mangrove forest, the world's longest unbroken beach, a unique ecosystem set in the world's largest alluvial delta with flora and fauna, and genuinely hospitable and pleasant people. Bangladesh has a wide range of natural variations (e.g. hilly areas, sea beaches, mangrove forest, and many historical places, which has a huge potential in the scenario of world tourism). Because of the tourism industry's interdependence, companies from various sectors (such as transportation, hotels, and tour operators) are exceedingly reliant on one another (Evans et al., 2003). In case one fails to deliver a service, the other suffers as a result. Tourism has long been recognized as a source of economic development and growth. Its role and significance in creating jobs and contributing to national income have been widely acknowledged around the world (Vethirajan & Nagavalli, 2014). Furthermore, tourism contributes significantly to a country's foreign exchange reserves (Fayissa et al., 2007; Iqbal & Sami, 2016; Mishra et al., 2011; Munshi & Mishra, 2016; Mishra & Rout, 2012; Rout et al., 2016).

Hossain (1999) found tourism to be a full-fledged industry, much like the rest. The role of marketing in the tourism industry, the current situation of the tourism industry in Bangladesh, and the value of the tourist industry in Bangladesh's economy are the three areas of this research. Hossain (1999) has also described the industry's major issues, which include a shortage of sufficient and exclusive tourist goods, inadequate access to destinations, and insufficient and efficient marketing promotion. Bangladesh receives the fewest foreign tourists in relation to its neighbors, and the scale of both the international and domestic tourism markets is negligible. Furthermore, the country is not well-known among international tourists as a favorite vacation spot. Bangladesh has been attracting a large number of domestic visitors since the nineteenth century. A significant number of international tourists visit Bangladesh almost every year, in addition to domestic tourists. The total number of international visitors in 1997 was 182,420, and this number is expected to rise in the coming year. In 2003, there were 244,509 visitors, a 2.92 percent increase over the previous year. In the years 2004 and 2005, approximately 12.75 percent of tourists visited Bangladesh for sight-seeing purposes. In 2006, there were 200,311 visitors. In 2006, the number of tourists arriving in Bangladesh decreased by 9.41 percent of the total number of visitors. Table 18.1 shows the number of tourists who visited Bangladesh from 1997 to 2006. It has been observed that the number of tourists visiting Bangladesh varies from year to year. It has been

Table 18.1 The SWOT analysis tool

Internal	Strengths (S)	Weakness (W)
External	Opportunities (O)	Threats (T)
	Positive	Negative

Source: SmartDraw (2020)

raised until the year 2004. However, after that time, the percentage of tourists arriving in the country has decreased once more (Ministry of Civil Aviation and Tourism, 2011). Bangladesh's tourism industry faces numerous challenges in its long-term growth. Infrastructure services are insufficient for this country's industry. This research proposes some strategies for resolving Bangladesh's fundamental tourism issue.

In our nations, tourism is a booming business. Income generation, jobs, and foreign exchange earnings are three high-need objectives for any nation. Ecotourism is currently a major issue for our country's growth. This classification accounts for a substantial portion of GDP and visitor differences. Bangladesh's tourism industry has enormous potential. If the market is done properly, this industry will be able to survive for a long time. For continuing the growth and sustainability of this industry, we must focus on its goals and outcomes, as well as its components and extensions, standards, and business objectives. Finally, we must consider the industry's capability by identifying the problems and addressing a few recommendations.

The chapter's aim is to help readers understand the tourism industry's potential and challenges in Bangladesh. However, particular goals include: first, exploring respondents' opinions on how to draw tourists; second, assessing the tourism industry's strengths, weaknesses, opportunities, and threats; and finally, proposing the best solution to solve the problems using the TOWS Matrix.

Literature Review

Natural Setting

India is a tourist paradise because of its natural surroundings, architectural masterpieces, music, dance, paintings, traditions, and languages (Bhatia, 2013). Romeril (1989) identified a number of potential environmental consequences of the tourism industry. Despite visitors' free time and thirst for growth, the scarcity of the natural resources we use for tourism reminds us that we must exploit them in a sustainable manner. Lim and McAleer (2005) examined management techniques and issues in the field of environmentally friendly tourism in depth. In order to build sustainable tourism, Eagles, McCool, and Haynes (2002) thought that local circumstances and communities should drive the planning and management of sustainable tourism in protected areas. Another critical aspect of resource and visitor management is to remember that the natural heritage area must be preserved for potential visitors' enjoyment. The sustainable management of protected areas will help to conserve natural resources and prevent the threat of natural beauty destruction. Mihalič (2001) emphasizes the need of a tourism-natural-environment collaboration, as well as the coordination of tourism's impact on the environment and the importance of a suitable physical environment for tourists. Despite the fact that they are equally important, the latter is frequently disregarded. According to Xu, Wan, and Yang (2007), different types of wetlands should adopt distinct ecotourism growth models, which should be derived from a study of the historical and cultural backdrops, the ecosystem's integrity, competitive market potential, and the regional economy.

Quality of Services

For a variety of reasons, the efficiency of a hotel can vary. Various studies have shown contradictory findings. Among the surrounding environmental variables, proximity to the ocean has the largest impact on performance, whereas historical structures and monuments have little impact (Shahroudi & Dery, 2011). According to study in Crete, nationally branded hotels are the most effective, followed by local brand and independent hotels, and internationally branded hotels are the least effective (Manasakis et al., 2013).

Co-ordination of Factors

Suppliers, manufacturers, rivals, public bodies, states, and other companies that perform complementary activities typically have close relationships with tourism organizations. Some of these ties are minor to the participants, while others are crucial. Collaboration with other organizations, for example, is increasingly acknowledged as a crucial aspect in organizational success and sustainability (Child & Faulkner, 1998). The multifaceted nature and variety of tourism packages (such as transportation, lodging, and sightseeing) necessitate inter-firm collaboration (Morrison et al., 2004); and (2) cooperating with competitors, suppliers, customers, and supply chain companies is critical for maintaining and developing sustainable destinations (Morrison et al., 2004). (Fadeeva, 2005a, 2005b; Ling et al., 2011; Kernel, 2005). For a long-term cooperating partner, a tour operator, Ling Guo and Liang (2009) study the optimum hotel price technique. Overall, there has been a lot of research on the coordination of many variables in the value chain, but this topic and its efficiency have received a lot of attention. In India, Sanjeev (2007) investigated the association between hotel efficiency and size but found no clear evidence. Assaf and Kneevi (2010) investigated the influence of the hotel's business year, star rating, and scale, and discovered that they are all interconnected. In Turkey, 4-star hotels are more efficient than 5-star hotels, according to Davutyan (2007). Large hotels and chain hotels produce more than tiny, independently owned hotels (Assaf et al., 2012). Before making decisions or introducing new processes and technology in hotels to increase performance, managers should refer to accounting data for advice.

Value Chain

Many scholars have examined the tourism industry holistically from a distribution channel viewpoint (Middleton & Clark, 2001; Halloway, 1998; Mill & Morrison, 2002; Laws, 1997). Buhalis and Laws (2001) focus on marketing, with little emphasis on value chain performance management and evaluation in the tourism industry. Porter's value chain idea was applied to the tourism industry by Poon (1993).

Innovation

Empirical research and official statistics show a relatively low degree of innovation in the tourism industry, and deliver a blurred picture of the possible explanations for that trend (Pivcevic & Petric 2011; Hjalager, 2002; Camison & Monfort-Mir, 2012). Literature on tourism innovation management replicate a patchwork of expertise on the topic, with many undeveloped areas that have yet to be thoroughly investigated empirically (Hjalager, 2010). Barriers to tourism innovation are one of those topics that is often discussed on a conceptual level but rarely investigated empirically (Hjalager, 2010).

Managerial Skills

Katz (1974) defines managerial skills the ability of a manager for transforming information and experience into action. Organizations today face difficult and unpredictable situations. As a consequence, more than ever before, a manager's ability to adapt their management style to modern business environments determines whether they succeed or fail (Bulog et al., 2016). While managerial skills are not firm or industry specific, they can appear as the outcome of the unique combination of managerial skills that each company possesses (Carmeli & Tishler, 2006). Only skilled managers have the potential to become successful managers, and effective managers are critical for organizational performance. Successful and efficient management necessitate managerial abilities (Mostafa et al., 2012). As a result, managerial ability is a critical factor in any sector. It is not an exception in the tourism industry.

Technology and Tourism

Information technology has played a big impact in the hotel and tourism industries over the last decade. Technology has contributed in cost reduction, operational performance improvement, and facility and customer experience improvements. The way we travel has changed as a result of technological advancements, and these latest technologies offer an even more engaging and exciting experience. According to a Google Travel Study, 74 percent of travelers plan their journeys online, with only 13 percent still using travel agents to organize their journeys (WAM, 2019).

Labor Productivity in Tourism

Long-term economic development requires a high level of labor productivity. In the tourist business, researchers have utilized data envelopment analysis (DEA) to analyze statistics and develop functional models that identify the most representative predictors of labor productivity (Barros & Alves, 2004; Hu & Cai, 2004; Sigala et al., 2005; Sigala, 2004). Labor productivity is needed in the tourism industry, as it is in other industries, though to a lesser extent. Labor may be in short supply, but if its productivity is high, that is, if the laborers are skilled, it will contribute to tourism growth and development.

Methodology

The research is focused on both primary and secondary data. During the time from November 30, 2019 to January 4, 2020, primary data was collected through a questionnaire filled out by a random sample of locals, domestic and foreign visitors, and people working in the tourism industry in the country. The data and information, on the other hand, were primarily gathered and analyzed from numerous research monographs, published articles and studies, Ministry of Civil Aviation and Tourism, and Bangladesh Parjatan Corporation (BPC), and various websites.

Sample Design

Information were gathered from the research areas through questionnaire surveys, in-depth interviews, and focus groups. Respondents were chosen at random from study areas using a basic random sampling process when it came to selecting individuals for the survey. The non-probabilistic sampling approach (purposive sampling) was applied to determine sample size based on respondents' categories and sampling strategy. Following extensive study, the authors identified three types of respondents as the most likely for conducting interviews: locals, visitors, and hotel and shop owners. Among the 360 total respondents, 120 were locals, 120 were visitors (both local and foreign), and 120 interviewees worked in the tourism industry as managers (Quota sampling). Again, the researcher used judgmental sampling to distribute the sample across the three study areas. An observational survey of respondents' presence in the three areas of Cox's Bazar beach, Foy's Lake, and Bandarban was conducted to focus on this. Secondary information was gathered from a variety of national and foreign studies, journals, books, magazines, and newspapers.

Survey Instrument

The questionnaire survey approach is used to gather primary data from respondents by distributing a 67-item questionnaire developed by the researchers, which includes questions about demographic characteristics, travel characteristics of respondents, and Strengths, Opportunities, Weaknesses, and Risks, as well as some questionnaires about Tourist Satisfaction. Questionnaires on satisfaction use two forms of 5-point Likert scales: (5=Strongly Agree...1=Strongly Disagree) and (5=Excellent...1=Very Bad).

Mode of Data Analysis

To achieve the goals, a variety of approaches were used. The following are the different types of data analysis:

Qualitative Analysis

Qualitative analysis is the study of qualitative data, such as text data from interview transcripts. Unlike quantitative analysis, which relies on statistics and is completely independent of the researcher, qualitative analysis relies heavily on the researcher's analytic and integrative talents, as well as personal knowledge of the social environment in which the data is obtained. Rather than predicting or describing, qualitative research focuses on "sense-making" or interpreting a phenomenon. For qualitative research, an innovative and investigative mentality is needed, as well as an ethically aware and participant-in-context attitude and a collection of analytic strategies (Lumen Learning, 2020).

Likert Method Analysis with Various Graphs

A psychometric scale known as the Likert scale is frequently used in surveys to pose questions. It's critical for determining a respondent's attitude or opinion on a certain topic. In general, a set of assertions is employed, each of which is intended to examine a concept from a somewhat different perspective. Typically, a five-, seven-, or nine-point scale is utilized (Questionpro, 2020).

Descriptive Statistics of the Factors

The core characteristics of a study's results are described using descriptive statistics. They provide quick summaries of the sample as well as the metrics. Descriptive statistics are used to organize and present numerical data in a logical and intelligible way. In a research report, there could be a lot of measurements. Alternatively, any technique can be used to assess a large number of people. Descriptive statistics help us make sense of massive amounts of data. Each descriptive statistic condenses a lot of information into a small amount of text (Trochim, 2020).

SWOT Analysis

A SWOT analysis is a high-level strategic planning approach that helps firms assess where they are doing well and where they may improve from both an internal and external perspective. The abbreviation for "Strengths, Weaknesses, Opportunities, and Threats" is "Strengths, Weaknesses, Opportunities, and Threats".

The steps of SWOT analysis are:

- Decide what your goal is. Choose a major project or plan to examine and place it at the top of the list.
- Decide what your goal is. Choose a major project or plan to examine and place it at the top of the list.
- Each box should be labelled. Fill in the words "Strengths," "Weaknesses," "Opportunities," and "Threats" in the top left, top right, bottom left, and bottom right boxes, respectively. Because these are names, they can be distinguished from the rest of the text by using a different font size or a different color. To help you get started quickly, SmartDraw includes a selection of SWOT diagram models.
- Make a list of your advantages and disadvantages. Fill in the project-related variables in the appropriate boxes. Qualitative and anecdotal information, as well as quantitative and analytical data, can be included in a SWOT analysis. The majority of the time, factors are listed in bullet form.
- Make a choice. Examine the SWOT diagram that has been completed. Keep track of whether the positive outcomes outnumber the unfavorable outcomes. If

they do, it's a good idea to go forward and achieve your goal. If they do not, changes will need to be made, or the plan may need to be scrapped entirely (SmartDraw, 2020).

TOWS Matrix by Using S-O, W-O, S-T, W-T Strategies

A TOWS analysis is a version of the SWOT analysis in which the Strengths, Weaknesses, Opportunities, and Threats are described, but the Threats and Weaknesses are then balanced. It is an excellent follow-up to SWOT since it allows you to act on your findings.

Because it considers the interaction between internal and external variables, TOWS is a far more useful matrix than a solitary SWOT. The main goal of a TOWS Study, according to (Lucidity, 2020), is to:

- Reduce threats
- Take advantage of opportunities
- Capitalize on strengths; and
- Eliminate flaws,

A company should use a TOWS study to balance its internal strengths with external opportunities (SO) to build "maxi-maxi" strategies – those with the best chance of succeeding. Strengths like high brand awareness or customer loyalty, for example, may be paired with the chance to launch a new product or service (Table 18.2).

In its most severe, it emphasizes the organization's susceptibility to threats based on its flaws and inspires the formation of strategies for mitigating and preventing threats (WT) – so-called "mini-mini" strategies. Developing strategic alliances, for example, or a more dramatic approach, withdrawing from a single market entirely, are examples of such strategies. The approaches of mini-maxi (WO) and maxi-mini (ST) are aimed to improve vulnerabilities by using opportunities and minimize risks by exploiting strengths. As an example mini-maxi strategy, a corporation may have discovered an opportunity to outsource specific portions of its business activities, overcoming the vulnerability of a lack of specialized competence within the organization (WO). It is vital to remember that a TOWS review won't tell you which strategy to take; instead, it will assist you focus on the areas where action is required and offer you an idea of what that action will be (Oxford College of Marketing, 2016).

Table 18.2 TOWS Matrix with four strategies

	Strength (S)	Weakness (W)
Opportunities (O)	S-O (maxi-maxi)	W-O (mini-maxi)
Threats (T)	S-T (maxi-mini)	W-T (mini-mini)

Source: Lucidity (2020)

Table 18.3 Identifications of variables

Variables	Internal/External
Natural settings	Internal
Quality services	External
Accommodation and food	Internal
Transportation	Internal
Culture and heritage	Internal
Festivals and events	Internal
Visa and immigration flexibility	Internal
Innovative tourism products	External
Resources	Internal

Source: Developed by the authors (2020)

Variables that Affect Tourist Satisfaction have been Identified

The following Table 18.3 shows the variables that influence tourist satisfaction with tourism products and services in Bangladesh:

Tourism and Bangladesh

Major Tourist Destinations in Bangladesh

Bangladesh is rich in tourist destinations, and different types of destinations are available for different types of tourists. Category-wise sites names are given below as per Bangladesh Parjatan Corporation (2020):

- *Archaeological sites*

  Mughal Eidgah, Ahsan Manjil, Sonargoan, Wari Bateshawar, Mainamati, Paharpur, Mahasthangar, Kantajew Temple, and the Sixty Dome Mosque etc.

- *Beaches*

  Patenga, Parki, Cox's Bazar, Kuakata, Teknaf, Kotka etc.

- *Religious places*

  Mosque, Hindu Temple, Churches, Buddhist Temples, etc.

- *Hills and islands*

  Rangamati as a lake district, Bandarban as Bangladesh's roof, Kaptai as a lake town, Khagrachhari as a hilltop town, Mymenshingh, Sylhet, Sonadia Island, Moheshkhali Island, and St. Martin's Island are some of Bangladesh's most well-known tourism attractions.

- *Historical places*

National Memorial, Martyred Intellectual Minar, Central Shahid Minar, National Poet's Grave, Mujibnagar Memorial, Baldha Garden, Bahadur Shah Park, Curzon Hall, Sohrawardi Park, Old High Court Building, Dighapatiya Palace, Shilaidaha Kuthibari, World War II Cemetery, Sagordari-Jessor, Trishal, Gandhi Asram are all well-known tourist attractions in Bangladesh.

- *Forest and swamp forest*

The Sundarbans, several eco-parks, the Ratargul Swamp Forest etc. are well known.

- *Others attractions*

Some of the tourist sites are Bangabhaban, Shankharibazar, Sadarghat, Ramna Park, National Botanical Garden, National Zoological Garden, National Park, Jamuna Bridge, DC Hill, Batali Hill, Rajshahi, Madhabkundu, National Assembly Building, and Jaflong.

UNESCO Recognized World Heritage Site of Bangladesh

There are three world heritage sites in Bangladesh that have been recognized by UNESCO as: first, Historic Mosque City of Bagerhat, Category: Cultural, Recognition year: 1985; second, Ruins of the Buddhist Vihara at Paharpur, Category: Cultural, Recognition year: 1985; and third, The Sundarbans, Category: Natural, Recognition year: 1997. Also, sites on the Tentative List are Mahansthangarh and its Environs – 1999; the Lalmai-Mainamati Group of monuments – 1999; Lalbagh Fort – 1999; Halud Vihara – 1999; and Jaggadala Vihara – 1999 (UNESCO, 2020).

- *Beaches and islands*

Bangladesh has a sea border that faces the Bay of Bengal. THz country has the world's longest sea beach that offer a view of both sunrise and sunset. This is Cox's Bazar that locates in the Cox's Bazar district. It is called the tourist capital of Bangladesh. Kuakata is the other famous sea beach. Kuakata sea beach is located in Patuakhali district of Bangladesh. The beautiful beach of sea is located in Kuakata, where only sunroof and sunset only can be seen in South Asia. Inani beach is located in Ukhia Upazilla in Cox's Bazar district, 35 km south of Cox's Bazar and the mountains on the east side of Inani Sea beach. Both Laboni and Inani sea beaches as the blue water beaches are extremely attractive for bathing. Patenga beach is located in Chittagong of Bangladesh. It is a sandy sandstone located in the mouth of the Karnafully River. Patenga Sea Beach is located on the side of the Bangladesh Naval Academy. Parki sea beach is a coastal sea beach in Bangladesh. At one time, as the sea beach in Bangladesh is considered to be only Cox's Bazar sea beach and Patenga Beach, but Parki sea beach is now becoming very popular to the tourists. Guliyakhali beach (Guliakhali Sea Beach) is located in Sitakunda Upazilla of

Chittagong district. This beach is also recognized as "Muradpur beach" for local people. The distance from Sitakunda Bazar to Guliyakhali Sea Beach is just 5 kilometers.

Bangladesh also has a unique coral sea beach, the St. Martin's Island. St. Martin's Island is located at Teknaf in Cox's Bazar district. However, it is located at Teknaf, 48 km from Teknaf and 115km from Cox's Bazar. St. Martin's is Bangladesh's only coral island. Nijhum Dwip is located in Hatia Upazilla of Noakhali district. The Island is About 14,050 acres. It has another four islands: Kamlar Char, Ballar Char, Char Usman, and Char Muri.

Distinction of Bangladeshi Tourism

According to Budgetyourtrip (2020), tourism in Bangladesh is budget-friendly and heaven for budget traveler where daily traveling price is BDT 4540 ($54), meal price is BDTK1114 ($14) per day, daily accommodation price is BDTK 5372 ($64) for typical double occupancy room and BDT 2686 ($32) for hotel or hostel for one person, daily transportation cost is BDTK 571 if the tourist travels in taxis, local buses, and subway, etc. The cost of traveling between cities is BDT 686, cost of entertainment such as entrance tickets, shows etc. is BDTK 1429, and the cost of an alcoholic drink for one day is BDTK 891.

- *Art and culture*

Bangladeshi art and culture encapsulate the country's and region's social and political transformations over time.

- *Art*

Bangladesh has a long and illustrious artistic history. Ancient pottery and terracotta specimens show implausible artistry. Artists as Zainul Abedin and Qamrul Hasan have pioneered modern painting. S. M. Sultan, Shafiuddin Ahmed, Rashid Chowdhury, Shafiqul Amin, Anwarul Haque, Qayyum Chowdhury, Abdur Razzak, Murtaza Baseer, Debdas Chakraborty, Aminul Islam, Kazi Abdul Baset, Mohammad Kibria, Syed Jahangir are a few of some other names to mention. Zainul Abedin, for example, became popular around the world in 1943 for his breathtaking sketches of the Bengal Famine.

- *Fairs and festivals*

The people of Bangladesh get a lot of pleasure, excitement, and color from the fairs and festivals. Although the majority of the festivities are based on religious traditions, the fairs have their roots in the hearts of the participants, regardless of faith, status, or ideology. Bashanta Utsab, or the first day of spring, is observed throughout the country on the first day of Falgun, the 11th month of the Bangla calendar, which corresponds to mid-February in the Gregorian calendar. People dress in bright colors for this day, and colorful fairs and cultural programs are held. Women usually dress in yellow "Sharis" (women's clothing) and accessorize with floral ornaments. On this day, greetings and gifts are commonly exchanged. The Bengali new year celebration day, known as Pahela Boishakh. Baishakhi Mela (Bengali New Year Fair), is facilitated with a cheering arrangement. The fairs accompany the mix of delight and art together. The Eid-ul-Fitr, Eid-ul-Azha, Durga Puja, Buddha Purnima, Christmas are celebrated widely throughout the whole country. Different special and delicious foods are prepared in that time with joy and delight. Although the people are from different religions, they celebrate the religious festivals altogether and share happiness with each other. On the 25th Baishakh (May), the noble laureate Rabindranath Tagore's birth and death anniversary, as well as that of the National Poet Kazi Nazrul Islam on the 11th Jaystha (May), are commemorated across the world. March 26 is the Independence Day, which is the biggest state festival. Countless flocks of parades can be seen celebrating the independence of the country and obviously remembering the martyrs. To celebrate this day, different popular Bangladeshi bands hold concerts in different prestigious venues in the country. Furthermore, the 21st of February is commemorated for paying tribute to the sacred souls of the martyrs of the 1952 Language Movement. It is a national mourning day and also recognized as International Mother Language Day all over the world.

- *Museum*

Bangladesh's museum houses a wide range of unique collections, including sculptures and paintings. National museum, Folk art museum, Ethnological museum, archeological museum, Varendra museum, Tribal museum, etc. are the renowned museums of Bangladesh, which focus on cultural and historical diversities of the country.

- *Bengali cuisine*

Curries are one of the most common types of Bangladeshi cuisine, as they are in many other Asian countries. Before eating, many popular ingredients in Bangladesh are incorporated into a curry in some way.

- *Amenities*

Tourist satisfaction necessitates the addition of value-added services. Swimming, boating, yachting, and surf-riding are all necessary activities for a seaside resort, as are other activities like dancing, recreation, and other amusements. Natural amenities (trekking, sandy beaches, fishing spots, sea-bathing, climbing opportunities, watching, and so on) and man-made facilities for entertainment (theatre, parasailing, paragliding, and so on) cater to the specific needs of visitors.

Results and Analysis

Based on the survey questionnaire, SWOT has been pointed out for tourism in Bangladesh. The major SWOTs mentioned by the respondents are given below:

Accommodation Facilities

The aim of this study was to learn about the respondents' impressions of the accommodation facilities to better understand the role of accommodation in tourism.

Table 18.4 and Fig. 18.1 is showing that more than half of the respondents' perception of accommodation is good. 14.44% of respondents perceive excellent accommodation facilities. 15.28% of respondents perceive quite bad accommodation facilities. Though the perception is good enough, more improvement is necessary. To improve in this industry, more developmental strategy is needed.

Food Facilities

Food facilities are very important as people stay outside of their residents most of the time. Especially tourists have very little time for cooking. Furthermore, they want to taste different foods while traveling. Also, they do not want to waste time on

Table 18.4 Opinions regarding accommodation

	Accommodation	%
Excellent	52	14.44
Good	198	55.00
Moderate	50	13.89
Not good	55	15.28
Very bad	5	1.39
Total	360	100

Source: Authors' calculation based on field study (2020)

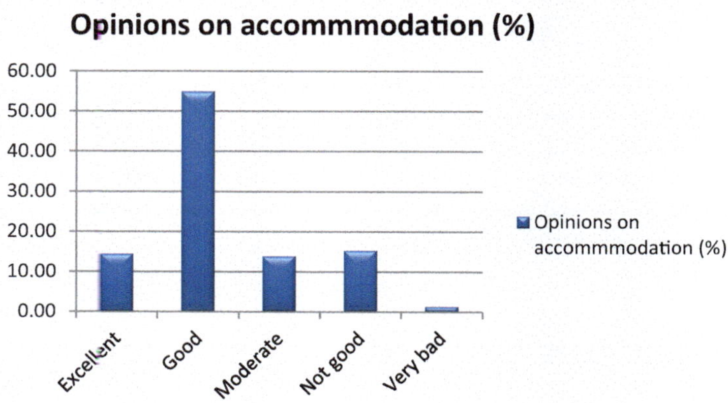

Fig. 18.1 The distribution of the opinions of respondents regarding accommodation facilities. (Source: Authors plot based on field survey, 2020)

Table 18.5 Opinions regarding food facilities

	Food	%
Excellent	65	18.06
Good	201	55.83
Moderate	56	15.56
Not good	29	8.06
Very bad	9	2.50
Total	360	100

Source: Authors' calculation based on field study (2020)

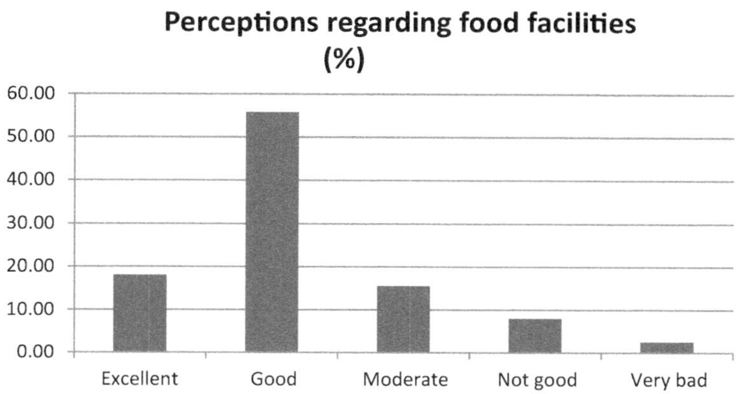

Fig. 18.2 The distribution of the opinions of respondents regarding food facilities. (Source: Authors plot based on field survey, 2020)

cooking when they come to a very new place. Restaurant facilities are very important for visitors (Rao, 1986).

Table 18.5 and Fig. 18.2 above show, only 8.06% of respondents opine that the food facilities in Bangladesh are not good, whereas 55.83% of respondents think that the food facilities in Bangladesh are good, and 18.06% opine that the food facilities are excellent.

Transport Facilities

Transportation is an essential element of the tourism industry. Good transportation is also a basic condition for the development of tourism (Hossain, 2001). Tourism destinations need to be well linked by road, air, and water (Young, 1992). Thus, a good transportation network can push forward a tourism industry (Theuns, 2002).

Table 18.6 and Fig. 18.3 above are showing that 17.22% and 68.61% of respondents believe that transport facilities are excellent and good, respectively. Few

Table 18.6 Opinions regarding transport facilities

	Transport	%
Excellent	62	17.22
Good	247	68.61
Moderate	45	12.50
Not good	3	0.83
Very bad	3	0.83
Total	360	100

Source: Authors' calculation based on field study (2020)

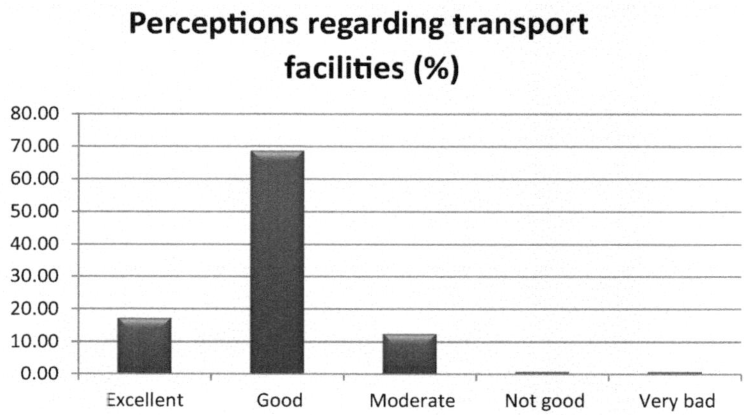

Fig. 18.3 The distribution of the opinions of respondents regarding transport facilities. (Source: Authors plot based on field survey, 2020)

people have negative perceptions. Though this result is good enough, Bangladesh needs to focus on the transport facilities in rural areas where many tourists' destinations exist.

Technological Support

In this competitive world, technology is a major concern now-a-days. Generally, tourists want to get good technological support as the world is converting to a digital world.

Table 18.7 and Fig. 18.4 above are showing that 18.61% and 16.11% of respondents believe that transport facilities are excellent and good, respectively. 57.78% believe that utilization of technology is at a moderate level that means not so good and not bad at all. Few people have negative perceptions. Though this result is good enough, Bangladesh needs to focus more on the technological facilities in remote areas where many tourists' destinations exist.

Table 18.7 Opinions regarding technological support

	Technology	%
Excellent	67	18.61
Good	58	16.11
Moderate	208	57.78
Not good	16	4.44
Very bad	11	3.06
Total	360	100

Source: Authors' calculation based on field study (2020)

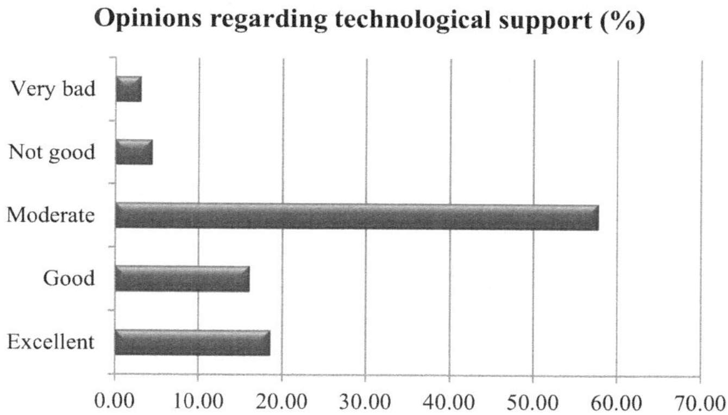

Fig. 18.4 The distribution of the opinions of respondents regarding technological support. (Source: Authors plot based on field survey, 2020)

Management

This research looked into tourist attraction management of in Bangladesh. The analysis gathered the views of visitors. The Table 18.8 depicts international tourists' views on the management of tourist attractions in Bangladesh (Fig. 18.5).

Just 15% of respondents thought Bangladesh's management system was fine; 0.56 percent thought it was really bad; and only 15.56 percent and 9.72 percent thought it was good and excellent, respectively. The management level is moderate, according to 59 percent of respondents. Overall management should be reformed for the sake of tourism development. The proper management of tourist attractions frequently contributes to the progression of tourism. Mishandling of many tourist destinations in Nepal is costing the country a lot of money. Bangladesh is also in a similar situation. As a result, this country should develop a sound management strategy.

Table 18.8 Management-related opinions

	Management	%
Excellent	35	9.72
Good	56	15.56
Moderate	213	59.17
Not good	54	15.00
Very bad	2	0.56
Total	360	100

Source: Authors plot based on field survey (2020)

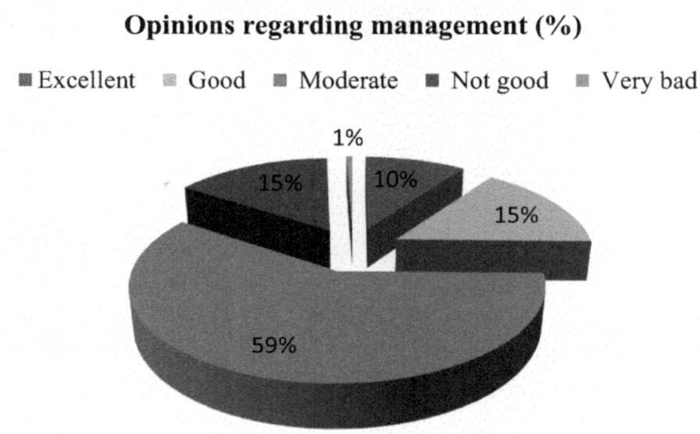

Fig. 18.5 The distribution of respondents' views on management is reflected in this graph. (Source: Authors plot based on field survey, 2020)

Table 18.9 Opinions on the collection of visas

	Visa collection	%
Excellent	56	15.56
Good	73	20.28
Moderate	213	59.17
Not good	15	4.17
Very bad	3	0.83
Total	360	100

Source: Authors plot based on field survey (2020)

Visa Collection

The Bangladesh visa collection scheme is the subject of a thorough investigation. The aim of this study was to gather tourist feedback on the visa application process. The following are some of the viewpoints:

Table 18.9 and Fig. 18.6 indicate, approximately 36% of respondents believe Bangladeshi visa collection is simple; 59.17 percent believe it is moderate; 4.17 percent believe it is complicated; and only 0.83 percent believe it is difficult. We were pleased to learn that information about the Bangladeshi visa collection system is very positive for attracting tourists to Bangladesh.

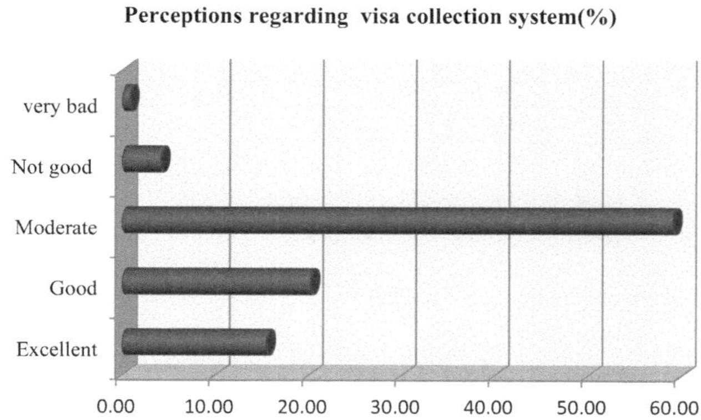

Fig. 18.6 The distribution of respondents' views on the visa collection scheme is seen. (Source: Authors plot based on field survey, 2020)

Immigration Formalities

This research looked into the immigration procedures in Bangladeshi airports. In this regard, the Table 18.10 and Pie Diagram present the responses of the respondents.

Table 18.10 and Fig. 18.7 show the opinion distribution of respondents about immigration formalities. 66% think that immigration formalities are good enough, meaning easy and flexible, whereas only 5% think that the formalities are not good and 12% think that the formalities are very bad.

Strengths

Geographical Location

Bangladesh enjoys a favorable geographic position. In most areas, tourist attractions are similar to one another. As a result, both the mountain range and the sea beaches can be visited in a single day. The presence of attractive tourist attractions such as St. Martin's Island, Cox's Bazar, Chittagong Hill tracts, and the Sundarbans, among others, contributes to the locations' high ranking.

Cultural Diversity

Bangladesh is a one-of-a-kind multicultural country. Along with the 98 percent Bengali community, at least 45 small ethnic groups exist, each with their own culture, language, food habits, clothing, music, and so on. The Peoples' Republic of Bangladesh's Constitution states that the state will protect all citizens in its territory's right to practice and cultivate all cultural patterns, regardless of caste, faith, origin, color, gender, or other factors (UNESCO, 2013).

Table 18.10 Opinions on the formalities of immigration

	Immigration formalities	%
Excellent	78	21.67
Good	159	44.17
Moderate	61	16.94
Not good	18	5.00
Very bad	44	12.22
Total	360	100

Source: Authors plot based on field survey (2020)

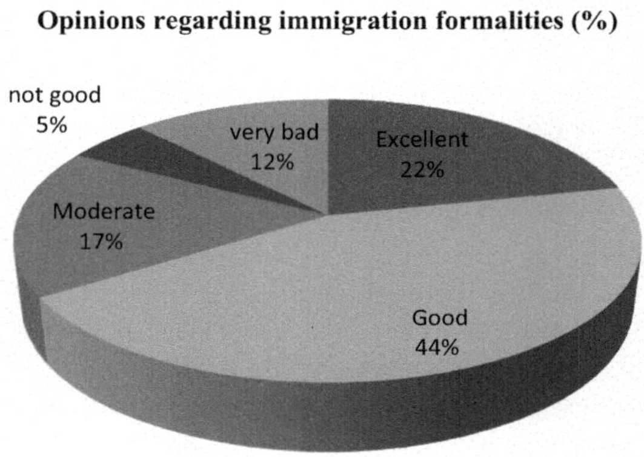

Fig. 18.7 The distribution of the opinions of respondents regarding immigration formalities. (Source: Authors plot based on field survey, 2020)

Tourism Infrastructures that have been Extensively Built

Bangladesh's tourism infrastructure is well-developed. Vision 2020 concentrates on the growth, conservation, and maintenance of tourist attractions in order to keep this facility sustainable. Paharpur, Mahasthangarh, and other cultural and archaeological sites must be properly preserved and conserved. The development of St. Martin's Island, Cox's Bazar, and other special tourist areas, along with adequate entertainment and infrastructure facilities, is highly recommended, as is ensuring safety and security. Ecotourism attractions such as the Chittagong Hill Tracts, Tanguar Haor, Sundarbans, and others require proper monitoring and security against illegal trespassers, which must be coordinated with the Department of Forest.

Strong Accommodation and Tour Operators are Available

Strong lodging facilities and tour operators are effective tools for advancing society. They will contribute to tourism growth by encouraging people to visit tourist attractions. In view of the environmentally friendly situation, tour operators and guides are qualified to operate ecotourism. They could ensure successful tourism and tourism education if they were properly educated.

Table 18.11 Opinions on the strengths of tourism in Bangladesh

Respondents	Agree (No. of people)	% of agreed people	Disagree (No. of people)	% of disagreed people
Local people	119	33.05556%	1	0.277778
Domestic tourists	110	30.55556%	0	0
Foreign tourists	10	2.777778%	0	0
Officials	119	33.05556%	1	0.277778
Total	358		2	

Source: Author's calculation based on field study (2020)

A Diverse Variety of Large and Small Festivals and Activities are held During the Year
Bangladesh is known for its vibrant festivities. Religious festivals, national gatherings, cultural festivals, and tribal festivals are the four main types of festivals in Bangladesh. Bangladeshis celebrate their religion, life, liberty, nature, joy, and accomplishments all year long through a variety of fairs and festivals, which are organized with unwavering enthusiasm and meticulous attention to detail. Some Bengali fairs and festivals have a history dating back over two thousand years.

Specialty Pieces, Particularly in the Areas of Heritage and Culture
Bangladesh is a country with a diverse community. The "Convention Concerning the Protection of World Cultural and Natural Heritage" has been ratified by 186 countries around the world, including Bangladesh. Bangladesh has three sites that have been designated as world heritage sites as of 2009. The Paharpur Buddhist Monastery of Noagaon, the Historic Mosque City of Bagerhat, and the Sundarbans are among these places. On the other hand, Bangladesh currently has four Intangible Cultural Heritages on the UNESCO list: Baul song in 2008, traditional Jamdani weaving in 2013, Mongol Shobhajatra in 2016, and Sylhet's traditional Shitol Pati weaving in 2017. Rickshaws and rickshaw painting in Dhaka are currently being considered for inclusion in the Representative List of Humanity's Intangible Cultural Heritage (Daily Star, 2015) (Table 18.11).

According to field survey, 33.06% local and 20.56% domestic tourists agreed with the factors of strengths, whereas only 0.28% local and 0% domestic tourists disagreed with the factors of strengths. No foreign tourists disagreed with the factors identified by the author.

Weakness

Promotional Events are Lacking
Due to policy directives, budget limitations, and a lack of marketing orientation, Bangladesh's tourism industry has been unable to conduct appropriate and successful promotional activities. Tourism goods, market analysis, and concerted

campaigns from both the public and private sectors should be the focus of tourism promotional activities. After conducting a market survey, Bangladesh Parjatan Corporation (BPC) should establish long and short-term marketing strategies aimed at specific tourists from specific countries. More promotional events, such as full-page advertising, television and radio advertisements, online promotion etc. can be promoted by the BPC.

Service Output is Inefficient as Compared to other Parts of the World
Quality is an important success factors for tourism service providers and the industry as a whole. In Bangladesh, the quality approach to tourism needs to evolve. Tourists would not visit tourist destinations due to inefficiency, despite the natural beauty of the locations. The tourism quality approach is a complex process that has developed over time as the tourism industry has grown. The tourism quality approach is a complex process that has developed over time as the tourism industry has grown. To achieve tourism development, Bangladesh must ensure high-quality services.

Inadequate Supply Chain Coordination
Value Chain Coordination. Coordination of the value chain is a popular interest amongst tourism service providers, such as car rental, airlines, leisure companies, and hotels, tour operators, and it allows tourists to have a more enjoyable, safe, and convenient vacation.

Low Creativity and a Scarcity of Reliable, High-quality Data
The tourism industry's decentralized nature, which is governed by small companies run by one individual or family, is generally correlated with the relatively modest innovativeness of tourism businesses (Pivcevic & Petric, 2011; Hjalager, 2002). Risk aversion, reluctance to change, a lack of understanding of the value of technology, and resource constraints, all of which are prevalent in small companies, become an industry-wide challenge.

Inadequate Management Abilities
To thrive in the current world, industries, like humans in every aspect of their lives, must grow and evolve. For businesses, this necessitates the mastery of a variety of management skills. Managerial abilities are a subset of managerial abilities (Smutny et al., 2016). However, in some areas of Bangladesh, management skills are lacking. This is an area where the country wants to change.

Inflexibility for Coping with Tourism Demand Changes
The world is rapidly changing over time. Technology training is key to efficiency increases. Industries and individuals must make a commitment to technology support and training. As technology advances, it becomes easier to use. But this country's technology is inflexible compared to other countries. To achieve optimal tourism growth, all the sub-industries in the tourism industry should be changed for coping with the tourism demand changes.

Low Labor Productivity
The tourism industry, in comparison to other economic sectors (such as manufacturing), has far less resources to substitute labor with capital and to introduce

Table 18.12 Opinions on the weakness of tourism in Bangladesh

Respondents	Agree (No. of people)	% of agreed people	Disagree (No. of people)	% of disagreed people
Local people	102	28.33333	2	0.555556
Domestic tourists	110	30.55556	5	1.388889
Foreign tourists	10	2.777778	0	0
Officials	109	30.27778	2	0.555556
Total	351		9	

Source: Author's calculation based on field study (2020)

labor-saving advanced technologies. Because of cheap labor, tourism industries in most of the developing countries can be operated at low marginal capital substitution rate than those in developed countries (Smeral, 2008) (Table 18.12).

According to field survey, 28.33% of local respondents agreed with the factors of weakness in the tourism industry, whereas only 0.56% disagreed. All foreign tourists agreed with the factors of weakness identified by the author.

Opportunities

According to Pannell Kerr Forster Associate's (1988) comprehensive report on the Strategic Master Plan for Tourism in Bangladesh, the government of Bangladesh has attempted to make substantial investments in national airlines, infrastructures, and general tourism in the country, which must be accurately maintained by effective management and activities of marketing. Hossain (2001) stressed the importance of implementing BPC marketing strategies for meeting the demands of visitors, achieve the BPC's goals, and contribute to Bangladesh's national economy. He has suggested that when developing and implementing marketing approaches, BPC has to carefully consider the factors that influence prospective tourists' decisions when choosing a destination and purchasing tourism items. The tourism industry offers the following opportunities:

- Potential tourism marketing budget increase
- Cultural tourism potential
- Growing income levels
- Create unconventional venues to draw new customers in the tourism industry
- Demographic shift
- Possible demand for creative goods (Table 18.13).

According to field survey, 33.33% of local respondents agreed with the factors of opportunities in the tourism industry, whereas only 0% disagreed. All foreign tourists agreed with the factors of opportunities identified by the author.

Table 18.13 Opinions on the opportunities of tourism in Bangladesh

Respondents	Agree (No. of people)	% of agreed people	Disagree (No. of people)	% of disagreed people
Local people	120	33.33333	0	0
Domestic tourists	110	30.55556	0	0
Foreign tourists	10	2.777778	0	0
Officials	120	33.33333	0	0
Total	360		0	

Source: Author's calculation based on field study

Threats

The global travel industry is rapidly approaching some of the most significant challenges it will face. Pollution by single-use plastic is a major concern for both coastal and inland destinations as a growing problem. Such threat is partly exacerbated by the ASEAN region. Almost half of its member countries are said to remain as the top ten countries of mismanaged plastic wastage. Improper solid waste disposal and the lack of appropriate sewage treatment, according to Professor Dr Amran Hamzah from Malaysia's University of Technology, are also threatening important marine destinations (The Asian Post, 2018). Security, according to some, is the most serious threat. Environmental deterioration, according to some experts, is one of the most serious challenges to the tourism industry. However, Bangladesh faces some unique tourism-related threats, which are mentioned below:

1. Marketing and product creation are not given enough attention
2. Implications of visa and immigration issues for visitor mobility
3. The tourism industry's fragmented existence
4. Global rivalry is fierce
5. Inadequacy of educational services
6. In many businesses, low-cost business models exist
7. Problems in the regulatory system due to bureaucratic issues
8. Inadequate financial resources
9. Tourism's delicate feature
10. Global warming and climate change (Table 18.14).

This study understands the satisfaction level of the tourism industry from the respondents. Table 18.15 is showing the distribution of opinions of the respondents regarding overall satisfaction level.

Table 18.14 Opinions on the threats of tourism in Bangladesh

Respondents	Agree (No. of people)	% of agreed people	Disagree (No. of people)	% of disagreed people
Local people	120	33.33333	0	0
Domestic tourists	109	30.27778	1	0.277778
Foreign tourists	9	2.5	1	0.277778
Officials	120	33.33333	0	0
Total	358		2	

Source: Author's calculation based on field study (2020)

Table 18.15 Opinions on overall

Opinions	Frequency	Percentage (%)
Excellent	98	27.22222
Good	203	56.38889
Not good	59	16.38889
Very bad	0	0
Total	360	

Source: Author's calculation based on field study (2020)

SWOT Analysis

A SWOT table has been developed, which represents a synthesis of key strategic factors influencing Bangladeshi tourism growth. The importance of key strengths and opportunities is highlighted. The author attempted to turn the identified vulnerabilities into strengths and the identified challenges into opportunities by implementing effective policies and activities. The key goal was to assess current tourism conditions in order to determine the best approach (Table 18.16).

Scoring on Some Factors of Tourism

Scoring of various factors helps to evaluate current state of the tourism industry of Bangladesh (Table 18.17).

Descriptive Statistics on the Strengths and Opportunities of Tourism Industry in Bangladesh

This table displays the summary statistics of selected three areas respondents' perception about strengths and opportunities statement of the tourism industry (Table 18.18).

Table 18.16 SWOT table. (Source: developed by author, 2020)

PURPOSE	
To evaluate the tourism market conditions so that we can fix up the best strategy	
S	INTERNAL **STRENGTHS**
1	Presence of attractive tourist destinations , for example, CoxsBazar, Saintmartin, Hill tracts, Sundarban
2	Cultural diversity
3	Sufficient natura resources to make top notch items
4	Wide range of large and small festivals and events
5	Special items especially heritage and culture.
W	INTERNAL **WEAKNESSSES**
1	Unattractive marketing strategies of tourist destinations
2	Inefficient quality of services compared to other regions in the world
3	Low innovation and lack of regular and high quality statistics
4	Insufficient managerial skills
5	Inflexibility to cope with changes in tourism demand
O	EXTERNAL **OPPORTUNITIES**
1	Potential tourism marketing budget increase
2	Cultural tourism potential
3	Potential demand for innovative products
4	Develop rural experiences and unusual venues to attract new audiences in leisure and business markets
5	Encourage more characterful serviced accommodation and a unique range of local dining facilities.
T	EXTERNAL **THREATS**
1	Not enough focus on marketing and product development
2	The effects of visa and immigration problems on touristic mobility
3	Bureaucratic difficulties in regulatory framework
4	Insufficient access to finance
5	Existence of low cost business models

TOWS Matrix

Despite the fact that SWOT analysis is a beneficial method for recognizing strengths, vulnerabilities, opportunities, and risks, it is stagnant and does not allow for the creation of novel strategies (Koontz & Weihrich, 2010).

Table 18.17 Scoring on tourism factors

Factors of tourism development	Cox's Bazar (score)	Foy's Lake (score)	Bandarban (score)
Tourism products and services	5	4	5
Cultural diversity	4	3	5
Natural resources	5	4	5
New/ innovative products potentials	4	5	4
Festivals and events	4	4	5
Increase budget	3	4	4
Cuisines	5	4	3
Quality accommodation	5	3	3
Mean score	4.29	3.86	4.14

Source: Author's calculation (2020)

Table 18.18 Summary statistics on the scoring of various factors of tourism

Cox's Bazar		Foy's lake		Bandarban	
Mean	4.285714	Mean	3.857143	Mean	4.142857
Standard error	0.285714	Standard error	0.26082	Standard error	0.340068
Median	4	Median	4	Median	4
Mode	4	Mode	4	Mode	5
Standard deviation	0.755929	Standard deviation	0.690066	Standard deviation	0.899735
Sample variance	0.571429	Sample variance	0.47619	Sample variance	0.809524
Kurtosis	−0.35	Kurtosis	0.336	Kurtosis	−1.81661
Skewness	−0.59529	Skewness	0.173897	Skewness	−0.35304
Range	2	Range	2	Range	2
Minimum	3	Minimum	3	Minimum	3
Maximum	5	Maximum	5	Maximum	5
Sum	30	Sum	27	Sum	29
Count	7	Count	7	Count	7
Confidence level (95.0%)	0.699118	Confidence level (95.0%)	0.638204	Confidence level (95.0%)	0.832116

Source: Author's calculation (2020)

	Strength	Weakness
Opportunities	***S-O strategy*** By increasing the potential marketing budget, attractive tourism destinations can be utilized. To utilize the cultural diversities, there is a need to take necessary steps of cultural tourism potentials. Sufficient natural resources can be used for fulfilling the demand for innovative products. Rural venues can be made attractive in various festivals and events to attract more new visitors. Tourism destinations with heritage and culture need to prioritize to provide quality service accommodation and traditional dining facilities.	***W-O strategy*** The best marketing strategy should be adopted with a potential marketing budget. To compete with other countries as tourism destinations, new innovative products can be launched with quality services. Cultural tourism can be exposed with regular quality assessment of the services. Develop managerial skills to organize events in the selected venues. Based on the dynamic change in tourism demand, heritage and culture-based products can be re-organized.
Threats	***S-T strategy*** New tourism products and services can be launched in popular tourist destinations. Touristic mobility should be increased by reducing visa and immigration restrictions to present cultural diversities to foreign visitors. Bureaucratic difficulties need to eradicate to get the proper outcomes of sufficient natural resources. Investment should be raised on large and small festivals and events. Conversion from the low cost tourism model to the highest possible cost tourism model is necessary to shed light on heritage and cultural products.	***W-T strategy*** Products and services development should be the main priority. Managerial skills need to increase to lessen visa and immigration problems. Reform the regulatory framework and arrange training programs to improve skills in managerial tasks to attract tourists. Finance sufficient quantity to innovate new products and services to survive in competition. Existing tourism models can be reformed according to dynamic tourism demand.

Conclusion

The tourism industry has the potential to help Bangladesh's economy develop. Earning hug foreign currency can be possible by ensuring the proper development of this industry, which will help the country's national economy expand. With cultural exchange and interaction of people of different races and nationalities, it can also play an important role in social and cultural growth.

The results of this study highlight many dimensions of strengths, weaknesses, opportunities, and threats. Based on SWOT analysis, the results of the study reveal the best strategy (S-O strategy) for Bangladesh through TOWS Matrix. If Bangladesh can adopt this strategy, an optimal level of tourism growth is possible. However, W-T strategy is such a strategy which is generally avoidable to policy planner. The most critical dimensions of tourist satisfaction, according to the findings, are the overall standard of tourism goods and services. This empirical study offers a novel perspective on the impact of expertise on product and service satisfaction in the tourism industry, and it will be a useful pedagogical and research resource for decision-makers, academics, and students involved in hospitality and tourism.

This chapter contributes to the study of natural environment, quality of services, co-ordination of factors, creativity, technology, and productivity of labour for tourism development by emphasizing the value of different strategies through an in-depth study of the Bangladeshi tourism industry. In addition, the value of the methods is examined in this chapter.

References

Assaf, A., & Knežević, C. L. (2010). The performance of the Slovenian hotel industry: Evaluation post-privatisation. *International Journal of Tourism Research, 12*, 462–471.

Assaf, A., Barros, C. P., & Josiassen, A. (2012). Hotel efficiency: A bootstrapped metafrontier approach. *International Journal of Hospitality Management, 31*, 621–629.

Bangladesh Parjatan Corporation (2020). *Home*. Retrieved from: http://www.parjatan.gov.bd/. Accessed 20 Apr 2020.

Barros, C., & Alves, P. F. (2004). Productivity in the tourism industry. *International Advances in Economic Research, 10*(3), 215–225.

Bhatia, A. (2013). SWOT analysis of Indian tourism industry. *International Journal of Application or Innovation in Engineering & Management, 2*(12), 44–49.

Budgetyourtrip (2020). *How much does it cost to travel to Bangladesh?* Retrieved from: https://www.budgetyourtrip.com/bangladesh. Accessed 20 Apr 2020.

Buhalis, D., & Laws, E. (2001). *Tourism distribution channels: Practices, issues and transformations*. Continuum.

Bulog, I., Jukić, I., & Kružić, D. (2016). Managerial skills: Does family ownership make a difference? In *Proceedings of the 5th international OFEL conference on governance* (pp. 346–360). Management and Entrepreneurship.

Camison, C., & Monfort-Mir, V. M. (2012). Measuring innovation in tourism from the Schumpeterian and the dynamic-capabilities perspectives. *Management in Tourism, 33*, 776–789.

Carmeli, A., & Tishler, A. (2006). The relative importance of the top management teams managerial skills. *International Journal of Manpower, 27*(1), 9–36.

Child, J., & Faulkner, D. (1998). *Strategies of co-operation: Managing alliances, networks, and joint ventures*. Oxford University Press.

Daily Star (2015). *The wish-list for our heritage*. Retrieved from: https://www.thedailystar.net/lifestyle/the-wishlist-our-heritage-1562821. Accessed 20 Apr 2020.

Davutyan, N. (2007). Measuring the quality of hospitality at Antalya. *International Journal of Tourism Research, 9*, 51–57.

Eagles, P. F. J., McCool, S. F., & Haynes, C. D. (2002). *Sustainable tourism in protected areas: Guidelines for planning and management*. World Commission on Protected Areas.

Evans, N., Campbell, D., & Stonehouse, G. (2003). *Strategic management for travel and tourism*. Butterworth-Heinemann.
Fadeeva, Z. (2005a). Translation of sustainability ideas in tourism networks: Some roles of cross-sectoral networks in change towards sustainable development. *Journal of Cleaner Production, 13*(2), 175–189.
Fadeeva, Z. (2005b). Development of the assessment framework for sustainability networking. *Journal of Cleaner Production, 13*(2), 191–205.
Fayissa, B., Nsiah, S., & Tadasse, B. (2007). *The impact of tourism on economic growth and development in Africa*. Middle Tennessee State University.
Halloway, J. C. (1998). *The business of tourism*. Addison Wesley.
Hjalager, A. M. (2002). Repairing innovation defectiveness in tourism. *Tourism Management, 23*, 465–474.
Hjalager, A. M. (2010). Progress in tourism management. A review of innovation research in tourism. *Tourism Management, 31*(1), 1–12.
Hossain, M. A. (2001). Marketing strategies of the tourism industry in Bangladesh: Implications for Bangladesh Parjatan corporation. *Dhaka University Journal of Marketing, 4*, 79–100.
Hossain, M.A. (1999). Marketing of tourism industry in Bangladesh: An empirical study of performance and strategies. In *Unpublished PhD Thesis*. University of Pune.
Hu, B. A., & Cai, L. A. (2004). Hotel labor productivity assessment. *Journal of Travel & Tourism Marketing, 16*(2), 27–38.
Iqbal, B. A., & Sami, S. (2016). Global and India's tourism scenario. *Journal of Tourism and Hospitality, 5*(1), 1–3.
Katz, R. L. (1974). Skills of an effective administrator. *Harvard Business Review, 52*, 90–102.
Kernel, P. (2005). Creating and implementing a model for sustainable development in tourism enterprises. *Journal of Cleaner Production, 13*(2), 151–164.
Koontz, H., & Weihrich, H. (2010). *Essentials of management* (8th ed.). New Delhi: Tata Mcgraw-Hill Publishing Company Limited.
Laws, E. (1997). *Managing packaged tourism*. International Thomson Business Press.
Lim, C., & McAleer, M. (2005). Ecologically sustainable tourism management. *Environmental Modelling and Software, 20*(11), 1431–1438.
Ling, L., Guo, X., & Liang, L. (2009). Optimal pricing strategy of hotel for cooperative travel agency. *2009 International Conference on Computational Intelligence and Software Engineering*. Wuhan, China: The 11th–13th December.
Ling, L., Guo, X., & Liang, L. (2011). Optimal pricing strategy of a small or medium-sized hotel in cooperation with a web site. *Journal of China Tourism Research, 7*(1), 20–41.
Lucidity (2020). *An introduction to TOWS Analysis*. Retrieved from: https://getlucidity.com/strategy-resources/an-introduction-to-tows-analysis. Accessed 20 Apr 2020.
Lumen Learning (2020). *Qualitative analysis*. Retrieved from: https://courses.lumenlearning.com/suny-hccc-research-methods/chapter/chapter-13-qualitative-analysis/. Accessed 20 Apr 2020.
Manasakis, C., Apostolakis, A., & Datseris, G. (2013). Using data envelopment analysis to measure hotel efficiency in Crete. *International Journal of Contemporary Hospitality Management, 25*, 510–535.
Middleton, V. T., & Clark, J. (2001). *Marketing in travel and tourism*. Butterworth-Heinneman.
Mihalič, T. (2001). Turizem kot dejavnik sonaravnega razvoja. In A. Lah (Ed.), *Turizem in okolje: Slovenija turistična dežela* (pp. 29–34). Svet za varstvo okolja Republike Slovenije.
Mill, R. C., & Morrison, A. M. (2002). *The tourism system*. Kendall/Hunt Publishing.
Ministry of Civil Aviation and Tourism, Bangladesh (2011). *Home*. Retrieved from: http://www.mocat.gov.bd/tourism_statistics.php. Accessed 20 Apr 2020.
Mishra, P. K., & Rout, H. B. (2012). Economic impact of tourism: A case study of Jharkhand. *International Journal of Economics and Managerial Thoughts*, 3(2), pp. 130-139.
Mishra, P. K., Rout, H. B., & Mohapatra, S. S. (2011). Causality between tourism and economic growth: Empirical evidence from India. *European Journal of Social Sciences, 18*(4), 518–527.

Morrison, A., Lynch, P., & Johns, N. (2004). International tourism networks. *International Journal of Contemporary Hospitality Management, 16*(3), 197–202.
Mostafa, A., Habib, H., Farzad, Q., & Nahid, J. (2012). Prioritizing managerial skills based on Katz's theory in physical education offices of universities in Iran. *World Applied Sciences Journal, 20*(3), 388–394.
Munshi, T., & Mishra, P. K. (2016). Can tourism a catalysts for economic growth of Haryana? *Paper presented at ICSSR Sponsored National Seminar on Make in Haryana: A Road Ahead.* Haryana: Central University of Haryana: The 17th–18th, March.
Oxford College of Marketing (2016). *TOWS analysis: A step by step guide.* Retrieved from: https://blog.oxfordcollegeofmarketing.com/2016/06/07/tows-analysis-guide/. Accessed 20 Apr 2020.
Pannell Kerr Forster Associates. (1988). *Strategic master plan for tourism, Bangladesh: Project findings and recommendations.* WTO and UNDP.
Pivcevic, S., & Petric, L. (2011). Empirical Evidence on Innovation Activity in Tourism: The Hotel Sector Perspective. *The Business Review*, 17(1), pp. 142–149.
Poon, A. (1993). *Tourism, technology and competitive strategies.* CAB International.
Questionpro (2020). *Likert scale: QuestionPro survey tools: Features.* Retrieved from: https://www.questionpro.com/features/likert-scale.html. Accessed 20 Apr 2020.
Rao, A. (1986). *Tourism and export instability in Fiji. Occasional Papers in Economic Development.* University of New England.
Romeril, M. (1989). Tourism and the environment -accord or discord? *Tourism Management, 10*(3), 204–208.
Rout, H. B., Mishra, P. K., & Pradhan, B. B. (2016). Nexus between tourism and economic growth: Empirical evidence from Odisha, India. *International Journal of Applied Business and Economic Research, 14*(11), 7491–7513.
Sanjeev, G. M. (2007). Measuring efficiency of the hotel and restaurant sector: The case of India. *International Journal of Contemporary Hospitality Management, 19*, 378–387.
Shahroudi, K., & Dery, M. (2011). Assessment of the efficiency of Guilan province's hotels using two-stage DEA method. *Australian Journal of Basic and Applied Sciences, 5*, 1495–1502.
Sigala, M. (2004). Using data envelopment analysis for measuring and benchmarking productivity in the hotel sector. *Journal of Travel & Tourism Marketing, 16*(2), 39–60.
Sigala, M., Jones, P., Lockwood, A., & Airey, D. (2005). Productivity in hotels: A stepwise data envelopment analysis of hotels' rooms division processes. *The Service Industries Journal, 25*(1), 61–81.
SmartDraw (2020). *SWOT analysis.* Retrieved from: https://www.smartdraw.com/swot-analysis/. Accessed 20 Apr 2020.
Smeral, E. (2008). *Growth accounting for tourism industries.* OECD.
Smutny, P., Prochazka, J., & Vaculik, M. (2016). The relationship between managerial skills and managerial effectiveness in managerial simulation game. *Revista Innovar Journal, 26*(62), 11–22.
The Asian Post (2018). *Identifying threats to tourism.* Retrieved from: https://theaseanpost.com/article/identifying-threats-tourism. Accessed 20 Apr 2020.
Theuns, H. L. (2002). International tourism in developing countries: Conditions and effects. *The Tourist Review, 28*, 109–115.
Trochim, W. M. K. (2020). *Descriptive statistics.* Retrieved from: https://conjointly.com/kb/descriptive-statistics/. Accessed 20 Apr 2020.
UNESCO (2013). *Quadrennial periodic report on measures to protect and promote the diversity of cultural expressions.* Retrieved from: https://en.unesco.org/creativity/governance/periodic-reports/2013/bangladesh. Accessed 20 Apr 2020.
UNESCO (2020). *Bangladesh.* Retrieved from: http://whc.unesco.org/en/statesparties/BD. Accessed 20 Apr 2020.
Vethirajan, C., & Nagavalli, P. (2014). Trends and growth of tourism sector in India – A research perspective. *Paripex: Indian Journal of Research, 3*(11), 9–11.

WAM (2019). *The new technology and travel revolution: Business intelligence and analytics*. Retrieved from: https://www.wearemarketing.com/blog/tourism-and-technology-how-tech-is-revolutionizing-travel.html. Accessed 20 Apr 2020.

Xu, F.-F., Wan, X.-C., & Yang, D.-Y. (2007). Development models of ecotourism in wetland areas. *Chinese Journal of Ecology, 26*(5), 771–774.

Young, R. C. (1992). The structural context of the Caribbean tourist industry. *Economic Development and Cultural Change, 25*, 57–72.

Chapter 19
Development and Investment Scenario in Tourism Education in Bangladesh

Yeasmin Islam and Nazia Akthar

Abstract Bangladesh is a land of natural treasures. It acquires the diverse nature of tourism prospects. The tourism industry of Bangladesh is yet to fully develop. One of the major deficiencies is identified here is the lack of appropriate tourism education. There is a dire need for tourism professionals in Bangladesh that can be prepared through proper tourism education. But most educational institutes are not equipped adequately and developed to provide an progressive level of tourism to cope with the current need of the tourism industry because of short of investment. This chapter mainly focuses on the development and investment scenario in tourism education in Bangladesh. This chapter finds out that the lack of coordination of different stakeholders of the industry is caused for not having a proper development plan for tourism education. Not only that, there is not enough investment in this sector from public and private organizations. It is recommended that this industry needs a fundamental transformation from the primary level to the tertiary level, which including curriculum, program offering, pedagogy, andragogy, research, training, and learning environment. The chapter concludes in the hope that these necessary measures will embellish this industry that will convert human forces to human capital and create a great impact on the country's GDP and the growth of the international market.

Keywords Tourism · Education · Development · Investment · Bangladesh

---

Y. Islam (✉)
School of Business and Economics, United International University, Dhaka, Bangladesh

N. Akthar
School of Business, Presidency University, Dhaka, Bangladesh

A. Hassan (ed.), *Tourism in Bangladesh: Investment and Development Perspectives*, https://doi.org/10.1007/978-981-16-1858-1_19

Introduction

Education is the primary building block of any industry. No industry can be developed and take a market position without it. Every industry needs a specialized, professional, and technical workforce to establish itself in the market. Most of the tourism economy based countries emphasize on advanced tourism education. In Bangladesh, the tourism industry is also in dire need of effective and appropriate tourism education. In the primary and secondary levels of education, there are no significant aspects of tourism education. Though there are enormous numbers of public and private universities at the tertiary level, only a few of them are providing tourism education. There are some institutions are working in that sector, but the contribution is not satisfactory. However, this industry needs various activities to develop an appropriate education system and a huge amount of investment in this purpose. For this reason, the primary objective of the chapter is to discuss the development and investment scenario of tourism education briefly. To accomplish this objective, data were collected through primary and secondary sources.

The Literary Investment and Development

The development of Tourism education is the best instrument to build up indigenous knowledge on tourism services and equip our people to excellence in the tourism business and tourism profession (Huda et al., 2012).

Malihah and Setiyorini (2014) defined education as a conscious and objective effort for developing human resources to be a better person, which cannot be separated by human life and their daily activities, including their activities in doing their leisure activities, recreation, and tourism. They further stated tourism education is the way tourists' destination prepare the human resources to work professionally in the development of the tourism industry.

Qian et al. (2019) have discussed the development and evolution of tourism education from vocational to university higher education levels in different countries. Apart from the tourism organizations (i.e. BTB, BPC, ATAB, TOAB, Civil Aviation of Bangladesh, DMO, BBA, NHTTI, etc.), there are different private and public universities (i.e. DU, RU, NSTU, PU, BSMRSTU, etc.) that offer hospitality degree to create workforce for this industry by providing theoretical and practical experiences (Ananya et al., 2019). A collaborative framework of these stakeholders can help to build the overall growth of this industry. NHTTI is working on tourism education that emphasizes training and different practical training even after getting graduation and post-graduation. They aim to develop manpower that may deliver reliant workforces to several hotels and tourism operating organizations. To ensure excellent customer service, its need to create an expert workforce. The private and public university is also making their effort in this regard. The Ministry of Education and UGC approve different tourism-related courses in secondary, higher secondary, and tertiary levels to create juvenile research in this field.

García and Ruiz (2020) and Penprase (2018) indicate that tourism is the third position as a worldwide export category and is the main export factor in many developing courtiers. They proposed an approach to the educational competencies in the training of tourism professionals for Industry 4.0, focusing on sustainable development and the exploration of the concepts of intellectual capital, tourism, and the fourth industrial revolution.

Ahmed (2019) discussed the bright prospects of the tourism industry. This industry is advancing its way by creating new jobs every year, expecting to increase its number. The development of tourism in Bangladesh mainly depends on the sustainable development of this sector, which can be achieved through infrastructure services and having an adequate marketing policy. This comprehensive development can increase the market, which creates new job opportunities that can only increase the different skills of human resources. Diversity of this developed skilled can develop different types of tourism.

According to Nipa et al. (2015), Bangladesh can be benefited from the prospects of public-private partnership (PPP) that leads the future economic growth in this industry. Tourism market is now identified through SCP (Structure, Conduct, Performance) framework. This can be created by PPP that drives different related issues for the growth of the tourism industry and generates a relationship between employment generation and revenue growth by providing employment opportunities to unemployed people. To support these unemployed people, PPP needs to work together to develop infrastructure and an oligopolistic market and it should ensure maximum outcome from this industry.

According to the USAID report regarding the potential tourism aspects of Bangladesh (Uddin, 2019) reviewed the report and discussed the focus of private sector contributions on countries economy through revenue and job creation. It identified that this industry's growth depends on having a skilled human resource in MMES (Micro, Small and Medium Enterprises) that impact the country's GDP. This also finds out the absence of fiscal policy in this sector, which needs to introduce to establish the proposed master plan and economic zones for tourism.

The Ministry of Education of Bangladesh has formulated a policy named National Education Policy 2010 as the directives provided in the country's constitution. This education policy aims at cultivating basic human values in people of every spare and pursues to suggest ways through which citizens can be groomed to become leaders in pro-people development programs and progress of the society (Ministry of Education, 2020). This policy plays an important role in paving the way for establishing tourism education at various education levels.

Bangladeshi policy planners are not giving enough importance to prepare advanced tourism policy for this sector rather. They are wasting their time by not being serious about it (Siddiq, 2019). For that reason, this industry is losing its share in the international market. There are not enough funds for infrastructure development and promotion of this country to disperse its negative image. Lack of commitment, outdated development concept, less effort to update, lack of professionalism delayed project of JICA and not enough latest effort to improve made it impossible to develop and improve this sector. BPC and BTB should emphasize creating skilled valued professionals rather than just changing the top.

Not only is tourism education important to develop this sector, but it also needs to increase and develop the country's overall literacy rate. Though it is a long process to develop the overall literacy rate to increase communication skills with foreigners, authorities can generate an emergency program to fulfill this industry's immense need. Most of the university in BD does not provide specialized formal education and training on tourism to develop skilled personnel in this regard. It has been suggested to plan accurate measures to rapidly train and develop good manpower for facing the current need.

Development and Investment Scenario of Tourism Education in Bangladesh: Agency Wise Roles

As changes occur in the global climate, globalization, political and legislative factors, technological factors, demographic factors, social and cultural factors, etc., those influence change in industries and businesses (Belcourt et al., 2010) the tourism industry. And these social, environmental, political, demographic, cultural, and economic reformations create change in the type of skills needed (Stanciulescu & Bulin, 2012) by the tourism industry. The development of tourism education is the best tool to build up indigenous knowledge on tourism services and equip our people to excellence in the tourism business and tourism profession (Huda et al., 2012). Not only that, a need-based academic curriculum and facilities both at the public and private stage of building up tourism entrepreneurs, professionals and most importantly, a well-groomed nation with the character and consciousness on hospitality will pave the way for Bangladesh to become an ideal tourist destination (Huda et al., 2012). We need to instill hospitality behavior and understanding in the nation that comprises people of all spares irrespective of age, gender, level of educational qualification, occupation, race, region, religion, etc. It is only possible if proper development and investment in tourism education is fostered in Bangladesh through primary & secondary, higher secondary, technical & Madrasa education, and tertiary level of education.

As opined by Qian et al. (2019), tourism education began as a development of technical/vocational schools in Europe. These schools emphasized training in core competencies such as hospitality, hotel management, and related business skills. In Bangladesh, this trend prevailed, and it too started with technical and vocational institutes. Later on, tourism education has been initiated by different public and private universities at the tertiary level also.

Development and investment in tourism education in Bangladesh can be discussed from the part of its stakeholders. Among many of them, the entities that are contributing to the development of tourism education are:

Bangladesh Parjatan Corporation (BPC)

Bangladesh Parjatan Corporation (BPC) was founded in 1973 to make Bangladesh one of the most attractive tourist cities in the world through the constitution of 1972, order no 143 (Bangladesh Parjatan Corporation, 2020). It is a National Tourism Organization (NTO) of Bangladesh and an autonomous organization under the Ministry of Civil Aviation and Tourism. Some of its missions, directly and indirectly, are related to tourism education. These are to develop and maintain international standard tourism products and ancillary facilities; to involve government and encourage the private sector in developing physical infrastructure (i.e. road, rail, air, and waterways for easy accessibility; to ensure security & safety of the tourists; to encourage pro-poor tourism industry for their econoic betterment and empowerment; to develop ecotourism based on nature and ethnic culture; enhance the balance of payment, create employment opportunities, alleviate poverty and create social harmony through tourism activities; to enhance the marketing of tourism products at home and aboard; to develop human resource in the tourism industry; promote and maintain close cooperation with regional and international counterparts of the industry, etc.. As stated by Ananya et al. (2019), BPC is trying to promote the heritage and tourist attractions, creating the bright image of Bangladesh to foreign and domestic countries, develop existing tourist sites, identifying new tourist zone and promotion, providing training to the workforce to make them skilled, in developing accommodation and quality of food, developing civic and official benefits, service development, etc.

National Hotel and Tourism Training Institute (NHTTI)

A Training institute named National Hotel and Tourism Training Institute (NHTTI) was established in 1974 under Bangladesh Parjatan Corporation (The National Tourism Organization). This institute offers various courses of different duration, with the purpose to develop professionalism and provide teams of skilled and trained manpower to meet the increasing demands of the trained personnel for the tourism and hospitality industry in the country. Besides offering regular courses, NHTTI has started a two-year duration diploma in Hotel Management in 1994. The courses offered by NHTTI are- diploma in Hotel management, diploma in culinary arts and catering management, national certificate course, professional chef course, diploma in tourism and hospitality management, etc. (National Hotel & Tourism Training Institute, 2020).

Pacific Asia Travel Association (PATA)

Pacific Asia Travel Association (PATA) is a voluntary association that was founded to stimulate and create awareness to ensure eco-friendly tourism in a certain territory. PATA is a strategic partner of UNWTO and a membership association that plays an active role in the responsible development of the Asia Pacific travel and tourism industry including Bangladesh (Pacific Asia Travel Association [PATA], 2020). They organize several seminars and conferences to ensure sustainable tourism by building effective human resources in Bangladesh for addressing climate change issues and Rohingya crisis in Cox's Bazar. It ensures a win-win situation among all the stakeholders of the industry by embedding ethical practices practically and manually. They also publish journals about the existing and upcoming tourism resources in the Pacific regions and thus help the member nations and Bangladesh to introduce new tourism products and services around the world. They conduct marketing researches to provide valuable insights, forecasts, and analysis to help members to make better business decisions.

Association of Travel Agents of Bangladesh (ATAB)

Association of Travel Agents of Bangladesh is the first registered association of travel agents operating in Bangladesh (Association of Travel Agents of Bangladesh [ATAB], 2020). It is one of the leading travel agents' association in the country, functioning with 3500 members. It has established the ATAB Tourism Training Institute (ATTI) in 2007 to provide skilled manpower for the travel and tourism industry of Bangladesh and other countries. About 2700 trainees have been trained and certified by ATTI who are already engaged at different organizations. The government has approved the ATAB Tourism Training Institute (ATTI) & thus, the institute is directing courses like "Ticketing & Reservation" by following the proper curriculum as NTVQF (National Technical and Vocational Qualifications Framework) Level-2.

Apart from these institutions, other institutions are also investing in offering different vocational courses. Some of these institutions are i. Bangladesh Hotel Management & Tourism Training Institute; ii. Bangladesh Skill Development Institute; iii. Bangladesh Tourism Guide; iv. Tommy Miah's Hospitality Management Institute; v. Meghna Shorobor Resort; vi. Seagull School of Hotel Management; vi. Western Institute of Business and Technology (WIBT) College; vii. Royal Park Institute of Hotel Management Tourism and Catering; viii. The City and Builders of London Institute; ix. Institute of Tourism and Catering Services; x. Cambrian College; xi. Hotel Razmoni Ishakha Hotel and Management Training; xii. The Guardian Hotel and Tourism Training Institute; xiii. Bangladesh Hotel and Tourism Training Institute; xiv. International Hotel and Tourism Academy; xv. The Regency Hospitality Training Institute (Pajatanbd.com, 2020).

Initiatives have been taken, and these organizations have made investments to provide tourism courses. They have developed scopes for skill-based course materials, labs, and practical exposure for ensuring quality education in the institutes.

Public and Private Universities

Later on, interest and demand from the public and private sectors encouraged rapid growth of tourism studies and the development, and establishment, of departments of travel and tourism at private and public universities in addition to technical schools. Stanciulescu and Bulin (2012) emphasized shaping the tourism higher education curriculum to make students achieve different technical, functional, and self-management.

The numbers of the institutions are providing tourism education is given in the following Table 19.1.

These institutions mainly emphasize the front desk, food and beverage related courses, core business, and some selected major related to tourism. But education in tourism is a vast area, and the institute or organization should cover as much area as possible.

Investment and development have been made at graduation level education to some extent also. Two public universities and 11 private universities are providing four-year degree (Parjatanbd.com, 2020). The name of these universities is given in Table 19.2. Some other universities are also applying to University Grant Commission (UGC) to open that course.

Ministry of Education, Bangladesh

Hospitality is an innate characteristic in Bangladeshi ethics, that the visitors appreciate the most during their visit in Bangladesh. But these values should be further instilled and enhanced using schooling and formal education from early age. It will evolve us as a host nation and instigate our youths to become tourism professionals, entrepreneurs, and researchers. Hence tourism education needs to be included in our primary and secondary, and higher secondary education curriculum. Ministry of

Table 19.1 Name of the institutions providing tourism education

Type of institution	Number
Public university	2
Private university	11
Public/private institutes/colleges	19
Consultancy firms	7

Source: Developed by the authors, 2020

Table 19.2 Name of the public and private universities that provide tourism education

Name of the university	Type of university
University of Dhaka	Public university
Rajshahi University	
International University of Business and Agriculture and Technology	Private university
American International University – Bangladesh	
Daffodil International University	
IBAIS University	
World University of Bangladesh	
Primeasia University	
Royal University of Dhaka	
Victoria University of Dhaka	
Victoria University of Bangladesh	
European University of Bangladesh	
Fareast International University	
Cox's Bazaar International University	

Source: Developed by the authors, 2019

Education of Bangladesh and their concern organization, the Bangladesh Education Board, is responsible for formulating these policies. They have formulated a nationwide education policy named "National Education Policy 2010" (Ministry of Education, 2020). The main aims of this policy are focused at the propagation of values, that tries to recommend ways through which people can be trained to become leaders in pro-people improvement agendas and society's development (Ministry of Education, 2020).

The education policy is a generalized one. This isn't easy to stem out information from it exclusively related to tourism education. Some significant points are highlighted here based on their relevance and significance in developing tourism education and developing a proper mindset among mass people, which will enable them to respond to changing dynamics of the global tourism industry.

There will be three sections in the secondary level of education- general, madrasa, and technical, and there will be some branches in each section. Yet, regularity will be maintained through some specified compulsory courses, such as Bangla, English, Bangladesh Studies, General Mathematics, and Information Technology for all sections (Ministry of Education, 2020). The syllabus of various classes will be organized keeping syllabus of other streams in mind comprising the compulsory subjects such as Bangla, English, Moral Science, Bangladesh Studies, General Mathematics, Social Environment, Environmental Science etc. that will include concepts of climate change and information technology (Ministry of Education, 2020).

This subject, in one way or another, fosters basic knowledge of tourism among students.

Morality is engrained and learnt from religion. Social and cultural ethics and regional veracities have significant contribution to it also. The objective of religious and ethical education is to create awareness about the respective religions of the learners, improve the pupils' behavioral patterns, establish moral qualities in life, and build the learners' ethical character (Ministry of Education, 2020).

This objective perfectly aligns with ensuring tourists' safety, getting cooperation, and positive treatment from local people.

This objective perfectly aligns with ensuring tourists' safety, getting cooperation, and positive treatment from local people. Subjects such as defense studies, comparative theology, peace and conflict, climate change, etc. will be included in the curriculum of higher education. The objective of this inclusion is aiding the students acquiring international standard education; creating keenness to learn, helping them grow up with human qualities; magnifying the prospect of knowledge; effectively familiarizing students with the knowledge of the up-to-date world; constructing citizens who will possess systematic, liberal, humane, progressive and forward-looking mindset; creating a community inspired by insight, originality, human values, and patriotism. (Ministry of Education, 2020).

As stated in National Education Policy 2010, the policy has treated business studies as a comprehensive establishment of essential divisions of knowledge that are necessary to operate industries, trade, and service organizations. One of its strategies clearly emphasizes the need for inclusion of tourism education in the curricula, which states: "It is necessary to identify the prospective sectors like tourism, hotel management and to expand them; sufficient and efficient manpower will be created. So, provision of education and training will be there for the sectors thus earmarked". Some other of its strategies to develop tourism education and human capital in this industry are first, measures are initiated to extend, revise and coordinate business studies depending on its local and international demand. Second, course syllabus and textbooks of business school will be designed keeping the need of industries, product and service based entities and trade organizations, along with their potential growth in mind will be introduced for all levels of commerce education. Third, it is necessary to have an exchange of views between universities and industries or business houses in designing syllabus and course curriculum at university and college levels. A combined committee will be made consisting university faculties, industrial professionals and representatives of the decision-making authorities to construct the syllabi and course materials to be exercised in all universities- which will ensure the sharing of ideas among industries, academia and government in preparing subject based syllabus and curriculum. Fourth, to ensure acquirement of industry based hands-on experience and knowledge, doing internships in different areas and divisions of business has been made mandatory for the students which is a basic requirement for completing their graduation and post-graduation degree. Fifth and finally, up to date skill based courses will be opened and required human resources will be appointed to ensure more effective utilization of the infrastructure of sixteen commercial institutions of Bangladesh.

Ministry of Civil Aviation and Tourism, Bangladesh

The Ministry of Civil Aviation and Tourism was formed in August 1975, with the objectives to facilitate safe, convenient and efficient civil aviation facilities as well as to ensure attraction of tourists by diversifying tourist products and improving associated services (Ministry of Civil Aviation and Tourism, 2020).

According to the Ministry of Civil Aviation and Tourism (2020), the ministry and its concerned organizations have undertaken investment and development initiatives in tourism education and training.

Civil Aviation Authority of Bangladesh has facilitated 10,992 man-hours of training to 2620 trainees under 195 courses at home and abroad. Other than that, 1162 employees have received on the job training for 15,942 man-hours.

Biman Bangladesh Airlines has initiated the Bangladesh Airlines Training Center to provide education and training to employees of Biman by six faculties as Operations Technical; Customer Service; Management Development; Sales and Marketing; Engineering (Aerospace); and Engineering (Avionic) (Ministry of Civil Aviation and Tourism, 2020).

The Ministry of Civil Aviation and Tourism (2020) plans to expand the scope of the training center by adding simulator, helicopter aerospace education, and transformation to type rated training organization.

As described by the Ministry of Civil Aviation and Tourism (2020), Biman Bangladesh Airlines also plans to initiate long –awaited Graduation Program named: BSC in Aircraft Maintenance Engineering (Aerospace); BSc in Aircraft Maintenance Engineering (Avionics); BBA in Aviation Management; and MBA in Aviation Management.

As per the Ministry of Civil Aviation and Tourism (2020), the ministry and Bangladesh Partatan Corporation (BPC) have been arranging courses through NHTTI on hospitality management and are contributing to the supply of an efficient workforce to the tourism industry. They are facilitating the flowing courses of different duration (Table 19.3).

BPC also initiated the following activities in 2018–2019 as mentioned in the Annual Report 2018–2019 (Ministry of Civil Aviation and Tourism, 2020, p. 49) (Table 19.4).

Bangladesh Tourism board, in 2018–2019 has arranged the following courses (Table 19.5).

Issues Need Attention to Minimize for Investment and Development of Tourism Education in Bangladesh

MSMEs (Micro, Small and Medium Enterprises) are the main driving forces for Bangladesh's tourism industry. This includes different restaurants and region-based hotels, tour operators and recreational initiative based businesses which are

Table 19.3 Courses offered by NHTTI

No	Course name	Duration
1	Diploma in Hotel Management	2 years
2	Diploma in Tourism and Hospitality Management	1 year
3	Diploma in Culinary Arts and Catering Management	1 year
4	Professional Chef Course	1 year
5	Professional Baking Course	10 months
6	National Certificate Course in Front office and Secretarial Operations	18 months
7	National Certificate Course in Food and Beverage Production	18 months
8	National Certificate Course in Food and Beverage Service	18 months
9	National Certificate Course in Housekeeping and Laundry	18 months
10	National Certificate Course in Bakery and Pastry Productions	18 months
11	National Certificate Course in Tour guide and Travel Agency operations	18 months
12	Special Fast-food, Snacks, and Dessert Bakery Course	5 months
13	Food, Hygiene, and Sanitation	3 months
14	Tin Chef Course	3 months

Source: Ministry of Civil Aviation and Tourism (2020, p. 48)

Table 19.4 Courses summary offered by NHTTI

Course	Number of courses	Number of students
National certification courses	18	921
1 and 2 years courses	5	303
Short-term courses	24	966
Total	27	2190

Source: Ministry of Civil Aviation and Tourism (2020, p. 49)

supporting this industry; though Uddin (2019) mentioned that this sector requires more skilled Human Resources trained by a suitable tourism education system. Human resources developed by tourism education can play a dynamic role by contributing to the country's GDP, which is expected to raise to US$7.5 billion by 2028. Uddin (2019) also mentioned some non-fiscal supports from Bangladesh Tourism Board (BTB) and Bangladesh Investment Development Authority (BIAD) for the the proposed tourism master plan and export zone for tourism that can only be achieved by developing an effective tourism education system.

The tourism industry has created a huge number of jobs every year, but the number is still not significant because of the lack of contribution to tourism education. The curriculums of tourism education are not focusing on the sustainable

Table 19.5 Courses offered by BTB

Course name	Number of courses	Number of students
New comer tour operator course	3	75
Tour guide course	3	90
Street food vendor course	5	218
Community based tourism course	5	104
Total	16	487

Source: Ministry of Civil Aviation and Tourism (2020, p. 61)

development of this industry. An advanced educational curriculum is absent and does not create expected jobs (Ahmed, 2019).

Many tertiary-level educational institutions are not providing formal education on tourism (Huda et al., 2012). Only very few numbers of private and public universities are offering tourism education. The number of professional institutes is also very minimal.

According to Huda et al. (2012), the public and private sectors need to take development initiatives to create a pro-tourism educational environment and to come up with effective directions to support hospitality education in Bangladesh. However, there is an insufficient investment from the public and private sectors to develop educational facilities in tourism that lack behind this industry to have skilled workforces.

This industry needs the development of professional competency. Siddiq (2019) acknowledges Bangladesh's two most responsible organizations BPC and BTB, for that. BPC and BTB are not giving sufficient effort to ensure development of professional capabilities that are necessary to support and develop this industry. The primary goals should retain not only the capability but also enhance it. For this purpose, these two organizations should emphasize developing advanced professional competency by ensuring appropriate tourism education.

The organization that provides tourism education are not able to meet the quality of international education. Because of inadequate advanced education facilities, employers do not get skilled employees for this industry. There is a lack of opportunity to develop skilled personnel through specialized training programs, and the number of qualified trainers is also very nominal. Foreign skilled labor takes a higher position in the industry because of a lack of skilled and education of local people in tourism.

Cross-cultural training is an important criterion for tourism education. Most universities teach their future human resources about the differences in culture and how to cope with them. Universities and institutions in Bangladesh should focus on cross-cultural training to serve this diverse industry.

In every country, government agencies have important contributions to promoting, developing, and monitoring hospitality and tourism education. So to ensure a prominent tourism industry it needs to make sure a well-structured tourism education system. The absence of coordination between different government agencies creates obstacles in developing an advanced tourism education system.

Conclusion

Now in Bangladesh tourism industry is just not a source of revenue rather, it becomes a sector to promote the future developments of this country. Through tourism, many other industries try to build their mark on the international market. The government is also taking different steps to make this industry more attractive, one of that is to develop educational prospects on tourism. Ministry of Education and UGC are working together to introduce and operate different courses on the tourism industry. They are increasing their attention and investment to make it successful. Different private and public universities and institutions are also making efforts to create skilled human resources in this industry. But comparatively to other tourism developed countries, the effort is not enough. It needs more strategically developed plans and investment in tourism education. For that reason, it should give more pay attention to private-public relationships. The responsible authority of this industry needs to be more proactive to work in advance to support tourism education. This industry requires more specialized training institutions as well as expert trainers.

References

Ahmed, S. (2019). *Tourism holds bright prospects*. Retrieved from: http://www.theindependentbd.com/post/219239. Accessed 10 Jan 2020.

Ananya, S. A., Islam, S., & Kuri, B. C. (2019). An exploratory study on the role of major tourism organizations in flourishing the tourism industry in Bangladesh. *IOSR Journal of Business and Management (IOSR-JBM), 21*(12), 35–43.

ATAB Tourism Training Institute. (2020). *About ATAB*. Retrieved from: http://www.atab.org.bd/Login/ATAB_Training_Institute. Accessed 23 Feb 2020.

Bangladesh Parjatan Corporation (BPC). (2020). *Home*. Retrieved from: http://www.parjatan.gov.bd/site/page/68c2d54a-112f-4804-8624-9180bf59df98/-. Accessed 21 Feb 2020.

Belcourt, M., McBey, K. J., Hong, Y., & Yap, M. (2010). *Strategic human resources planning* (pp. 5–59). Nelson Education.

Cecilia, G., Stanciulescu, G. C. J., & Bulin, D. (2012). Shaping tourism higher education curriculum-strategy to develop skills for tomorrow's jobs. *Procedia Economics and Finance, 3*, 1202–1207.

García, R. M. R., & Ruiz, J. A. C. (2020). Professional training in tourism for the fourth industrial revolution. In E. O. C. Espinosa (Ed.), *Management Training Programs in Higher Education for the Fourth Industrial Revolution: Emerging Research and Opportunities* (pp. 62–78). IGI Global.

Huda, K., Haque, A., & Jahan, I. (2012). Importance of tourism education in Bangladesh: A diagnostic appraisal. *International Academic Research Journal of Business and Management, 1*(2), 5.

Malihah, E., & Setiyorini, H. P. D. (2014). Tourism education and edu-tourism development: Sustainable tourism development perspective in education. In *The 1st International Seminar on Tourism (ISOT)-"Eco-Resort and Destination Sustainability: Planning, Impact, and Development"*. Bandung: the 27–28 October, pp. 1–7.

Ministry of Civil Aviation and Tourism (MoCAT). (2020). *Annual report 2018–2019*. Retrieved from: https://mocat.portal.gov.bd/sites/default/files/files/mocat.portal.gov.bd/page/

b0c352d4_03b2_45fd_8f2a_d0541fd6ce75/Annual%20Report%202018-19.pdf. Accessed 25 Feb 2020.

Ministry of Education. (2020). *National education policy 2010*. Retrieved from: https://rhtibd.com/program-fees/. Accessed 22 Feb 2020.

National Hotel & Tourism Training Institute (NHTTI). (2020). *About NHTTI*. Retrieved from: https://www.nhtti.org/page_detailsf019.html?type=page&id=6. Accessed 22 Feb 2020.

Nipa, N. J., Sultana, J., & Rahman, M. H. (2015). Prospect of private-public partnership in tourism of Bangladesh. *Journal of Investment and Management, 4*(3), 73–83.

Pacific Asia Travel Association (PATA). (2020). *About PATA*. Retrieved from: https://www.pata.org/sic/. Accessed 22 Feb 2020.

Parjatanbd.com. (2020). *Higher study and short course of tourism subject in Bangladesh*. Retrieved from: http://www.parjatanbd.com/Education/Study.htm. Accessed 22 Feb 2020.

Penprase, B. E. (2018). The fourth industrial revolution and higher education. In N. W. Gleason (Ed.), *Higher education in the era of the fourth industrial revolution* (pp. 207–229). Palgrave Macmillan.

Qian, J., Law, R., & Li, X. (2019). Education research in tourism: A longitudinal study of 77 articles between 2008 and 2017. *Journal of Hospitality, Leisure, Sport & Tourism Education, 24*, 120–129.

Siddiq, R. (2019). *Bd to wait longer for new development master plan*. Retrieved from: https://menafn.com/1098896310/Bangladesh-BD-to-wait-longer-for-new-tourism-development-master-plan. Acessed 10 Jan 2020.

Uddin, S. (2019). *Tapping tourism potential crucial to BD economy: USAID*. Retrieved from: https://thefinancialexpress.com.bd/economy/bangladesh/tapping-tourism-potential-crucial-to-bd-economy-usaid-1574483512. Accessed 10 Jan 2020.

Chapter 20
Investment and Development for Tourism and Hospitality Revenue Collection in Bangladesh

Jamee Ahmad

Abstract This review paper consists of details about Bangladesh's total revenue from tourism with the actual figure, investment amount, employment status, and roles of government and private organizations in the tourism and hospitality industry. The chapter also shows data from different sources like Bangladesh Tourism Board (BTB), Bangladesh Parjatan Corporation (BPC), World Travel & Tourism Council (WTTC), United States Agency International Development (USAID), National Board of Revenue (NBR) and different ministries of the Government of Bangladesh (GOB). The chapter finds new steps that are taken to make the tourism industry more attractive. New infrastructure, projects, and marketing activities that are currently done by government and private organizations are also mentioned. Effects of COVID-19 on Bangladesh's tourism was also focused with detailed statistics about total job loss, and industry-wise revenue loss. Other barriers and shortcomings that stop international tourists from coming to the country and resulting in low revenue collection were also identified. Recommendations were made based on former research results and observations from news portals, journals, and other sources.

Keywords Tourism · Hospitality · Revenue · Investment · Bangladesh

Introduction

Currently, Bangladesh is becoming a destination for tourists from across the globe. It is believed that the tourism industry will become a big revenue generator in a short time. This chapter sees the hospitality and tourism industry's investments to generate revenue. An analysis is made that shows the different sections of this

---

J. Ahmad (✉)
College of Business, IUBAT-International University of Business Agriculture & Technology, Dhaka, Bangladesh

A. Hassan (ed.), *Tourism in Bangladesh: Investment and Development Perspectives*, https://doi.org/10.1007/978-981-16-1858-1_20

industry where those investments are made. According to a report published by the Bangladesh Ministry of Civil Aviation & Tourism (MoCAT), some of the current famous tourist attractions in Bangladesh are the world's largest undivided sea beach in Cox's bazar, Kuakata sea beach, the world's single largest mangrove forest, the Sundarbans, hills, tea gardens, etc. (Ministry of Civil Aviation & Tourism, 2010). Tourism refers to the activities commenced by visitors. It is also called the "visitor economy". The industry of tourism includes different activities that occurs within the "visitor economy" (Stainton, 2019). Tourism revenues are a measure of the economic impact of tourism. In easy words, it is the earning from the tourism industry after deducting costs. It includes all sorts of tourism-related expenditure within a region by non-resident visitors and locals traveling within the province/district. It includes accommodation, transportation, food and drinks, cultural sites, recreation and entertainment, travel agency, and other reservation services (Tourism Nova Scotia, 2020). This chapter deals with the investment and development for tourism and hospitality revenue collection in Bangladesh.

Present Scenario

Status of Revenue from Tourism

Bangladesh has been earning revenue from the tourism and hospitality industry. This part of this chapter shows related statistical data regarding earning of revenue. Following the United States Agency International Development (USAID) (2019:23), the tourism industry, which reported USD 5.3 bn annual revenue last year, and is currently growing around 7 million domestic travelers per year. Despite the fact that it has enormous potential in terms of GDP contribution, Bangladesh has yet to establish itself as a popular tourism destination. In 2017, the Bangladesh tourist business was estimated to be worth USD 5.3 billion, contributing roughly 2.2 percent to the country's GDP. Local tourists contribute 98% of the revenue, and 2% comes from foreign tourists, for both business and traveling purposes.

Visitor exports generated USD 228 million in 2017, and are predicted to expand at a 6.20 percent annual pace from 2018 to 2028, resulting in USD 445 million in visitor exports. In 2017, the tourist industry contributed around 1.17 million jobs to the economy. Between 2017 and 2018, 100,000 foreign visitors (90 percent business travelers, 5 percent vacationers) visited Bangladesh, a 22 percent increase (USAID, 2019:1).

As per the World Travel & Tourism Council's (2020), annual research report of 2019, tourism contributes 3% of the entire economy, which totaled to T&T GDP = BDTk 772.7BN (USD 9,113.2MN) and has created 185,89 jobs (2.9% of total employment). The International visitor impact is BDTk 28.3bn as visitor spends (0.7% of total exports). Another report on "Global Economic Impact & Trends 2019" from the World Travel & Tourism Council's (2020), shows that Bangladesh is among the TOP 20 countries for travel & tourism GDP growth in 2018 by holding the 11th position with 11.6%. the World Travel & Tourism Council's (2019) also shows that Bangladesh is estimated to be in the 9th position by contributing to 6.1% growth to its GDP by 2029.

Financial Investment Details

Government of Bangladesh (GOB) has allocated BDTk 34 billion in FY2019-20 for the Ministry of Civil Aviation and Tourism. This was more than double of the last fiscal year's allocation, which shows that the government is giving importance to the tourism industry (Ministry of Finance (2019).

Governmental Initiatives to Promote Tourism in Bangladesh

To become a revenue-generating industry, requires good publicity and advertising. So, the GoB has done a lot for Bangladesh's tourism publicity via different mediums and programs in different times.

Promotion and Marketing

Bangladesh Tourism Board (BTB) promoted the positive image of Bangladesh to the world via advertisements in July 2010. A promotional video was seen during the ICC Cricket World Cup 2011 named "School of Life: Beautiful Bangladesh". This television commercial had obtained 3rd position at the prestigious Zagreb Film Festival for TVC.

Digital Marketing Campaign

BTB has created its website www.tourismboard.gov.bd for a digital marketing campaign (Bangladesh Tourism Board, 2020). The parent organization Bangladesh Parjatan Corporation also featured a lot of details. Unfortunately, this website is not working at the moment. Another website is www.visitbangladesg.gov.bd. This was launched during the World Cup Cricket 2011, and was re-vamped for the purpose of digital marketing campaign.

Participation in Foreign Tourism Exhibitions

In order to attract more tourists to Bangladesh and to promote its tourism, BTB launched the "Visit Bangladesh" Campaign. GoB had allocated a total of BDTk 66 crore to facilitate the project. BTB took several steps for this massive project.

Bangladesh Premier League 2011

Bangladesh Cricket Board and BTB co-sponsored and collaborated in arranging the opening ceremony of the Bangladesh Premier League. The Bangladesh Premier League is an International Standard T20 Cricket Tournament that has lots of well-known foreign and local players as participants. This event is now taking place every year.

Asian Tourism Fair 2011

An event named as "Asian Tourism Fair 2011" was arranged in Dhaka from September-October. It was a three-day regional tourism fair, and organized to make common tourism-promotion strategies for the Asian countries. This fair highlighted their uniqueness in tourism in particular regions, and focused on attracting international tourists for those particular regions (Tourism BD, 2018).

Public-Private Partnership (PPP) Projects

According to the "Public-Private Partnership (PPP)" Authority under the Prime Minister's Office, several tourism enhancement projects are being initiated under the PPP.

The PPP is another way to enrich the tourism industry jointly. Many private companies are willing to invest in tourism but cannot get much access due to not having government support, infrastructure, and finance. So, the term of PPP was introduced where government and private industry can work in partnership to develop faster and in a competitive manner (Nipa et al., 2015). A project under the PPP was initiated titled "Development of Integrated Tourism and Entertainment Village at Parjatan Holiday Complex, Bangladesh Parjatan Corporation, Cox's Bazar Under PPP" with private partner Orion Group. To date, its status is "contract to be signed". The total estimated cost was fixed at USD80–200mn (PPPA, 2020a).

Another project under the PPP named the "Development of Five Star Hotel and Zakir Hossen Road in Chittagong" was planned by Bangladesh Railway and due to insufficient fund it was done on PPP, with a budget at USD80–200 mn and the current status is "no bids received" (PPPA, 2020b).

Joint Cruise Service

Dhaka Dinner Cruise with Bangladesh Parjatan Corporation jointly took the initiative to create awareness to conserve the country's perpetual rivers (Dhaka Dinner Cruise, 2020).

Project with China

China and Bangladesh agreed to expand cultural exchange, education, tourism, etc., during the last visit of Chinese President Xi in 2016. They also agreed to promote interactions among media, think tanks, youth activities, and women entrepreneurship for non-government groups and local authorities. Both Bangladesh and China view the tourism industry's potentials via the Bay of Bengal Initiative for Multi-Sectoral Technical & Economic Co-operation (BIMSTEC). The member countries in BIMSTEC agree to the concrete steps to promote intra- BIMSTEC tourism (The Independent, 2020).

Institute for Training on Tourism Management

The Bangladesh Parjatan Corporation has National Hotel and Tourism Training Institute (NHTTI) for professional tourism management training. NHTTI was established in 1974 under Bangladesh Parjatan Corporation. Also a good number of private universities are offering courses on hotel tourism management now a day (Roy and Roy, 2015).

New Tourist Destinations for Prospective Revenue Collection

Investments are made by individual institutions and the GoB jointly to bring more professionalism, competitiveness, and meet up to date client demands. The world's longest marine drive of 80-kilometer road from Cox's Bazar to Teknaf along the Bay of Bengal is famous and attractive for tourists. The Bangladesh Prime Minister inaugurated the coastal road at the 28th kilometer point of the road near Inani beach. It was constructed by the Bangladesh Army Engineering core under the supervision of the Roads and Highways Department. The total cost was around BDTk1050 crores (1.2billion approximately) (Dhaka Tribune, 2020a, b).

After assessing a famous travel website Booking.com, it was found that more than 15 resorts and big hotels are built to create tourism and entertainment in different parts of Bangladesh in the last 15 years. Many of them focus on eco-tourism, tea resorts, natural views, and water parks, cottage and rural tourism beside rivers, mountain, and hill tracks, island tours, etc. (Booking.com, 2020).

Mithamain and Nikli haor, Austogram (inland water reservoir) are famous tourist spots in Kishoreganj. These attractions are creating interests among many new tourists. Several new roads are constructed to connect the three villages; the recent one is 29.73 km with BDTk 874.08 crore cost. It has created an extraordinary scenario and tourist spot where people come for a one-day tour in between the busy life. It has become a tourist spot and has made the lives of people living in the Haor (water sources) area easy to commute and earn revenue from tourists (The Daily Star, 2020a).

BiriShiri is another beautiful place in Netrokona, Durgapur having a canal with blue water, and small restaurants and motels (Journey for Better Life, 2020). People usually go for a day-long trip.

The first cruiser in Bangladesh has set its journey in the Bay of Bengal. The cruise ship arrived at Chittagong port on Sunday (the 20th September). It is currently located there. It is expected to be connected by sea from Cox's Bazar to St. Martin's island. The ship's previous name was "Salvia Maru". The ship was bought from Japan to provide an international quality travel experience to the tourists. After being brought to Bangladesh, it was re-named "One Bay" (Somoy News, 2020).

Sajek Valley is a mountain area in the hill tracts Chittagong and an emerging tourist spot. Several tourism agencies are operating tours in this area where tourists can feel the cloud in their arms. Even in the Corona pandemic, it was opened in September, and the cottages are finding a large number of travel lovers (Dhaka Tribune, 2020a, b).

Drawbacks of Tourism in Bangladesh

Lack of Tourists from the Western Countries

Bangladesh gets half of its international tourists from India and has failed to draw tourists' attention from the western countries. According to the World Tourism Organization (2020), the Americans spent USD334 billion, and the Europeans spent USD570 billion for tourism while the global expenditure was tourism is USD1.7 trillion. For Bangladesh, only 5% of the tourists come from the US, including the Bangladeshi origin Americans, and 7% from the rest of the world, which is pretty low compared to neighboring countries. It shows that Bangladesh is not attracting enough high spending tourists.

Lack of Security and Public Safety

Security can be a huge issue for foreign tourists making them unwilling to travel to Bangladesh. The US Department of State had issued a warning against traveling to Dhaka and the Southeast of Bangladesh, citing crime, terrorism, kidnapping as the main reasons and Bangladesh being at level 2 (exercise increased caution). The Safety and Security Index of 2019 by the World Economic Forum shows that Bangladesh has improved its rankings to 105th from 123rd in the list of unsafe countries to travel (U.S. Department of State-Bureau of Consular Affairs, 2020).

Insufficient Local Transports

The number of local transports going to popular tourist destinations is not enough. Also, facilities and quality of these transports fail to attract foreign tourists, and to the locals, these seem to be a bit expensive.

Lack of Proper Policy

In 1992, the National Tourism Policy (NTP) was formulated, as a Strategic Master Plan for tourism development. It was prepared by UNDP/WTO in 1990 and updated by WTO. Bangladesh Parjatan Corporation also took many initiatives but still, efforts are not enough (Afroz & Hasanuzzaman, 2012).

Lack of Adequate Infrastructures

Bangladesh is unable to build enough infrastructure that can help the tourism industry grow. Recently, the private sector is doing a lot but still GoB support is inadequate regarding hotels, motels, and transportation (Moniruzzaman & Abedin, 2015).

Political Unrest and Decisions

Political unrest that is known as "Hartal" (blockade) was a big reason for tourism to go downwards from the year 2008-2014. Also, there was not enough budget allocated to develop public hotels in those years (Chowdhury & Chowdhury, 2015).

Inactive Worker Unions

Several private agencies are there to negotiate with the government regarding tourism development (National Board of Revenue, 2020). But it seems that these worker unions are mostly quite inactive.

The Rohingya Crisis

The Rohingya crisis is a big issue now, fleeing from Myanmar as refugees (UN News, 2020). As per UN mandates, Bangladesh is letting them in. Most of them are staying at Cox's Bazar, the largest undivided sea beach in the world, which is also the biggest tourist destination in Bangladesh. Their presence makes tourism difficult in the region.

COVID-19's Effects on the Industry of Tourism in Bangladesh

According to a report of the Pacific Asia Travel Associations (PATA Bangladesh Chapter) (2020), in the aviation sector, more than 2000 lost their jobs and had a financial loss of 6 billion. More than 100,000 employees from across Bangladesh in different hotels/motels/inns and resorts lost their jobs, a financial loss of 1500 crore BDTk (15 billion). Also in Bangladesh, these losses accounted to 150,00 jobs and 300,00 crore BDTk (30 billion) among travel agents; over 100,00 jobs and 100 crores BDTk (1 billion) among inbound tourism, and for both inbound and outbound 150,00 people lost their jobs and faced a loss of 3000 crores of BDTk (25 billion). For Umrah (Hajj), 5000 people became jobless. Fast food, restaurant, bar, coffee shops, and other small food preparation points have lost 150,000 jobs and a loss of 4.5 billion BDTk. Among vehicle operators in tourism, the losses were 1000 jobs and a financial loss of 4 billion BDTk., and in tourists' vessels, the loss was 15 billion BDTk. The total losses until June 2020, were 9705 crore BDTk. (97.5 billion) approximately, and a total of 309,500 jobs.

Recommendations to Collect More Revenues from the Tourism and Hospitality Industry

Benchmarking with Other Countries

Neighboring countries as India and Sri Lanka are well ahead of Bangladesh in tourism. Despite having great potential and resources, Bangladesh lacks behind. So the idea to benchmark neighbors who are competitors can help to know their people, nature, food, and culture to follow their success steps. Eco-tourism, water life tourism, and wildlife tourism can be good sources to develop tourism revenue and interests about Bangladesh. Some neighboring countries are practicing them and getting great attention (Bandara & Tisdell, 2004). Sri Lanka has good places for wildlife tourism, and these are practiced at the local and regional levels. Bangladesh has the largest mangrove forest that is globally known as the Sundarbans with full of wild animals, especially the Royal Bengal Tiger, deer, and expensive woods can be great sources to earn foreign currency. If tourism companies are given permission, and proper roads, quality resorts and hotels are built, Ecotourism can be done in locations like Sylhet that is famous for growing tea and exports. The golden fiber of Bangladesh "Jute" and eco-friendly products made from it can also be a source of ecotourism. Water life tourism can be arranged on the largest coral island of Bangladesh, St. Martin's Island. For a long time, international football, tennis, athletics etc. tournaments are not arranged. Hosting international sports tournaments can also be a great way to attract foreign guests for revenue earning (Arulmozhi & Gomathinayagam, 2019).

Other Recommendations

Exploring new tourist spots or building them can be a way to attract more tourist to generate revenue. Already the GoB has taken initiatives through different ministries to enrich the tourism industry, but it is believed that it has to be more. While collecting the data, it was found that several government tourism websites do not open or have not been updated for a long time. Upgrading the government tourism website, providing proper feedback, online advertisements can make good impacts. Bangladesh Parjatan Corporation Sponsorships can be introduced for domestic tourism for educational institutes and private tourism agencies so that study tours can be arranged regularly that will help to enhance concern and awareness about tourism among the youth (Hossain & Wadood, 2020). Tourism and hospitality infrastructural development is needed Bangladesh. These need to be the most modern and technologically sophisticated. Also, their updating and proper maintenance are essential. Sponsoring festival-based tourism can be another effective way to promote Bangladeshi culture and tourist destinations. The "Pahela Baishakh" (Bengali new-year), Bengali autumn festival, folk music festivals can be some of the primary tourism products. VAT-free concept of tourism for a foreigner can be marketed more. The NBR website mentions that tourist who stays less than 180 days and buys from specific outlets marked as "VAT Refund for Tourists". Showing the refund certificate issued by those outlets, the VAT money can be refunded and made cash from the Tourist Return Desk at Airport Customs Office (National Board of Revenue, 2020). It is essential to make the most prominent tourist destination (Cox's Bazar) of Bangladesh safe. Considering the risk of natural disasters, security concerns and taking cognizance of tourism interests in the area, the GoB has taken the initiative to transfer the Rohingya camps from Cox's Bazar to Bhashan Char (an island),

Noakhali. Bangladesh Navy has implemented the project costing USD280 million to construct shelters for Rohingyas on the island (Center for Policy Dialog, Bangladesh, 2020). Khan et al. (2020) states in their study that many hotels are under performing compared to the private hotels due to lesser investment and no modern revenue management system. However, Ivanov and Zhechev (2011) suggested hospitality industry will have to cope with online distribution and more intelligent approaches through technology in order to sustain.

Future Possibilities

Bangladesh's tourism is well flourished compared to its economic development in the past decade, as a good number of the population is overcoming poverty. According to The Daily Star (2020b), in the past decade, her economic growth was 6.5% on an average; but in the last three years, it had more than 7% growth. As a result, peoples' buying capacity and ability to afford tours to different destinations raised rapidly. Also, resorts and tourist spots opened up for business and created thousands of jobs for residents. At the end of 2017 (the last fiscal year), the per capita income raised to USD1,751 from USD703. Bangladesh does not get many foreign tourists, but it has huge number of local tourists who go out for vacations very often. According to different tour operators, the number of domestic tourists rose to 7 million in 2017 from 6 million from the previous year.

Conclusion

This chapter aims to outline the investment and development aspects of tourism and hospitality revenue collection in Bangladesh. The study finds that the number of both domestic and international tourists and revenue collection from the tourism industry in Bangladesh are increasing. Proper initiatives of the GoB and private sector investments can support the generation of more revenues. Also, this industry needs more professionals and client supportive services. Thus, future studies can cover post-COVID-19 context of the tourism industry.

References

Afroz, N., & Hasanuzzaman, M. (2012). Problems and prospects of tourism in Bangladesh Bandarban District case. *Global Journal of Management and Business Research, 12*(23), 22–30.

Arulmozhi, J., & Gomathinayagam, V. (2019). Marketing sport tourism in India. *International Journal of Advance and Innovative Research, 6*(2).

Bandara, R., & Tisdell, C. A. (2004). *Use and Non-use Values of Wild Asian Elephants: A Total Economic Valuation Approach.* University of Queensland.

Bangladesh Tourism Board (BTB). (2020). *Promotion and marketing*. Retrieved from: https://bit.ly/2RJVp6t. Accessed 28 June 2020.

Booking.com. (2020). *Resorts in Bangladesh*. Retrieved from: https://www.booking.com/resorts/country/bd.html. Accessed 26 July 2020.

Center for Policy Dialog, Bangladesh. (2020). *CPD's budget recommendations for FY2018–19* (p. 37). Dhaka.

Chowdhury, T., & Chowdhury, M. (2015). An analysis of guest occupancy and profit of private and public hotels in Cox's bazar. *Asian Business Review, 4*(3), 50.

Dhaka Dinner Cruise (2020). *Dhaka Dinner Cruise: Exploring Riverine Bangladesh*. Retrieved from: https://dhakadinnercruise.com. Accessed 15 Oct 2020.

Dhaka Tribune (2020a). *Sajek Valley to reopen in September*. Retrieved from: https://www.dhakatribune.com/bangladesh/2020/08/28/sajek-valley-to-reopen-in-september. Accessed 15 Oct 2020.

Dhaka Tribune (2020b). *World's Longest Marine Drive Opens in Bangladesh*. Retrieved from: https://www.dhakatribune.com/bangladesh/2017/05/06/bangladesh-gets-worlds-largest-marine-drive/. Accessed 19 July 2020.

Hossain, B., & Wadood, S. (2020). Potential unexplored? Tourism and economic growth of Bangladesh. *Journal of Tourismology, 6*(1).

Ivanov S. H., & Zhechev, V. (2011). Hotel revenue management - a critical review. *Tourism: An International Interdisciplinary Journal, 60*(2), 175–197.

Journey for Better Life (2020). *Travel to Birishiri, Netrokona*. Retrieved from: http://journey-for-better-life-bd.blogspot.com/2014/12/travel-to-birishiri-netrokona.html#.X4gAXtAzbIU. Accessed 27 June 2020.

Khan, M.Y.H., Islam, S.T., & Hassan, A. (2020). Revenue management in the tourism and hospitality industry with special reference to Bangladesh. In A. Hassan (Ed.), *Tourism Marketing in Bangladesh* (pp. 257–270). Routledge.

Ministry of Civil Aviation & Tourism (2010). *National Tourism Policy*. Retrieved from: https://bit.ly/3ktWvj5. Accessed 27 June 2020.

Ministry of Finance. (2019). *Budget Speech 2019–20* (p. 109). Dhaka.

Moniruzzaman, M., & Abedin, M. (2015). Tourism promotional dimension on Bangladesh prospective. *Global Journal of Management and Business Research: F Real Estate, Event & Tourism Management, 15*(1), 16–22.

National Board of Revenue (NBR). (2020). Business organization. Retrieved from: http://nbr.gov.bd/business-org/eng. Accessed 29 July 2020.

Nipa, J., Sultana, J., & Rahman, M. H. (2015). Prospect of private-public partnership in tourism of Bangladesh. *Journal of Investment and Management, 4*(3).

Public Private Partnership Authority (PPPA). (2020a). *Development of Integrated Tourism and Entertainment Village at Parjatan Holiday Complex, Bangladesh Parjatan Corporation, Cox's Bazar Under PPP*. Retrieved from: https://www.pppo.gov.bd/projects-tourism-entertainment-village.php. Accessed 13 July 2020.

Public Private Partnership Authority (PPPA) (2020b). *Development of Five Star Hotel and Zakir Hossen Road in Chittagong*. Retrieved from: https://www.pppo.gov.bd/projects-five-star-hotel-at-zakir-hossen-road-chittagong.php. Accessed 13 July 2020.

Roy, C. S., & Roy, M. (2015). Tourism in Bangladesh: Present status and future prospects. *The International Journal of Management Science and Business Administration, 1*(8), 53–61.

Somoy News (2020). *The country's first luxury cruise ship is going to sea*. Retrieved from: https://bit.ly/2HdF3kg. Accessed 15 Oct 2020.

Stainton, H. (2019). *TEFL Tourism: Principles, Commodification and the Sustainability of Teaching English as a Foreign Language*. CABI.

The Daily Star (2020a). *PM opens 'Haor Road' in Kishoreganj*. Retrieved from: https://www.thedailystar.net/country/news/pm-opens-haor-road-kishoreganj-1974525. Accessed 13 Oct 2020.

The Daily Star (2020b). *Tourism booming with economy*. Retrieved from: https://www.thedailystar.net/business/news/tourism-booming-economy-1673479. Accessed 15 Oct 2020.

The Independent (2020). *5th BIMSTEC Summit in Jan; Bhutan nominates Lekphell as next SG.* Retrieved from: http://www.theindependentbd.com/post/252647. Accessed 15 Oct 2020.

The Pacific Asia Travel Association (PATA). (2020). *PATA Bangladesh Chapter: COVID-19 Impact on the Tourism Industry in Bangladesh – PATA*. Retrieved from: https://www.pata.org/pata-bangladesh-chapter-covid-19-impact-on-the-tourism-industry-in-bangladesh/. Accessed 31 June 2020.

Tourism Nova Scotia (2020). *Tourism revenues*. Retrieved from: https://tourismns.ca/research/tourism-statistics/tourism-revenues. Accessed 1 Sep 2020.

U.S. Department of State-Bureau of Consular Affairs (2020). *Bangladesh travel advisory*. Retrieved from: https://travel.state.gov/content/travel/en/traveladvisories/traveladvisories/bangladesh-travel-advisory.html. Accessed 3 July 2020.

UN News (2020). *Rohingya refugee crisis*. Retrieved from: https://news.un.org/en/focus/rohingya-refugee-crisis. Accessed 29 July 2020.

United States Agency International Development (USAID). (2019). *Comprehensive private sector assessment*. Retrieved from: https://pdf.usaid.gov/pdf_docs/PA00TWMH.pdf. Accessed 27 June 2020, pp. 1–23.

World Tourism Organization (UNWT0). (2020). *International tourism highlights, 2019*. Retrieved from: https://www.e-unwto.org/doi/pdf/10.18111/9789284421152. Accessed 16 July 2020.

World Travel & Tourism Council (WTTC). (2019). *Travel & tourism global economic impact & trends 2019*. Retrieved from: http://ambassade-ethiopie.fr/onewebmedia/Tourism-WTTC-Global-Economic-Impact-Trends-2019.pdf. Accessed 27 June 2020.

World Travel & Tourism Council (WTTC). (2020). *Bangladesh 2020 annual research: Key highlights*. WTTC.

Part VI
Sustainability Practices

Chapter 21
Green Marketing, Investment and Sustainable Development for Green Tourism

Md. Nekmahmud and Maria Fekete-Farkas

Abstract Green tourism, a rapidly rising popular concept among tourists, investors, and other stakeholders in both developing and developed countries, offers sustainable, environmentally friendly, and socially responsible tourism services. This study aims to discuss the present scenario of the investment on green tourism and explains important driving forces for sustainable development of the tourism industry in the context of developing country's concept such as Bangladesh. It also describes green marketing, promotional mix, investment policy in the green tourism industry, and sustainable tourism development (STD) in Bangladesh. Besides, this chapter highlights the ways and the reasons that government reacts to actual or potential investments on green tourism development in terms of environmental and sustainable development. In methodology, a qualitative sampling study by a systematic literature review analysis was conducted. More particularly, the PRISMA statement is used to comprehensively evaluate green marketing, green tourism, and sustainable tourism development in developing countries, reviewing 70 research papers in Scopus and Web of Science database. The systematic literature review and findings of the chapter are expected to encourage new investors, tourist agencies, financial institutions, and governments to raise investments on green tourism aiming towards providing customized green tourism service. Green tourism contributes to economic growth and protecting the environment and culture of tourist areas. Finally, the practical significance and recommendations are discussed in the concluding section.

Md. Nekmahmud (✉)
Management and Business Administration Sciences, Doctoral School of Economic and Regional Sciences, Hungarian University of Agriculture and Life Sciences, Gödöllő, Hungary
e-mail: nekmahmud.argon@gmail.com

M. Fekete-Farkas, PhD
Institute of Economic Sciences, Hungarian University of Agriculture and Life Sciences, Gödöllő, Hungary

A. Hassan (ed.), *Tourism in Bangladesh: Investment and Development Perspectives*, https://doi.org/10.1007/978-981-16-1858-1_21

Keywords Green marketing · Green tourism · Green investment · Sustainable tourism development · Bangladesh

Introduction

Green marketing offers eco-friendly items such as refillable, ozone-friendly, nutritious foods, phosphate-free and recyclable products. In addition, green marketing may be used in environmentally friendly ways to meet consumers' needs, wants, and requirements while also protecting the environment and society (Nekmahmud & Fekete-Farkas, 2020). Likewise, green tourism is dynamically growing globally, offering tourism services based on sustainable, environment friendly and responsible tourism. Nowadays, both government and private sector are financing the green tourism industry for sustainable development. However, global warming, climate change, emissions, pollution, diminution of natural resources, food deficiencies, erosion, and other threats are the key challenges of current societies (Grosseck et al., 2019). Tourism has increased significantly over the last 70 years, owing to technological advancements in travel and transportation, economic development, and an increase in disposable income (Scott & Cooper, 2010). Bangladesh is one of the fastest-growing economies globally, offering a diverse range of tourist attractions (e.g. resorts, beaches, historical monuments, forests and tribal people, picnic spots, wildlife of various species) (Nekmahmud & Hassan, 2021). In 2017, the travel and tourism sector contributed 10.4 per cent of global GDP and 9.9 per cent of global employment. In Bangladesh, the tourism sector contributes 4.4 per cent of the total GDP and 3.8 per cent of total employment (2,414.4) (WTTC, 2019). While this scenario is a step forward compared to the past, the global picture indicates that the country has not yet reached its full potential. Tourism is one of the growing sectors in Bangladesh and international visitors praise its rich cultural heritage, beautiful natural scenery and people's hospitality (Ali, 2004). The marketing concept may be used in tourism by utilizing various marketing tools, strategies, and approaches (Popadopoulos, 1989; Calantone & Mazanec, 1991).

Environmental marketing or green marketing strategy can influence positively to achieve green tourism goals. For example, in Bangladesh, many tourism agencies, hotels, resurgents using green marketing tools to attract both foreign and local travelers (Nekmahmud et al., 2020). The Bangladesh government is also focusing on green investment in the tourism industry and making a sustainable tourism industry to attract foreign and local tourists. Tourism may assist a country in immediately create the required infrastructure to support not just tourists but also local communities (Johannesburg Summit, 2002). Besides, tourism may be the main factor for developing countries to alleviate poverty. Tourism may provide significant economic, social and environmental advantages, especially in rural and developing countries. Although mass tourism is connected with negative consequences and various tourist destinations, tourism policy may benefit the economy

(Roy & Roy, 2015). Thus, the tourism industry contributes to the achievement of the Sustainable Development Goal (SDGs). The United Nations (UN) (2020), announced the 17 Sustainable Development Goals (SDGs) in September 2015, a worldwide plan for ensuring a resilient future for all stakeholders. The 12th SDG focuses on maintaining sustainable consumption and production patterns. The aim and indicator are to establish and implement mechanisms to measure the impact of sustainable tourism on employment creation and promotion of indigenous culture and products (United Nations Development Program, 2015). The target of the 11th SDGs also describes strengthening conservation of the world's cultural and natural heritage (Simon et al., 2016). Climate change response in cities focused on poverty and gender sensitivity and mitigation contribute to SDGs achievement (Reckien et al., 2017). As a result, this chapter aims to describe both theoretical and practical consequences in green tourism and sustainable development. The chapter highlights the present scenario of the investment in green tourism. It explains which factors are essential for the development of the green tourism industry in Bangladesh. Furthermore, it explains green marketing strategy and promotional mix (7Ps) on the green tourism industry using a systematic review. At the same time, it bridges the gap between green tourism and sustainable development.

We also followed structured ways to present this systematic review chapter. The first section describes the methodology, research setting and PRISMA statement. The second section presents an overview and the current status of green tourism products and services, green marketing, and green marketing mix on tourism. The third section addresses investment in green tourism, investment policy in Bangladesh, sustainable tourism development (STD). Finally, the chapter highlights the key drivers of sustainable tourism business and makes recommendations for the development of green tourism in Bangladesh. The finding of the chapter supports the tourist agency, tourism industry and marketer to improve the value and consider the current wants or expectations of tourists to deliver more environmentally friendly products and services.

Methodology

This chapter reviews the literature on green marketing, investment, and sustainable development in the tourist sector, as published in reputable journals. The PRISMA model (Preferred Reporting Items for Systematic Reviews and Meta-Analyses) was used to perform a systematic review of green tourism in Bangladesh (Liberati et al., 2009; Moher et al., 2009). The primary goal of a systematic review is to give summaries of prior research in a specific field (Budgen & Brereton, 2006). The PRISMA paradigm assists researchers and academics in comprehending and perfecting the literature review (Liberati et al., 2009). There is no detailed timetable for the publishing date of the article in this chapter. We considered not only green tourism practices but also other aspects of sustainable tourism separately (e.g. sustainable tourism, sustainable development of tourism, green tourism investment). A series of research review approaches are incorporated on green tourism, and sustainable tourism includes keywords "green tourism", "environmental marketing or green marketing", "green marketing mix", "green marketing mixes on green tourism", "investment", "green investment", "sustainable tourism", "sustainable development", "sustainable development on tourism", and so on. The majority of the research articles were gathered from leading publishers such as Springer, Springer Nature, Elsevier, Emerald, Wiley, Sages, Taylor & Francis, and others. Along with the documents provided by those publishers, we gathered articles from other publications to compile the essential information (Ahi & Searcy, 2015).

To complete the research objectives, we also evaluated certain country and global annual reports on the tourism industry published by Bangladesh Parjatan Corporation (BPC), United Nations (UN) report, a statutory board under the Ministry of Civil Aviation & Tourism of Bangladesh, world tourism and World Travel & Tourism Council (WTTC). Nonetheless, our chapter was followed by a systematic review technique, which aided us in systematically concluding our research project. Figure 21.1 exhibits a systematic research framework to explain the research procedure.

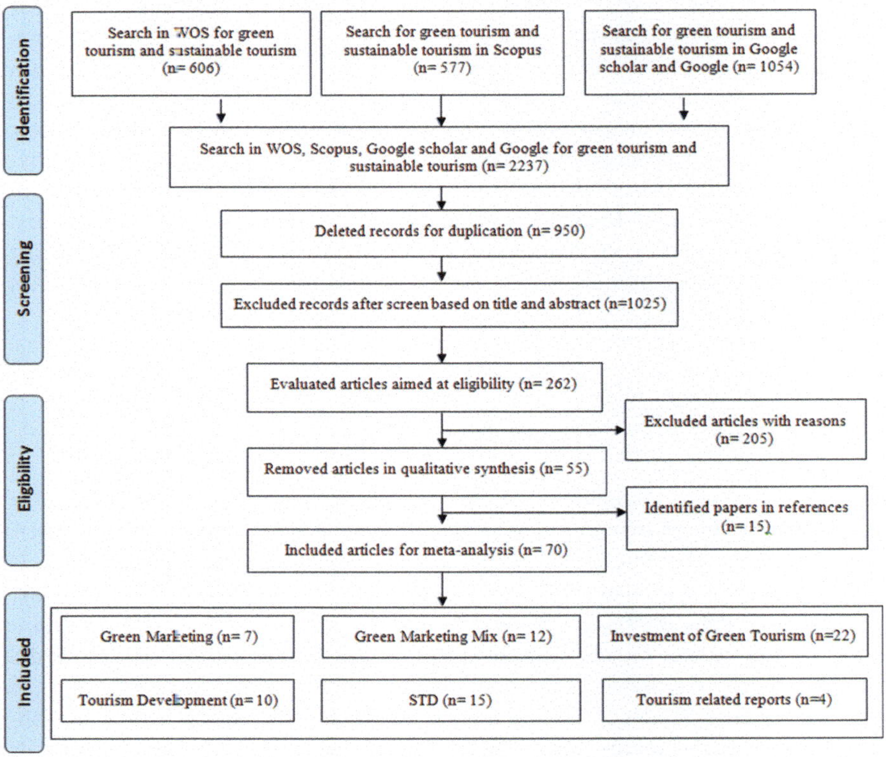

Fig. 21.1 Flowchart of study based on PRISMA statement. (Source: developed by the author's)

Tourism Places of Bangladesh

Bangladesh is naturally beautiful in South Asia, with stunning tourist attractions such as historic sites, picnic spots, resorts, beaches, forests, and tribal people, as well as diverse wildlife (Nekmahmud & Hassan, 2021). Tourists may enjoy water skiing, river cruise, fishing, hiking, sailing, rowing and sea bathing (Bangladesh has several tourist attractions, including Cox's bazaar (the world's longest uninterrupted stretch of clean sand), Dhaka (the capital known as the city of mosques and muslin), the Sundarbans (home to the majestic Royal Bengal tiger and spotted deer), Sylhet (home of famous religious saints' holy shrines and fascinating hills and tea gardens), Rangamati (the heart of the panoramic lake district), and Chittagong (the country's largest port city known as the city of holy shrines), Mohasthangarh, Mainamati, and Paharpur are the country's archaeological monuments from 300 BC to 1200 AD. The awe-inspiring landscape, vibrant tribal life, and simple rural life of the friendly millions are the primary motivators for potential travelers.

Green Tourism

'Green Tourism' refers to any tourism activities in a natural location where the focus is on the natural resource or tourism that is deemed ecologically friendly (Henion, 1976). An important role of green tourism is to conserve large and small places and their accompanying wildlife for future generations (Meler & Ham, 2012). Therefore, green tourism, a subset of ecotourism, is low-impact tourism aimed at preserving an area's ecology and culture. Although the United Nations (2020) has established certain requirements for ecotourism, green tourism can encompass a wide variety of norms and circumstances, ranging from totally compliant to less harmful to the environment than conventional tourism.

Green tourism is a reflection of sustainable tourism practices that balances the demands of the ecology and environment, residents, businesses, and visitors (Azam and Sarker, 2011; Ibnou-Laaroussi, Rjoub and Wong, 2020). It is committed to minimizing the harmful environmental and social consequences of its tourist activities. Green tourism is a way of travelling that reduces your harm to the environment while you travel. Green travel practices include:

- To choose eco-friendly hotels
- To choose eco-friendly modes of transportation
- To limit or even eliminating flights
- Not over-ordering in restaurants to prevent waste
- To eat local and organic foods
- To reduce your use of plastic and other garbage while you travel

Green Marketing

Green marketing can be defined as a holistic management process that emerges from societal marketing (Peattie and Crane, 2005). Green marketing is responsible for profitably and sustainably identifying, predicting, and serving the needs of consumers and society (Peattie, 1995). Polonsky and Rosenberger III (2001) defines "green marketing as all activities designed to generate and facilitate any exchanges intended to satisfy human needs or wants, such that the satisfaction of these needs and wants occurs, with a minimal detrimental impact on the natural environment". It became increasingly popular due to the rising importance of environmental sustainability (Dangelico and Vocalelli, 2017; Nekmahmud & Fekete-Farkas, 2020). Additionally, green marketing is a progression of thoughts toward environmental concern in decision-making, philosophy, ecological responsibility, and general organizational tools (Sharma et al., 2010). Green marketing brings ecosystem compatibility and achieves the goals of the consumer and organizational satisfaction than traditional marketing.

Green Marketing for Green Tourism

The application of green marketing in tourism is a proactive approach to stakeholders and customers. Green marketing now applies in the tourist business, which considers the tourism product providing based on the idea of aggregate supply relating to specific partial goods. Green marketing offers environmentally friendly products and services which is reusable, recycling, and remanufacturing (Hasan et al., 2019). Likewise, green tourism is low-impact tourism that aims to conserve an area's ecology and culture (Nekmahmud, 2020). Green marketing is an essential precondition for the operation of green tourism and is interconnected, having reciprocal and mutual effects (Meler & Ham, 2012). New concepts like social media or electronic word of mouth continue to develop in green tourism marketing. Green marketing and green tourism in Bangladesh are yet badly undeveloped (Nekmahmud et al., 2021). But now, the government and private tourism companies (e.g. hotels, restaurants, and tourist agencies) are focusing green marketing or environmental marketing strategy on green tourism.

Green Marketing Mix on Tourism

Kotler and Keller (2018) define a variety of fundamental marketing terminology, including marketing mix (product, price, place, and promotion), marketing strategy (segmentation, differentiation, targeting, and positioning), marketing communication mix, and branding. Several marketing academics (e.g. Polonsky & Rosenberger

III, 2001; Davari & Strutton, 2014; Hasan et al., 2019) focus on studying the classic green marketing mix. The classical marketing mix model is known as the 4Ps (price, product, place, and promotion) which were structured by McCarthy in 1975 (McCarthy, 1975). The marketing mix refers to the "seller's view of the marketing tools available for influencing buyers" (Kotler & Keller, 2006: p. 19). Tourism marketing takes use of the adaptation of the novel marketing mix (Pomeringa et al., 2011). Morrison (2009) added eight elements to the tourism context, which is a popular tourism marketing paradigm. Additionally, Bao (2018) incorporated 10 elements into the tourism marketing mix: product, price, promotion, and place (McCarthy, 1960); participants, process, and physical evidence (Booms & Bitner, 1981); and partnership, packaging, and programming (Morrison, 2009). Nevertheless, Polonsky and Rosenberger III (2001) introduced green 4Ps for the implementation of green marketing strategies. We apply an enlarged marketing mix (7Ps) to the green tourism industry, which is applicable for hospitality service providers, green tourism organizations, and other environmentally friendly travel-related firms.

Product

Production is "anything that can be offered to a market to satisfy a want or need" (Kotler & Keller, 2006: p. 372). Green products and services are defined by their capacity to have less negative environmental consequences than comparable products and services (iSustainableearth, 2014). The destination may be the final product in tourism, comprised of both private and public products, or it may be an "industry" component, such as an attraction, restaurant, accommodation facility, or tour operator (Pomering et al., 2011). Green tourism products need the commitment of all relevant stakeholders and strong government commitment to guarantee that tourism products meet their economic, environmental and socio-cultural goals. Different sorts of green tourism products along the "tourism supply chain" participate in the green travel market such as green destinations, protected areas, organic food or local vegetarian/vegan foods, green hotel and services, green restaurant, environmental-friendly travelling products, environmental-friendly travel agency, environmental friendly incoming tour operators and transport services, recyclable, reusable, Bio-products and local green shop and its products (Nekmahmud et al., 2021).

Price

The pricing of tourism services is a very key factor compared to other elements of the marketing mix. The price is the amount of money paid by visitors for a product or service. The price must cover the production cost and make a profit (Bowie &

Buttle, 2004). While specialist sustainable tourism goods may command premium pricing, the price rarely reflects the entire cost of the product's negative externalities. The primary focus of tourism pricing has been on cheap prices to attract large volume and hence large profit margins for businesses (Swarbrooke, 1999). Sustainability principles also entail ensuring that tourists feel they have got value for money, rather than leaving them with the notion of being exploited (Batra, 2006).

Place

The term "place" refers to the 'direct or indirect channels to market, geographical distribution, territorial coverage, retail outlet, market location, catalogues, inventory, logistics and order fulfilment' (Blythe, 2009: p. 146). The place where tourists purchase tourism products varies significantly. For example, travel agents, tour wholesalers, and tour operators are the distribution channels for tourism products. Product or service must be offered in the appropriate place, at the right time and in the right quantity to serve the demands of tourists. Convenience, easy information search about the product, and related elements are factors of critical importance in placing and distributing business offerings into the tourism markets (Bao, 2018).

Promotion

Promotion is "how firms attempt to inform, persuade, and remind consumers – directly or indirectly – about the products and brands that they sell" (Kotler & Keller, 2006: p. 536). Promotional activities may also be utilized to influence tourists purchasing decisions. For instance, it aims to promote more environmentally friendly modes of transport, such as train travel, via its websites and marketing campaigns in addition to reducing the carbon footprint of Scotland's 75% of visitors who arrive by car or air (Lane, 2009). One effective promotional instrument for green marketing strategy is green promotion. There are numerous ways to express green practices to consumers, including green product labels or eco-labels, green packaging, advertising through trade publications and newsletters, flyers, leaflets, and social media. Social media and eco-labels are the two most effective platforms for promoting green marketing activities in the tourist and hospitality industries. Ecolabels are defined as the hotel industry's assurances of high-quality green operations. A tourism eco-label is a quality certification that minimizes environmental adverse consequences and increases standards of environmental quality. Numerous ecolabels are used in the hospitality sector, including ISO 14001. Nowadays, green hotels and restaurants are responsible for energy & water saving, using greenhouse gas, waste management, reusing, and recycling strategy.

Physical Evidence

Physical evidence consists of "the environment in which the service is delivered and where the firm and customer interact, and any tangible components that facilitate performance or communication of the service" (Zeithaml et al., 2006: p. 27). Physical evidence is critical in the tourist environment, both at the firm level and at the destination level (Urry, 1992). Tourism firms must continue to spend in refurbishing to maintain high levels of repeat and recommended sales and to foster consumer loyalty (Bowie & Buttle, 2004; Morrison, 2002). Tourism businesses that fail to maintain their physical environments will ultimately get fatigued and will be forced to compete on price to attract tourists, resulting in lower profitability.

People

People are vital to promoting any product or service. In the service industries context in general and in the tourism industry in particular, people are not producers, but rather the products themselves (McLean, 2002). The individual who provides the services are the ones who represent the company and a key to the success of the experience. People deliver services in different circumstances, including travel agencies, tourist guides, hotels and restaurants, and transport companies. The physical appearance, dressing code, attitude, behaviour, knowledge/competencies, green knowledge, and skills greatly impact customers' perception of the tourism offering and experience.

Process

Process is described as how the service is accumulated, the "actual procedures, mechanisms, and flow of activities by which the service is delivered – the service delivery and operating systems" (Zeithaml et al., 2006: p. 27). The term "process" refers to "the combination of actions that results in the delivering of product advantages" (Rafiq and Ahmed, 1995). In the tourist industry, the process may refer to the transition from paper-based to digital booking and account management systems, as well as the management of the food and beverage supply chain and labour, the use of low-carbon transportation modes, such as hybrid or electric vehicles, to transport visitors to, within, and from a destination; and the implementation of effective waste reduction and recycling systems, particularly in environmentally vulnerable locations, for example, national parks, islands, and protected areas (Bohdanowicz et al., 2011).

Investment in Green Tourism in Bangladesh

Green investment is the expenditure needed to minimize greenhouse gas emissions and air pollutants without considerably affecting non-energy products' production and consumption. In the tourist sector, investments are crucial for the tourist business to flourish. Both government and private companies are involved in the development of the tourist sector in Bangladesh. Governments and private owners should invest some of their capital in the growth of tourism (Shamsuddoha & Nedelea, 2008). The government should implement a long-term strategy for ecotourism and tourism sustainable development. Figure 21.2 illustrates the investment in the green tourism industry by various stakeholders.

Nowadays, the government is investing in the green tourism industry, for example, eco transportation, ecotourism zoon, marine drive road, green infrastructure development, conservation of biodiversity, marine, natural resources, and land ecosystems, etc. The private sector in tourism is being provided with a variety of concessionary investment incentives. Commercial banks provide major loan for financing to the tourist industry. Bangladesh has a limited foreign exchange market (Shamsuddoha & Nedelea, 2008). The private sector is invested in organic food & restaurant, eco-friendly tourism products & services, green hotel, recycling tourism products, energy efficiency, green transportation, etc. On the other hand, Public-Private Partnerships (PPPs) is an essential instrument that governments may use to provide vital infrastructure services. PPPs are a method of contracting for services that makes use of private sector experience and innovation. They

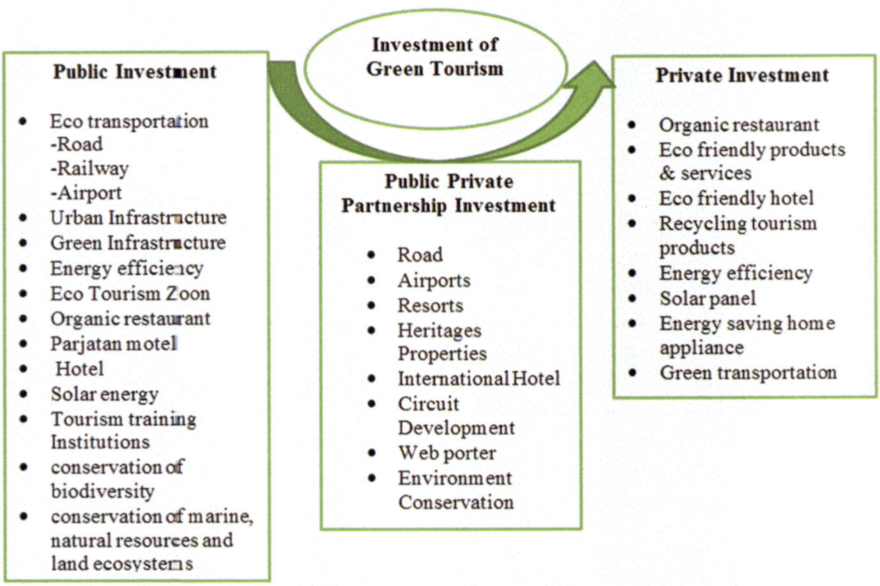

Fig. 21.2 Investment of Green tourism industry. (sources: illustrated by author's, 2021)

regularly force private finance in development in the tourism industry for development in road, airports, resorts, heritage properties, international hotel, circuit development, web porter, and environmental conservation. PPPs implements under the correct circumstances enhance economic growth, enhance service delivery, and improve tourism in Bangladesh.

Tourism Developments and Present Status of Tourism in Bangladesh

The tourism business in Bangladesh has enlarged significantly over the previous decade, resulting in the establishment of new travel agencies and hotels, as well as increased rivalry among them. According to Bangladesh Parjatan Corporation (2019), the Bangladesh government developed the tourism industry by reconstructing (e.g. new Parjatan holiday complex, kids zone, money exchange, tourist ship, international hotel, children amassment park, dusty free shop, ecotourism park). Private sectors build many 5-star and 4-star hotels and restaurants for tourists. In 2019, the total budget of the tourism industry was 6169.00 (Lakh BDT) for development, and total expenditure was 3848.43 (Lakh BDT). The Table 21.1 summarizes Bangladesh Parjatan Corporation's revenue, expenditure, and profit for the fiscal year 2019-2020.

Green Tourism and Investment Policy in Bangladesh

Bangladesh government approved the first tourist policy in 1992. On 14 December 2009, Bangladesh released its National Tourism Policy in response to global demand. It includes 31 objectives and goals, as well as short policy actions. The

Table 21.1 Income, expenditure, and profit of Bangladesh Parjatan Corporation for 2019–2020

Fiscal year	Total income (Lakh Taka)	Total expenditure (Lakh Taka)	Profit before tax (Lakh Taka)
2015–2016	3403.32	2602.82	788.98
2016–2017	3888.76	2896.86	976.76
2017–2018	4485.04	3191.78	1273.82
2018–2019	3433.67	2593.82	820.81
Total	15210.79	11284.88	3859.97

Source: Bangladesh Parjatan Corporation (2019)

most fundamental issue on which the Bangladesh tourism policy might be criticized is that the policy is published in the Bengali language on the official website of the Ministry of Tourism and Civil Aviation (2012). The tourism policy has no rules and regulations related to green investment and sustainable development. But it has one objective of policy 2010 that "enhancement and maintenance of tourism resource through the development of ecotourism while protecting the balance of environment and neighbourhood". Besides, the primary objectives of Bangladesh's National Tourism Policy 2010 are: establishing a strong and secure foothold for Bangladeshi tourists through coordination with relevant government organizations; creating a tourism-friendly environment, and marketing its tourism potential both domestically and internationally; and raising public awareness about protection, development, and promotion in tourism.

In 1992 the Bangladeshi National Tourist Policy aimed at investing, conserving, developing and maintaining tourism resources and opening up a recognized industry for private capital investment. The policy is ambiguous, and it should be informed about global trends and based on the most up-to-date sociopolitical/economic facts available (Hassan & Burns, 2014).

Bangladesh government should enact new legislation and regulations for the growth of sustainable tourism. But it has taken the initiative to introduce ecotourism zoon for protecting the tourism business. However, the existence of a tourism strategy that acknowledges the value of tourists, the tourist trade, investors, and interest groups increases and strengthens a tourist destination's legitimacy and appeal. In the absence of such a tourist strategy, there is typically insufficient knowledge regarding the destination's current condition of tourism, attitudes toward tourist, and future orientations for tourism. At the national level, policy formulation involves professional knowledge of global shifting forces as well as the availability of local resources and voices. An efficient tourist policy plan should incorporate theoretical and thematic considerations (Hassan & Burns, 2014).

Sustainable Development

Mostly, development implies a process that increases the living conditions of the human being (Bartelmus, 1986). "Development is not just about increased wealth. It means change; changes in behaviour, aspirations, and in the way which one understands the world around one" (Dudley, 1993: p. 165). Even, it encompasses larger concerns of the quality of life (e.g. infant mortality, life expectancy and educational attainment, access to basic freedoms, spiritual welfare, and nutritional status) (Pearce et al., 2013). In this point of view, the United Nations World Commission on Environment and Development (WCED) defines sustainable development as "a development that meets the needs of the present without compromising the ability of future generations to meet their own needs" (WCED, 1987: p. 42). Thus, sustainable development is a type of development plan that maximizes long-term wealth and welfare by managing all-natural resources, assets, and physical assets, as well as humane sources.

Sustainable Development for Tourism (STD)

The concept of "sustainable tourism" is derived from the more general concept of "sustainable development". STD is a concept that describes balanced economic, social, and cultural development without harming the environment, allowing for sustained growth. Sharpley (2000) defined the concept of sustainable tourism in two ways: as an economic activity and as a component of wider sustainable development. Figure 21.3 presents the conceptual model of STD proposed by (Grah et al., 2020), where sustainable urban tourism development merges. Tourism is a socio-cultural and economic phenomenon having extensive economic, social, cultural and environmental consequences (Padin, 2012). Sustainable tourism also depends on economic, ecological, and social & cultural sustainability (WTTC, 1995).

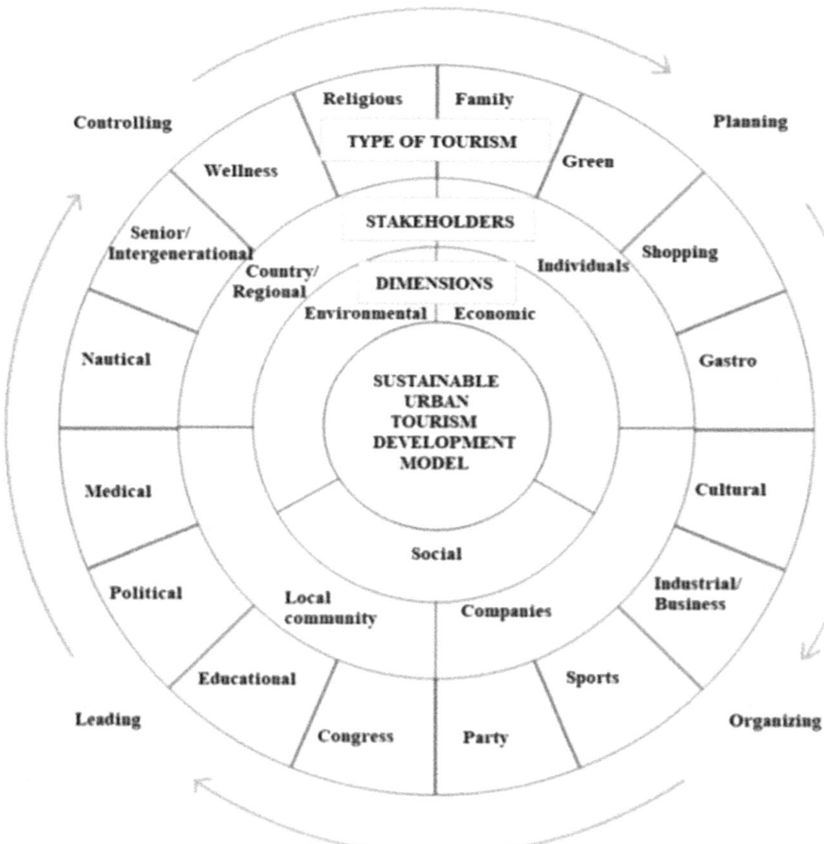

Fig. 21.3 Sustainable urban tourism development model. (Sources: adapted from Grah et al., 2020)

Social and Cultural Sustainability

Sustainable social and cultural development ensures compatibility, which is accomplished via a community's traditional values while also enhancing its identity (Angelevska-Najdeska & Rakicevik, 2012; WTTC, 1995). It demonstrates the host communities' sociocultural authenticity, traditional values, and the preservation of their architectural and cultural heritage, while also contributing to multicultural understanding and tolerance (Pomering et al., 2011).

Environmental Sustainability

Environmental or Ecological sustainability refers to the practice of certifying development that is sustainable with the preservation of critical ecological processes, biological diversity, and biological resources. Thus, sustainable tourism development should make use of environmental resources that support important ecological processes and contribute to the conservation of natural heritage and biodiversity.

Economic Sustainability

Economic sustainability enables cost-effective development and resource management to support future generations. It proves feasibility by providing socioeconomic benefits, long-term economic operations that are equally distributed to all stakeholders, including stable employment and income-generating opportunities, as well as social services to host communities, and contributing to poverty reduction (Pomering et al., 2011).

Nonetheless, different local, national, and global stakeholders are active in the development of sustainable tourism, including governments, non-governmental organizations, tourism industry sectors, host communities, travelers, academics, and the media (Crah et al., 2020; Ellis & Sheridan, 2014; Swarbrooke, 1999). Thus, regardless of their role in tourism, all stakeholders may express their concerns and contribute to sustainable growth. In Bangladesh, there are several categories of tourism, such as cultural, historical, natural, sports, religious, nautical, industrial, and so on. To achieve STD, the implementation of SDGs and stakeholder involvement and interaction are very mandatory. By contrast, creating sustainable tourism should consider a wide range of consequences and activities described by Horner and Swarbrooke (2009):

- Resource usage – water and land usage.
- Wildlife – Safaris, hunting, zoos, natural environments.
- The host community – allocation of advantages from tourism, degree of control.
- Tourist – cultural issues, behaviour abroad.

- Operational practices of tourist enterprises (hotels, resorts, points of attraction) – recycling, buying, waste disposal, energy efficiency
- Public sector policies – financing, legislation, planning.
- The industry – self-regulation, norms of practice, engagement with the host community, employment policies, environmental measures.
- Conservation policies and practices – landscapes, townscapes, animals.
- Pollution – air, water, noise, visual.

Linking Green Tourism and Sustainable Development

Green tourism and sustainable development are interconnected with each other. Green tourism is the key indicator for sustainable tourism development. Sustainability can be considered as an integrated paradigm when evaluating links between tourism and the natural environment (Throsby, 2016). Tourism must be rebuilt in this century with consideration to changing global linkages and social structures, technological innovations, increased spatial awareness and environmental issues (Pigram & Wahab, 2005). The combination of environmental efforts within the larger development framework will allow achieving the SDGs. Since the Earth Summit in 1992, demand has risen to strengthen the environmental performance of the tourism industry alongside other economic sectors (Pigram & Wahab, 1997; 2005), the movement to create green tourism across the globe. Besides three sustainability concepts such as ecological, economic and social, cultural sustainability are an important issue, especially when tourist operations are based on local or indigenous culture and tradition. Globally sustainable tourism development remains a telic idea. The path to sustainability is vibrant for present and future economic, ecological and socio-cultural well-being (Murphy & Price, 2005) and should be connected with community-based economics, equity conservation and environmental integration (Fig. 21.4).

Driving Forces of Sustainable Tourism Development in Bangladesh

There are several driving forces for sustainable tourism development (STD) in Bangladesh, especially accommodation, new rules and regulations, transportation development, infrastructure development, political, cultural, community development, culture protection and tourism taxes, and economic development and so on. The notable key forces of the tourism industry depend on different kinds of infrastructure improvements such as new roads, improvement in telecommunications and electrical power plants to meet the needs of the tourism industry together as well as offering benefit to all members of the host community (Singh & Mishra, 2004). Figure 21.5 exhibits a different kind of driving forces of sustainable tourism development in Bangladesh.

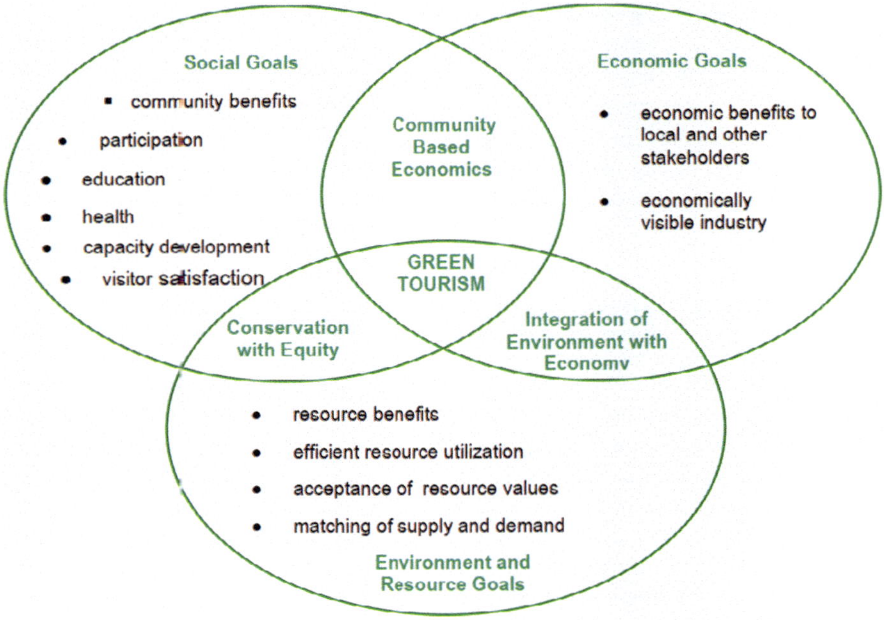

Fig. 21.4 A model of Green tourism values and principles. (Source: Adapted from Hall et al. 1997)

Fig. 21.5 Driving forces of sustainable tourism development (STD) in Bangladesh. (Source: explanations by author's, 2021)

Accommodation

Accommodation is the vital driving force of tourism development in Bangladesh, which directly links with tourism service (e.g. hotels, restaurants, lodges, guest-houses, etc.). Firstly, it promotes creating additional job opportunities to serve visitors requirements and, secondly, it supports trade and commerce (Singh & Mishra, 2004). New hotels, motels, guest houses, and apartments are built based on the tourism industry, which is financed by both public and private. At the present, tourist spots in Bangladesh have available accommodations such as private hotels, state tourism Parjatan motels, huts/cottages, tourist hotels, forest department restrooms etc. There are available accommodation facilities in Cox's Bazar, Bandarban, Sylhet, and others tourist destination.

Transportation

Local transportation provides an immediate benefit in terms of accessibility and significantly reduces travel time for visitors and members of the local community. Local transportation connects all tourist fascinations directly to the destination and indirectly generates job opportunity for private tour operators, auto-rickshaw drivers, and taxi drivers, among others (Singh & Mishra, 2004). There are not available local city bus services and adequate space for parking in Bangladesh. Even the public bus service is insufficient to view local attractions, and alternative modes of transportation are extremely expensive in comparison to the public bus.

Tourism, Taxes, and Economic Development

Taxes imposed on tourism development can contribute to local governments reduce their tax burdens. Tourism is frequently considered an economic driver for regional development (Singh & Mishra, 2004). Economic systems should operate under specified cultural parameters. Local administrative tasks include issuing permits and operating times, providing road maintenance, controlling architectural designs, promoting public financing and taking other actions that are affecting business in the tourist industry. The economy is a major factor that affects the tourism industry. For example, when the economic condition is so bad, people do not wish to travel as much. According to econometric models, the GDP growth rate in Bangladesh was 7.8 in 2018 and is expected to reach around 7.20 percent in 2021 and 7.30 percent in 2022. So day by day, it is increasing, and it also a key-driven force of sustainable tourism in Bangladesh.

Infrastructure Development

There are several issues with basic infrastructure facilities that are monitored and controlled by many authorities and departments. In Bangladesh, governments, private and public-private partnerships (PPP) are investing and budgeting for the development of infrastructure. When the infrastructure is developed, thus it helps to increase the sustainable tourism industry. Furthermore, It contributes to the growth of GDP. The main features of green tourism include environmental conservation and education, as well as income distribution to indigenous people through significant partnerships

Political Stability

Political stability is an important issue of the macro environment for achieving sustainable development. If a country is free of wars, conflicts, and terrorism, visitors are willing to visit.

Achieving Sustainable Development Goals (SDGs)

Bangladesh is a fast-growing developing country and trying to achieve 17 sustainable development goals (SDGs) and its target. Sustainable tourism or green tourism is part of SDG. Therefore, the tourism industry is contributing to achieving SDGs effectively. For instance, the 12th SDG focuses on ensuring sustainable consumption and production patterns, with the target and indicator aimed at developing and implementing instruments to assess the impact of sustainable tourism on sustainable development. Sustainable tourism generates jobs, promotes indigenous culture and products, and implements a variety of sustainable tourism strategies, policies, and action plans with adequate monitoring and assessment. Bangladesh government takes initiatives to fill up the SDGs within 2030. That is why SDGs is also the key driving force of sustainable tourism development.

Resources and Renewable Energy

Bangladesh has different types of tourism resources such as resorts, beaches, historical monuments, urban tourism, tribal people, forests, and wildlife of various species. Utilizing these resources are the key driving for sustainable tourism development. The government of Bangladesh should also invest in these resources to attract tourists and create employment opportunities. Nonetheless, renewable energy in Bangladesh refers to the generation of electricity from renewable energy sources such as biogas, biomass, hydropower (Islam, 2013), solar, and wind (Akter, 1997). Renewable energy also cuts greenhouse gas emissions by up to 20% and creates full-time equivalent jobs. So it is also driving forces of sustainable tourism development.

Employment and Cultural Development

The job crisis is one of the world's most serious problems. In Bangladesh, a key source of concern is the youth unemployment rate. Therefore, the tourism industry can help to create huge job opportunities. Bangladesh's government is now investing in the tourist sector in order to expand employment opportunities. On the other hand, conserving, and protecting all cultural and natural treasures are the key drivers of sustainable tourism development. The 11th SDG aim further reinforces the conservation of world cultural and natural heritage (Simon et al., 2016).

Conclusion and Recommendations

Greening is a crucial issue in the tourism industry. People are more conscious of environmental issues and seek environment-friendly services. The implication of environmental marketing or green marketing is essential for achieving green tourism and sustainable tourism. On the other hand, Bangladesh is a third-world country with limited financial resources, but it may boost GDP by focusing a reliance on tourism. Bangladesh Parjatan Corporation (BPC) is the country's only government-owned tourism operator that should employ a modern marketing strategy. Bangladesh has several potential tourist destinations, but it needs to increase its visibility, and it has several prospects to gain international and domestic revenue from tourists. Foreign tourists are concerned about environmental issues. Therefore, if you want to attract foreign tourists, you have to invest more in the green tourism industry. Green tourism and sustainable tourism development are parts of the Sustainable Development Goals (SDGs) that will aid in their achievement.

In this chapter, we have put forward some suggestions and recommendations for the development of the sustainable tourism industry. Suppose the Bangladesh Parjatan Corporation (BPC) follows on from these long-term suggestions. In that case, it will become a revenue-generating organization, and the Bangladesh tourism industry will be the main earning source from foreigners. Bangladesh government should formulate an environmentally friendly or green tourism policy for sustainable development. In addition, the private sector should push the government to cooperate in the development of the tourist business.

After reviewing the related published papers and Bangladesh government policy 2010, it has found many complications regarding green marketing, investment, and sustainable development in the tourism industry. Following actions are important for the growth of the Bangladesh green tourism sector: first, there is no international standard tourism education to develop marketing or salespeople to effectively offer the tourism product in local and worldwide perspectives. Tourist-related disciplines like hotel and tourism management, ecotourism and sustainable tourism, tour guide training, leisure and recreation studies should be introduced at all universities and colleges. Besides, it should begin training programs and workshops related to tourism and tourism marketing. Second, the Bangladesh government should open an international tourism information desk in train stations, bus stations, and airports to give foreign tourists facilities. Third, Bangladesh Parjatan Corporation (BPC) should put pressure on government authority to pursue its sustainable strategy more carefully to explore the place, investment and growth of tourism. The government should encourage private investors to green investment in the proper place. Fourth, the government should make environment friendly and sustainable tourism policy for developing green tourism. Fifth, the government should encourage reducing the tax on environment friendly and green tourism products and services. Sixth, the government should invest in renewable energy in the tourism industry. Seventh, the government should take action about the security system at the tourist place. Eighth, infrastructure for tourism should be developed and maintained. There is a demand for a rail link between Cox's Bazar and Chattogram because this communication infrastructure alone is insufficient to ensure visitor attraction. Ninth, maintaining political stability to attract international tourism. Ninth, to attract international tourists, political stability must be maintained. Tenth, different types of tourist vehicles (i.e. buses, taxis, rental cars, boats, etc.) should be well maintained so as not to cause unnecessary air and noise pollution and to employ non-polluting vehicles such as electric automobiles, eco-cars, energy-saving or shuttle buses in resorts (Altinay & Hussain, 2005). Eleventh, road and other transportation systems should develop adequately in order to stop traffic jamming, and ensuring the use of mass transit and pedestrian systems. Twelfth, applying appropriate architectural design standards and using construction materials for all structures in tourism regions to make them architecturally and environmentally friendly and applying energy conservation design strategies (Altinay & Hussain, 2005). Finally, solid waste disposal systems should be implemented appropriately, with waste tourist products recycled to the maximum extent feasible.

References

Ahi, P., & Searcy, C. (2015). An analysis of metrics used to measure performance in green and sustainable supply chains. *Journal of Cleaner Production, 86*, 360–377.

Akter, N. (1997). Alternative energy situation in Bangladesh: A country review. In *Regional Training Orientation Course on Alternative Energy Technologies*. Approtech Asia, Philippine Social Development Centre.

Ali, M. (2004). *Bangladesher sandhanay*. Student Ways. (in Bengali).

Altinay, M., & Hussain, K. (2005). Sustainable tourism development: A case study of North Cyprus. *International Journal of Contemporary Hospitality Management, 17*(3), 272–280.

Angelevska-Najdeska, K., & Rakicevik, G. (2012). Planning of sustainable tourism development. *Procedia-Social and Behavioral Sciences, 44*, 210–220.

Azam, M., & Sarker, T. (2011). Green tourism in the context of climate change towards sustainable economic development in the South Asian region. *Journal of Environmental Management and Tourism, 1*(3), 6–15.

Bangladesh Parjatan Corporation (BPC) (2019). *Home*. Retrieved from: http://www.parjatan.gov.bd/. Accessed 1 Dec 2019.

Bao, H. (2018). Marketing of tourism services/experiences. In M. Sotiriadis (Ed.), *The Emerald handbook of entrepreneurship in tourism, travel and hospitality: Skills for successful ventures* (pp. 261–276). Emerald Publishing.

Bartelmus, P. (1986). *Environment and development*. Allen & Unwin Australia.

Batra, A. (2006). Tourism marketing for sustainable development. *ABAC Journal, 26*(1), 59–65.

Blythe, J. (2009). *Key concepts in marketing*. Sage.

Bohdanowicz, P., Zientara, P., & Novotna, E. (2011). International hotel chains and environmental protection: An analysis of Hilton's *we care!* Programme (Europe, 2006–2008). *Journal of Sustainable Tourism, 19*(7), 797–816.

Booms, B. H., & Bitner, M. J. (1981). Marketing strategies and organizational structures for service firms. In J. H. Donnelly & W. R. George (Eds.), *Marketing of services* (pp. 47–51). Chicago, IL.

Bowie, D., & Buttle, F. (2004). *Hospitality Management an Introduction*. Retrieved from: https://images.template.net/wp-content/uploads/2015/10/01204854/Free-Hotel-Marketing-Plan-Template.pdf. Accessed 29 Dec 2019.

Budgen, D., & Brereton, P. (2006). Performing systematic literature reviews in software engineering. In *Proceedings of the 28th international conference on Software engineering* (pp. 1051–1052). ACM.

Calantone, R. J., & Mazanec, J. A. (1991). Marketing management and tourism. *Annals of Tourism Research, 18*(1), 101–119.

Dangelico, R. M., & Vocalelli, D. (2017). "Green marketing": An analysis of definitions, strategy steps, and tools through a systematic review of the literature. *Journal of Cleaner Production, 165*, 1263–1279.

Davari, A., & Strutton, D. (2014). Marketing mix strategies for closing the gap between green consumers' pro-environmental beliefs and behaviors. *Journal of Strategic Marketing, 22*, 563–586.

Dudley, E. (1993). *The critical villager: Beyond community participation*. Routledge.

Ellis, S., & Sheridan, L. (2014). A critical reflection on the role of stakeholders in sustainable tourism development in least-developed countries. *Tourism Planning & Development, 11*(4), 467–471.

Grah, B., Dimovski, V., & Peterlin, J. (2020). Managing sustainable urban tourism development: The case of Ljubljana. *Sustainability, 12*(3), 792.

Grosseck, G., Țîru, L. G., & Bran, R. A. (2019). Education for sustainable development: Evolution and perspectives: A bibliometric review of research, 1992–2018. *Sustainability, 11*(21), 6136.

Hassan, A., & Burns, P. (2014). Tourism policies of Bangladesh—A contextual analysis. *Tourism Planning & Development, 11*(4), 463–466.

Henion, K. E. (1976). Ecological marketing. Grid.

Huq, S. M., Nekmahmud, M., & Aktar, M. S. (2016). Unethical practices of advertising in Bangladesh: A case study on some selective products. *International Journal of Economics, Finance and Management Science, 4*(1), 10–19.

Islam, M. (2013). *Renewable energy prospects & trends in Bangladesh*. Bangladesh Power Development Board.

Kotler, P., & Keller, K. L. (2006). *Marketing management* (12th ed., p. 536). Upper Saddle River, NJ.

Kotler, P., & Keller, K. L. (2018). *Marketing management*. Pearson Education Limited.

Lane, B. (2009). Implementing sustainable tourism in Scotland: An interview. *Journal of Sustainable Tourism, 17*(6), 747–752.

Liberati, A., Altman, D. G., Tetzlaff, J., Mulrow, C., Gøtzsche, P. C., Ioannidis, J. P., Clarke, M., Devereaux, P. J., Kleijnen, J., & Moher, D. (2009). The PRISMA statement for reporting systematic reviews and meta-analyses of studies that evaluate healthcare interventions: Explanation and elaboration. *BMJ (Clinical Research Ed.), 339*, b2700.

McCarthy, E. J. (1960). *Basic marketing: A managerial approach*. R.D. Irwin.

McCarthy, E. J. (1975). *Basic marketing: A managerial approach*. R.D. Irwin.

McLean, R. (2002). *The 4 C's versus the 4P's of marketing. Custom fit online*. Retrieved from: http://www.customfitonline.com/news/2012/10/19/4-cs-versus-the-4-ps-of-marketing. Accessed 30 Sept 2017.

Meler, M., & Ham, M. (2012). Green marketing for Green tourism. In *Biennial International Congress. Tourism & hospitality industry* (p. 130). University of Rijeka.

Moher, D., Liberati, A., Tetzlaff, J., Altman, D. G., & The PRISMA Group. (2009). Preferred reporting items for systematic reviews and meta-analyses: The PRISMA statement. *PLoS Medicine, 6*(7), e1000097.

Morrison, A. (2009). *Hospitality and travel marketing* (4th ed.). Delmar Cengage Learning.

Murphy, P. E., & Price, G. G. (2005). Tourism and sustainable development. *Global Tourism, 3*, 167–193.

Nekmahmud, M., & Rahman, S. (2018). *Measuring the competitiveness factors in telecommunication markets*. Springer.

Padin, C. (2012). A sustainable tourism planning model: Components and relationships. *European Business Review, 24*(6), 510–518.

Pearce, D., Barbier, E., & Markandya, A. (2013). *Sustainable development: Economics and environment in the Third World*. Routledge.

Peattie, K. (1995). *Environmental marketing management: Meeting the green challenge*. Pitman Publishing.

Peattie, K., & Crane, A. (2005). Green marketing: Legend, myth, farce or prophesy? *Qualitative Market Research, 8*, 357–370.

Pigram, J. J. J., & Wahab, S. (1997). *Tourism, development and growth: the challenge of sustainability*. Routledge

Pigram, J. J., & Wahab, S. (2005). Sustainable tourism in a changing world. In J. J. Pigram & S. Wahab (Eds.). *Tourism, development and growth* (pp. 27–41). Routledge.

Polonsky, M. J., & Rosenberger, P. J., III. (2001). Reevaluating green marketing: A strategic approach. *Business Horizons, 44*(5), 21–30.

Pomeringa, A., Noblea, G., & Johnson, L. W. (2011). Conceptualising a contemporary marketing mix for sustainable tourism. *Journal of Sustainable Tourism, 19*(8), 953–969.

Popadopoulos, S. (1989). A conceptual tourism marketing planning model: Part 1. *European Journal of Marketing, 23*, 31–40.

Rafiq, M., & Ahmed, P. K. (1995). Using the 7Ps as a generic marketing mix an exploratory survey of UK and European marketing academics. *Marketing Intelligence & Planning, 13*(9), 415.

Reckien, D., Creutzig, F., Fernandez, B., Lwasa, S., Tovar-Restrepo, M., McEvoy, D., & Satterthwaite, D. (2017). Climate change, equity and the sustainable development goals: An urban perspective. *Environment and Urbanization, 29*(1), 159–182.

Roy, S. C., & Roy, M. (2015). Tourism in Bangladesh: Present status and future prospects. *International Journal of Management Science and Business Administration, 1*(8), 53–61.

Scott, N., & Cooper, C. (2010). Innovation for sustainable urban tourism: Some thoughts on best practice. *Revista de Administração Pública, 44*(5), 1171–1190.

Shamsuddoha, M., & Nedelea, M. O. (2008). Exploring tourist spots for promoting Investments in Tourism Industry of Bangladesh. *Revista de turism-studii si cercetari in turism, 6*, 5–9.

Sharma, A., Iyer, G. R., Mehrotra, A., & Krishnan, R. (2010). Sustainability and business-to-business marketing: A framework and implications. *Industrial Marketing Management, 39*(2), 330–341.

Simon, D., Arfvidsson, H., Anand, G., Bazaz, A., Fenna, G., Foster, K., & Nyambuga, C. (2016). Developing and testing the urban sustainable development Goal's targets and indicators – A five-city study. *Environment and Urbanization, 28*(1), 49–63.

Singh, R. B., & Mishra, D. K. (2004). Green tourism in mountain regions-reducing vulnerability and promoting people and place centric development in the Himalayas. *Journal of Mountain Science, 1*(1), 57–64.

Swarbrooke, J. (1999). *Sustainable tourism management*. CABI.

Throsby, D. (2016). Tourism, heritage and cultural sustainability: Three 'golden rules'. In L. F. Girard & P. Nijkamp (Eds.), *Cultural tourism and sustainable local development* (pp. 31–48). Ashgate.

United Nations (2020). *Sustainable development goals*. Retrieved from: https://sustainabledevelopment.un.org/?menu=1300. Accessed 17 Jan 2020.

United Nations World Commission on the Environment and Development (WCED). (1987). *United Nations World Commission On Environment And Development (WCED) Our Common Future Report (1987)* (p. 42). New York, NY.

United Nations World tourism Organization (2017). *UNWTO tourism highlights: 2017 Edition*. Retrieved from: https://www.e-unwto.org/doi/pdf/10.18111/9789284419029. Accessed 27 Dec 2019.

Urry, J. (1992). The tourist gaze "revisited". *American Behavioral Scientist, 36*(2), 172–186.

World Travel and Tourism Council (WTTC). (1995). *Travel and tourism in the world economy*. WTTC.

World Travel and Tourism Council (WTTC). (2019). *Annual report-2019: The economic impact of travel and tourism 2019*. WTTC.

Zeithaml, V., Bitner, M. J., & Gremler, D. D. (2006). *Services marketing, integrating customer focus across the firm*. McGraw-Hill.

Chapter 22
Sustainable Tourism in Bangladesh: The Demand for Investment and Development

Md. Asad Iqbal Chowdhury, Md. Alauddin, and Mohammad Rahim Uddin

Abstract This research places particular emphasis on sustainable tourism in Bangladesh. Theoretical understanding with some facts and figures are used to understand the different aspects of sustainable tourism in Bangladesh perspective. This research outlines the scope, importance, challenges, and suggestions to promote and ensure sustainable tourism in Bangladesh. Bangladesh is observing a growing trend in the tourism industry. This research discusses the scope of sustainable tourism from multiple dimensions from Bangladesh's perspective. Then it briefly shows the importance of ensuring sustainability in the tourism industry of Bangladesh. After that, it identifies some challenges that the tourism industry facing and can face implementing sustainable tourism in Bangladesh. On the basis of the challenges, some propositions are also given to improve the process of achieving sustainable tourism. Moreover, worldwide experiences are brought into the debate, making it more significance-focused to offer some recommendations that could be helpful to Bangladesh's promotion of sustainable tourism. This research views sustainable tourism not only for increasing the awareness of the tourists but also for improving the policies related to the tourism industry. This research considers the past experiences from around the world and even the probable future consequences. In addition, this research helps future researchers as guidance to do their extensive research.

Md. A. I. Chowdhury
Department of Economics & Banking, International Islamic University Chittagong (IIUC), Chattogram, Bangladesh

Md. Alauddin (✉)
Department of Marketing, City University of Hong Kong, Kowloon Tong, Hong Kong

Department of Business Administration, International Islamic University Chittagong (IIUC), Chattogram, Bangladesh
e-mail: alauddin4693@gmail.com

M. R. Uddin
Department of Business Administration, International Islamic University Chittagong (IIUC), Chattogram, Bangladesh

A. Hassan (ed.), *Tourism in Bangladesh: Investment and Development Perspectives*, https://doi.org/10.1007/978-981-16-1858-1_22

Keywords Sustainable tourism · Bangladesh · Environment · Wildlife · Society

Introduction

The acceleration of the tourists around the world is increasing due to the information circulation among the people with the use of social media and as well as the improvement in the communication facilities (Fernandes & Fernandes, 2017). Such excessive growth of the tourists in the natural areas where residents and environment are suffering from the consequences of the seasonal and temporary tourists pressure, resulting in a permanent change in the residents' lifestyle and damages in the natural environment (The Conversation, 2018). Such consequences have proven that the development of tourism has both adverse and optimistic impact on the destinations (Shakil, 2016). The concept of sustainable tourism emerges to identify and reduce the adverse effect of tourism development (Forsyth et al., 1995).

Sustainable tourism is a multidimensional concept. Simply, sustainable tourism is also known as "responsible tourism" is to make the destination a better place for both the hosts and visitors by visiting, and it considers the sustainability of the economy, environment, society, and animal life. United Nations World Tourism Organization (UNWTO) stated sustainable tourism as the combination of the three main aspects of tourism, socio-cultural, economic, and environmental, and made balance among these three dimensions for the assurance of long-term sustainability (UNWTO, n.d.a).

In the last few decades, sustainable tourism has taken on a global perspective. According to World Commission on Environment and Development (WCED), the sustainable tourism concept came forward at a global pace in 1987 alongside the idea of sustainable development, which considers the development that meets imminent needs without undermining future generations' resources to meet their demands (WCED, 1987). The concept of sustainable tourism has gradually become a decisive factor for destinations whose main tourist attraction relies on nature (Larson & Herr, 2008). Recently Bangladesh has also prioritized the tourism industry, and the government has taken multiple policies to ensure sustainable tourism. The National Tourism Policy of 2010 and the Seventh Five Year Plan (2016-2020) can increase the contribution of this industry in the national GDP (Mondal & Haque, 2018). This research work recognizes the growing trend in sustainable tourism with a cross-country view of Asian countries that relies mainly on natural tourism spots, considering the scope, challenges with recommendations. They are then relating the output of the research with some specific suggestions for Bangladesh.

This research initially outlines the evolving scope of sustainable tourism over time. Then it discusses the growing importance of sustainable tourism. After that, emerging challenges towards sustainable tourism are discussed. Finally, some suggestions are outlined to ensure sustainable tourism.

Scope of Sustainable Tourism

The concept of sustainable tourism is not developed over the night; it has been evolving over the decades. With the change in the delimitation, the scope of sustainable tourism also broadening. In the early stage (the late 1980s) sustainable tourism concepts developed with the concentration on the link between the activities of tourism and the environment (Aall, 2014). The 1997 Berlin Declaration added biological diversity with sustainable tourism. In 1999, the World Travel & Tourism Council (WTTC) added workers and trade unions, local authorities, and NGO perspectives with sustainable tourism. In the same year, UNWTO suggested ethics for tourism. Later in 2002, UNWTO and the United Nations (UN) incorporated ecotourism and responsible tourism with sustainable tourism. Again in 2007, WTO added climate change perspective with sustainable tourism. After the Rio + 20 conference in 2012, importance has been given to the transition of sustainable tourism to a green economy and also on poverty eradication. In 2014, WTO suggested accessible tourism for sustainable tourism. Currently, the scope of sustainable tourism is limited by multiple stakeholders connected with tourism (Buckley, 2012) (Fig. 22.1).

UNWTO recommended a number of goals and principles concerning the broader 2030 development agenda and SDGs (Sustainable development goals). The first pillar is connected with the SDGs 8,9,10, and 17, prioritize economic growth inclusively and maintaining sustainability. The second pillar is concerned with the society where employment, social inclusiveness, and poverty reduction are highlighted, linked with SDGs 1, 3,4,5,8, and 10. The third pillar emphasizes environmental protection, climate change, and resource efficiency and is connected with SDGs 6, 7,8,11,12,1314, and 15. The fourth pillar is based on cultural values, heritage, and

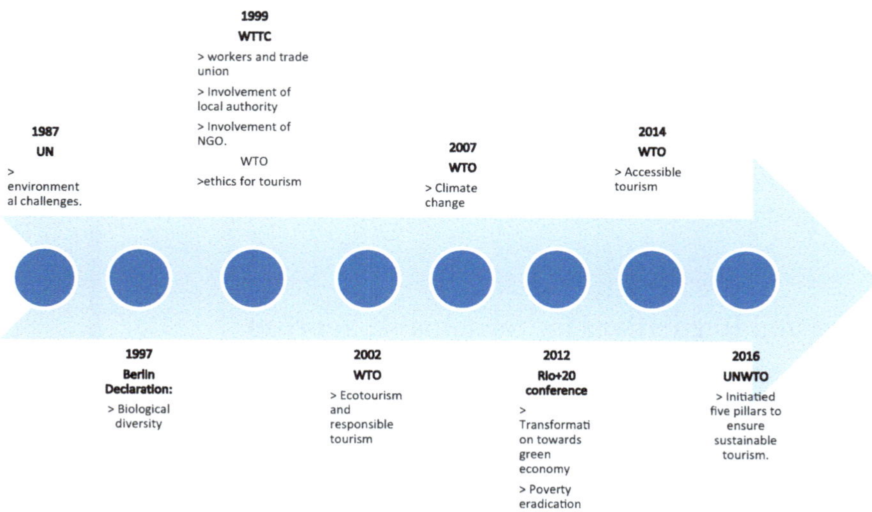

Fig. 22.1 Continuous evolving of the scope of sustainable tourism. (Source: Pan et al., 2018)

diversity, targeting SDGs 8, 11, and 12. The fifth pillar suggested peace, mutual understanding, and security following the SDGs 18 (UNWTO, n.d.b).

Considering the above changing nature of the scope of sustainable tourism following figure has been developed to show the overall range (Fig. 22.2).

Wildlife and Sustainable Tourism

Wildlife tourism is connected with the intervention in wildlife using multiple activities. The experiences in wildlife tourism are diverse that include scuba dive, safari, whale watching from speedy boats, scuba dive, bird watching, staying in resort neighboring the forest, fishing, zoo visiting, etc. All of these activities involve the life of the animals directly or indirectly (Shackley, 2006). To promote wildlife tourism in Bangladesh, the government has taken multiple steps. Wildlife tourism in Bangladesh includes watching birds and wild animals at the reserved forest, national park, Eco Park, botanical garden.

Relevant Scopes of Sustainable Tourism

Food and beverage, clothing, and housing are the induced contributors to tourism. Food tourism has emerged as a popular theme among tourism researchers in recent years (Ellis et al., 2018). Travelers are motivating by food choices while selecting destinations, and they are spending more money and time on having quality foods. As per recent research, visitors spend around 25–35% of their travel budget on food (World Food Travel Association, 2020).

Infrastructure facility is another essential part of sustainable tourism development. Tourism infrastructure includes the services required to satisfy the tourist during their travel and stay in the tourist destination (Jovanović & Ivana, 2016). Sustainable tourism is also directly and tightly connected with easy mobility towards the tourist destination. It has been found that around 72% of CO2 emissions generates from the transport facilities used during tourism (Pan et al., 2018).

Rules and Regulations for Sustainable Tourism

To ensure sustainable tourism development, effective policies are a pre-requisite. Such policies can be related to conservation policies that can protect the landscape, wildlife, and townscapes. The policy also required to generate funds and planning under specific legislation to protect the tourism destination. Self-regulations, codes of practice, and employment-related policies are also needed in the tourism industry (Pan et al., 2018).

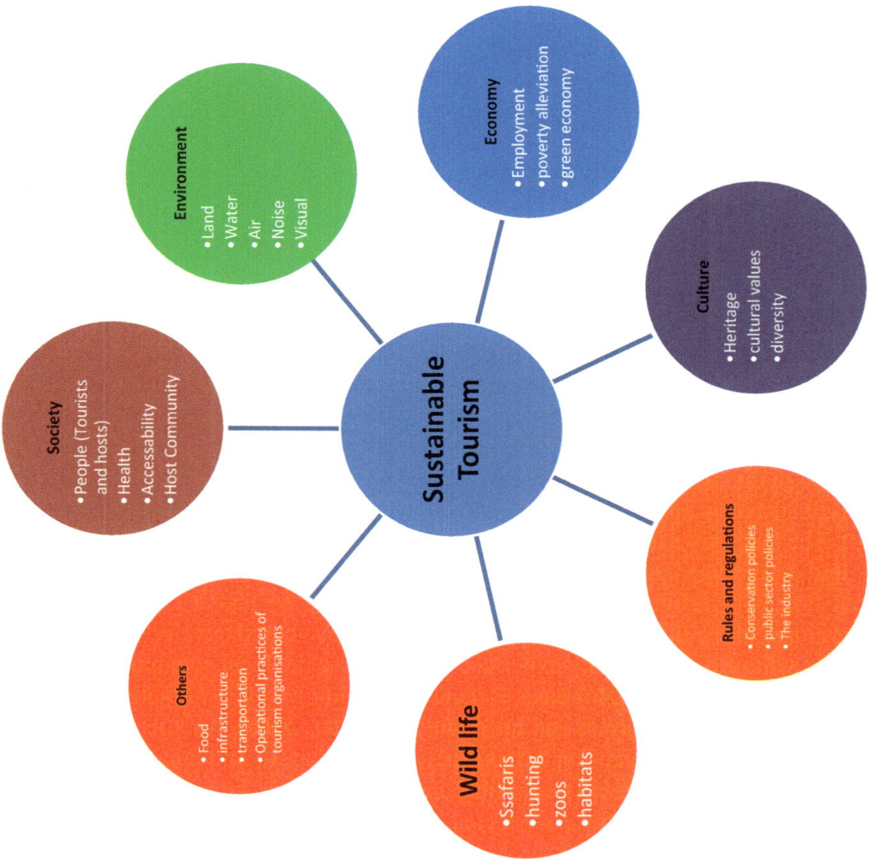

Fig. 22.2 Scope of sustainable tourism. (Source: Pan et al., 2018; Swarbrooke, 1999)

Environment and Sustainable Tourism

The concept of sustainable tourism came forward with environmental deterioration due to the rapid growth in the tourism industry. To protect the original site, the term ecotourism emerged in the late 1970s. Ecotourism gives special concentration on the sustainability of the environment and ecology of the tourist destination (Pan et al., 2018). Such tourism can empower the host community to optimize their natural resources that will ensure the sustainability of the environment as well as generates employment opportunities (Viljoen, 2012). Bangladesh has immense potentiality in the nature tourism industry with natural variations, sea beaches, hill tracks and forest areas (Ahmed et al., 2010). WTTC (2018) also recognized Bangladesh having plenty of opportunities in nature-based tourism. One of the recent studies on the nature-based tourism of Bangladesh depicted that protecting the nature-based tourism communication and infrastructure facilities is essential with the Government's initiatives (Islam et al., 2017).

Society and Sustainable Tourism

Society is one of the essential stakeholders in ensuring sustainable tourism. The society incorporates mainly the host community (i.e. the local people of the tourist destination). The host community embodied with the people directly and indirectly employed in the local tourism business, and the benefits from the tourism activities must be distributed among the host people following their participation (Swarbrooke, 1999). The hosts have the controlling power to or responsible for developing itineraries and development actions regarding the tourism activities (Aall, 2014). Society should take care about the accessibility of the tourists so that it can ensure the improvement in the life quality of the host people. They should develop an environment that will provide a quality experience to optimize the local economic benefits (Park & Yoon, 2009). In such a case, Bangladesh is doing well in attracting tourists with their exemplary behavior and hospitality (Sultana, 2016). Another researcher did his research around the Sundarbans area of Bangladesh and summarized that community-based tourism will be feasible and can ensure the host people's alternative employment opportunities (Islam et al., 2013).

Economic Contribution and Sustainable Tourism

Economic benefit from tourism formally came to sight by WTTC with the incorporation of workers and trade unions. In 2012, the Rio +20 suggested the transformation of tourism towards a green economy and poverty alleviation. The travel and tourism industry is now considering one of the largest economic areas as it generates prosperity worldwide, drives exports, and creates employment. WTTC summarized the impact of tourism on 25 regions and 185 countries as this sector contributes 10.4% of world GDP and generates 319 employment or 10% of total jobs in 2018 (WTTC, 2019). According to the global and regional tourism performance statistics published by UNWTO from 2010 to 2018, the receipts from the tourism are around 5% with a minimum of 3% in 2016, a maximum of 6% in 2013, and 5% in 2018. Europe has the highest amount of receipts (39%) where Asia and the pacific second-highest amount of revenues (30%). In terms of tourism expenditure, Asia and the Pacific has the highest position with 36% of world tourism expenditure and Europe in the second position with 35%. Around 58% of people travel over the air, which needs to be improved for grabbing the income from that sector (UNWTO, n.d.c). In alleviating poverty, tourism can develop and diversify the skill of the poor and also can create ownership sense.

Although tourism is one of the profitable sectors, Bangladesh's share in the South Asia region in terms of tourist arrivals is small (Pennington & Thomsen, 2010), and as per WTTC report its position was 150th out of 180 countries in 2017 (contribution in GDP). According to the WTTC report on Bangladesh in 2017, tourism had 4.3% contribution in total GDP, employment contribution was 3.8%, and investment contribution is only 1.4%.

Culture and Sustainable Tourism

Cultural tourism, also known as heritage tourism, is consistently developed to protect the tourist destination's heritage (Light, 2017). Cultural heritage can be divided mainly into two parts the monuments and the living culture. Memorials are of two types one is immobile, and another one is movable. Stationary monuments incorporate buildings, archaeological sites, gardens, and parks. The movable monument comprises handicrafts, arts, paintings, agricultural tools, documents, industrial machines, and artworks. Living culture includes religion, festivals, education, behavior, legends, music, dress, rites, culinary culture, and habits.

Bangladesh has much diversification in its culture as it is ruled by multiple rulers in different periods, including Paul kings, Sen Kings, Sultans, and British. These diversifications enrich the cultural tourism scope, including archaeological attraction, culture, religious attraction, tribal tourism, and multiple festivals (Rahman, 2012). Many kings and sultans ruled Bangladesh, and they leave many marks like monuments and magnificent cities, such as monasteries, mosques, tombs, temples, buildings, and so on. History and traditions are also timely intermingled with the landmass and its distinct customs and cultural characteristics (Rahman, 2012). Although the majority of the people are Muslims, religious festivals of all religions are observed with joy.

For the development of sustainable tourism, the Bangladesh government has established multiple policies and governing bodies. Some of these are (a) Bangladesh tourism board act-2010; (b) "Ministry of civil aviation and tourism, Medium-term Budget framework"; c) "Implementation strategy of national tourism policy".

Importance of Sustainable Tourism in Bangladesh

Many researchers have suggested that tourism has a tremendous and increasing contribution to world's GDP (Boley et al., 2017; Dwyer et al., 2016; Pan et al., 2018). It also contributes to generating job opportunities directly and indirectly (UNWTO, n.d.c). However, alongside economic development, it also harms the ecosystem and environment (Peeters & Dubois, 2010). The effect on climate change due to the tourists' increasing growth rates is not so small (Bows et al., 2009). Only a single factor or sector cannot ensure sustainable tourism, travel system needs a radical shift to ensure sustainable tourism. Considering the faster-growing economic sector, it should be managed to provide benefits to the related environment, culture, and local communities.

Bangladesh's tourism industry is growing at a faster rate due to its natural beauty with encompasses with beautiful tropical rain forest, evergreen hills, cultural heritage, longest natural sea beach, ethnic diversity, and small islands (Islam & Nath, 2014). However, rapid growth in the tourism industry also creates significant pressure on natural resources and the environment. This is particularly true in developing nations where tourism is expanding rapidly in ecologically vulnerable areas with modest environmental concerns (Aminu et al., 2013). Bangladesh is at high risk due to the rapid growth in the tourism industry with unplanned development and a fragile ecosystem (Mondal & Haque, 2017). To reduce the problems arising from the tourism industry's rapid growth, the term sustainable tourism has emerged (Kasimoglu, 2012). Sustainable tourism can be used to protect the environment, culture, and biodiversity of the tourist destination along with the generation of employment and alleviating poverty (UNEP & UNWTO, 2005). The previous study on sustainable tourism in Bangladesh suggested its importance in the light of economic benefit, environmental and cultural sustainability.

Economic Sustainability

The tourism industry of Bangladesh has not yet flourished. The total arrival of tourists in Bangladesh is small comparing to other countries of Asia (Pennington & Thomsen, 2010). However, the overall contribution of this sector in economic development is increasing. Current status of this sector shown in Fig. 22.3:

It is essential to develop sustainable tourism so that the earnings from this sector will increase. It has been revealed that revenues from this sector can be improved with proper promotional activities as well as employing a creative person and ensuring quality services (Al-Masud, 2015).

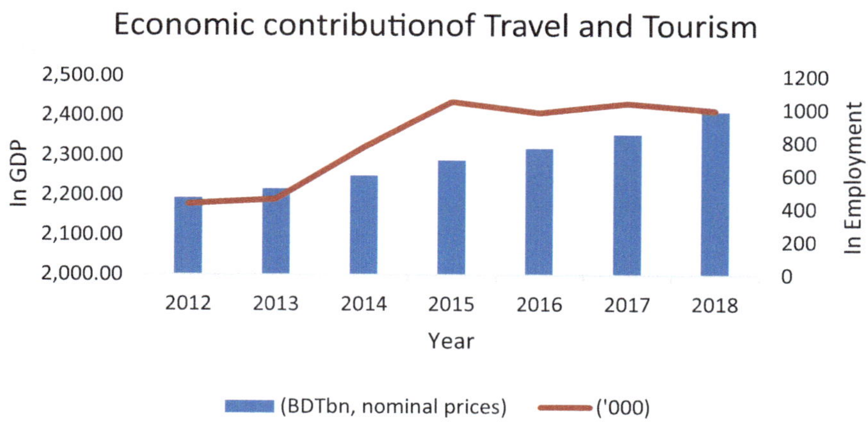

Fig. 22.3 Contribution of travel and tourism in the economy of Bangladesh (source: WTTC, 2018)

Socio-Cultural Sustainability

In addition to the economic benefit, tourism activity can also improve the identity of the local community in front of the whole world. It helps the local community to motivate their own culture and traditions towards the world (Roy & Roy, 2015). While tourist visits a new place, they also interact with the local cultures and customs of the tourist destination (Landlopers, 2011). Sustainable tourism is essential to prevent local heritage and culture, ensuring the social wellbeing of the host community and promoting cross-cultural understanding (Smith, 2015). Bangladesh is full of heritages left by multi rulers (Haque, 2003). It has a diversity of cultures, including cultural attractions, religious attractions, tribal attractions, and festival attractions (Rahman, 2012). It is therefore essential to develop sustainable tourism to protect the local attitudes, cultures and heritage with proper initiatives (Williams & Lawson, 2001).

Environmental Sustainability

With the increasing number of tourists, the economy of the host country is growing at the expense of ecological degradation. All the direct, indirect, and induced activities related to tourism have a severe impact on the environment. For instance, oil spell from the water transport used in tourism causes damage to the aquatic and marine ecosystem, wildlife tourism impacting the natural regeneration process, wildlife breeding, and extinction of the rare species (Higginbottom, 2004). Construction works and transportation facilities are increasing air pollution in natural tourism destinations, causing severe damage to the natural environment of the host area (Sorupia, 2005). Recent research on nature-based tourism in Bangladesh identified some significant factors influencing the environment, such as waste disposal could hamper the species breeding and create both air and water pollution, scarcity of drinking water in the coastal areas can impact future tourism and create pressure on the water supply in the local area and visitors are destroying young plants during their forest visits (Islam, 2015). To ensure the protection of the natural environment and a better place for the future, it is necessary to have sustainable tourism development.

Challenges of Sustainable Tourism in Bangladesh

The development and management of tourism will face a multitude of significant problems for policymakers and planners related to sustainability. A plethora of research depicted the interdisciplinary relationship between tourism and sustainability to highlight the importance of redefining sustainability challenges so that tourism development can ensure the benefits for society and the environment (Kristjánsdóttir et al., 2018). The major problem of sustainable tourism is to minimize the negative impacts and to maximize the economic benefit from the tourism industry (Tourism Notes, 2020). Ecologically sustainable and economically profitable tourism can ensure the satisfaction of the visitors and the well-being of the local community (Lim & McAleer, 2005). Considering the tourism chain showing in Fig. 22.4, some other influencing challenges are identified; lack of skilled labor, unplanned business development, food waste, energy consumption, waste management, lack of planned transport management, lack of financing and investment. These challenges are faced in multiple stages of the tourism chain.

Excessive Use of Energy and CO^2 Emission

With the extreme pressure by the increasing rate of the tourism industry, greenhouse gas emissions also increase. Tourism starts and ends with the use of transportation where need excessive use of energy. Energy use can be divided into three parts: transportation, accommodation and activities (Pan et al., 2018). One of the recent research of 160 countries found that from 2009 to 2013 the carbon emission increased from 3.9 to 4.5 GtCO2e. Major contributors are high-earning countries (Lenzen et al., 2018). Visitors from the top income countries are responsible for generating elevated CO2 from excessive energy and visitors from low-income countries responsible for CO2 emission by using road transport and non-energy CO2 emission (Lenzen et al., 2018).

As per the WTTC report (2018), in Bangladesh, around 85.2% of tourism spending is done for leisure purposes where only 14.8% are for business purposes. Among the visitors, only 2.6% of the expenditure is done by international visitors, where

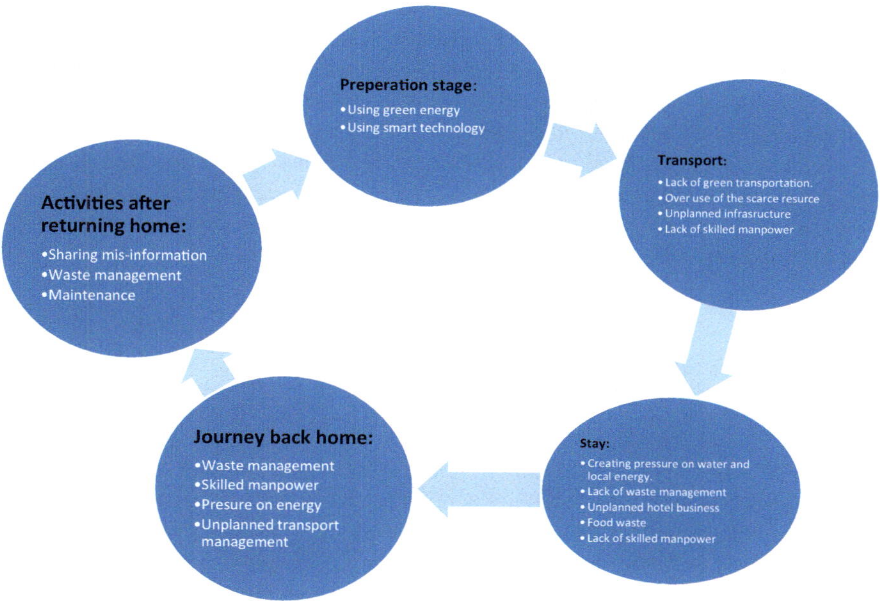

Fig. 22.4 Tourism chain and its connection with the challenges to sustainable tourism. (Source: Pan et al., 2018)

domestic visitors do 97.4% of spending in 2017 (WTTC, 2018). Since Bangladesh cannot provide enough air communication facilities, the number of international tourists is not increasing. As a result, energy pressure by road transport is rising. The transport sector uses fossil fuel (comprises oil, petroleum, natural gas, and coal) as energy. Among the 209 countries, Bangladesh's position is 48 in 2016, with a contribution of 14% by the transport sector (Worldometer, 2016). Accommodation facilities are another source of excessive use of energy. Around 50% of the energy is used for air conditioning or room heating followed by making hot water and cooking (Beccali et al., 2009). Due to the increasing demand for tourism in Bangladesh number of hotels with luxurious facilities is also increasing. According to the tourism ministry, there are 13 five stars, four four-star, and 17 three-star hotels are operating in Bangladesh with high-class accommodation facilities (Dhaka Tribune, 2017).

Excessive Use of Groundwater

Information regarding water usage by hotels is not so much available. In research, it has been found that most of the water is used in tourism area are by the accommodation facility (Gössling, 2005). According to UNEP and UNWTO (2005), direct water use varies from 100 to 2000 L per guest night. Another study on hotels in Tanzania identified that around 50% of water is used for the gardening purpose, 20% used for toilets, showers, and tap water (Gössling, 2001). Coastal zones are most attractive to tourists, especially in Bangladesh. Although agriculture has the highest need for groundwater in coastal areas, increasing water demand due to tourism development creates additional pressure on the underground (Mas-Pla et al., 2014). A recent study on the groundwater level at Cox's Bazar area depicted that the groundwater level has already gone below the sea level due to excessive abstraction of the groundwater in some tourist zone (Fatema et al., 2018).

Inappropriate Management and Disposal of Waste

The overwhelming number of tourism industries has led to more waste production (Jeftic et al., 2006). Tourism can directly affect water quality by discharging untreated wastewater (sewage) and abstracting freshwater (Pan et al., 2018). One recent primary research on "identifying the factors influencing the sustainable tourism in Bangladesh" identified that 72% of the participants in the survey agreed that waste disposal has a longer negative impact on the environment followed by pollution of water, air and disturbing the animal breeding process (Islam, 2015). The coastal areas covering Kuakata, Cox's Bazar-Teknaf, Sonadia along with St. Martin's Island are profoundly affected by waste disposal by the tourists. Cox's Bazar has the most popular tourist area in Bangladesh, around 300,000 visited in 2018 and responsible for generating a considerable amount of waste. Tourism business people just collect this enormous amount of garbage and dump it to the nearest river, Bakkhali. Municipal Corporation reported that around 120 tons of waste are deposited in the river each day by the beach umbrella owners and other local tourism business owners (New Age, 2019). Alongside environmental concerns, such as mismanagement in waste disposal can severely impact the tourism industry's growth.

Biodiversity Loss and Habitat Degradation

Large-scale tourism could also have damaging effects on the natural landscape, biodiversity, and habitats such as coral reefs, rainforests, mountains, arid, wetlands, and semi-arid ecosystems. For instance, due to unplanned resorts in the coastal areas, wetlands are destroying seriously (UNEP & UNWTO, 2005). In 2011, a court ordered the authorities to dismantle all unauthorized beach structures to stop more pollution, but the request went unheeded (New Age, 2019). St. Martin's Island is one of the major tourist attractions, and around 3000 people visit this small island during peak season. Sea turtles and corals are central biodiversities of this island. Due to the overnight running off generators and scuba diving severely hampering this island's biodiversity (Upal, 2015).

Threats to Cultural Integrity and Heritage Management

Heritage represents the host area's historical background or region, which incorporates cultural, natural, and built heritage (Pan et al., 2018). Cultural instability is a critical issue for worldwide indigenous cultures and inclusive societies. Also, the construction and development of infrastructure may be a threat to the integrity of heritage. Bangladesh is a land of multi-culture diversity, as so many rulers ruled it over time (Rahman, 2012). Bangladesh is facing many challenges to maintain its heritage. Many heritage spots are not adequately explored and managed, lack of investment to retain the structures, lack efficient service from the tourism provider, lack of proper policy to interact with the tribal and their villages (Rahman, 2012).

Lack of Information Platforms and Communication Channels

Lack of direct communication between the local tourism industry and the visitors can create conflicts between them (Pan et al., 2018). The communication system around Bangladesh's tourist spots is still not enough to attract tourists and maintain sustainability in the tourism industry (Roy & Roy, 2015). Lack of information platforms regarding the policies, vulnerability of the host area, the tourist spot's readiness, and security can result in a concern of awareness, which is a unique challenge for implementing a sustainable tourism system (Khandakar, 2014).

Climate Change

Climate change also has a severe impact on the sustainability of tourism in Bangladesh. A recent report published by Ministry of Environment and Forests, Bangladesh identified some significant risks of the climate change on touris spots due to increased scale of coastal disaster, loss of shorelines, increasing number of cyclones, loss of mangrove kinds, increased flash flood, increased mudslides, etc. All of these have a direct impact on the sustainability of tourism (Ministry of Environment and Forests, 2009).

Suggestions to Ensure Sustainable Tourism

Considering the tourism chain shown in Fig. 22.4, the researcherers suggested implementing multiple greening elements related to tourism, including using renewable energy, building transportation, infrastructure, agriculture, and smart technology (Pan et al., 2018). In this section, the role of these elements towards sustainable tourism will be discussed.

Using Green Energy

The energy need is one of the significant challenges in tourism areas. For instance, the tourism industry can adopt Green energy, the power generated by wind, solar, hydro, ocean, geothermal, and bioenergy. Solar energy can be used to generate electricity with the use of photovoltaic cells (Shih et al., 2016). However, another research finds that renewable energy also has a negative impact on biodiversity, and it must be considered while developing green energy policies (Gasparatos et al., 2017).

Greening Transportation

For greening the transportation facilities, modified transport can be used where energy source will be renewable energies as solar, hydro and so on. Air space management, using Green energy in the aviation sector, and efficient travel can ensure air transport greening (World Economic Forum, 2009). Economic instruments such as PMC (Passenger movement charge), environmental taxation can also be used to manage Green transportation (Tourism & Transport Forum Australia, 2012).

Greening Accommodation

From the accommodation perspective, many green practices can be implemented; such as encouraging visitors to re-use towels, restaurant menus and posters can use recycled paper, visitors can be charged for over usage of the power or resources. (D'Alessandro et al., 2016) demonstrated agri-tourism as an example of maintaining the green building in the tourist areas, where multiple green instruments are used to lessen the use of power as composite windows, diode lamp, PV panels on the roof, and so on. Research also suggests that higher consumption of green products can increase tourist attraction (Kim et al., 2016).

Greening Infrastructures

The concept of green infrastructure is connected with the use of both new and existing elements to ensure the benefits of air quality, sufficient lighting, and open space (Setyaningsih et al., 2015). Implementing green roads (full of trees and plants beside the streets), vegetated surfaces, green corridors etc. can ensure the green infrastructure (Serra-Llobet & Hermida, 2017).

Using Smart Technologies

The use of ICT in the tourism industry continuously increasing, for instance, collecting feedback from the visitors, reservations, travels, managing tourism firms, and so on (Leung et al., 2013). Some benefits from one of the recent smart technology are; real-time monitoring of wind speed, humidity, temperature, CO^2 concentration, probability of rainfall, weather forecasting, and counting the presence of the visitors (Novas et al., 2017). Such technology can help policymakers getting the results of the increasing number of tourists in the environment.

Conclusion

The transformation toward the sustainability of tourism requires a multi-dimensional approach as an investment in smart technologies, Green transportation, Green accommodation, integrative policies, strong bonding between tourism firms, and the government, creating awareness about Green tourism among the multiple stakeholders of tourism, etc. Knowledge should be designed in such a way that when people start to make travel, they will give preference for the use of the green product and complete their tour through the green system. Sustainable tourism challenges begin from the initial stage of the tourism chain, so initiatives should be taken from that stage and will continue until the end of the chain, and then sustainable tourism can be ensured.

References

Aall, C. (2014). Sustainable tourism in practice: Promoting or perverting the quest for a sustainable development? *Sustainability, 6*(5), 2562–2583.

Ahmed, F., Azam, M. S., & Bose, T. K. (2010). Factors affecting the selection of tour destination in Bangladesh: An empirical analysis. *International Journal of Business and Management, 5*(3), 52–61.

Al-Masud, T. M. M. (2015). Tourism marketing in Bangladesh: What, why and how. *Asian Business Review, 5*(1), 13–19.

Aminu, M., Ludin, A. N. B. M., Matori, A.-N., Yusof, K. W., Dano, L. U., & Chandio, I. A. (2013). A spatial decision support system (SDSS) for sustainable tourism planning in Johor Ramsar sites, Malaysia. *Environmental Earth Sciences, 70*(3), 1113–1124.

Beccali, M., La Gennusa, M., Coco, L. L., & Rizzo, G. (2009). An empirical approach for ranking environmental and energy saving measures in the hotel sector. *Renewable Energy, 34*(1), 82–90.

Boley, B. B., McGehee, N. G., & Hammett, A. T. (2017). Importance-performance analysis (IPA) of sustainable tourism initiatives: The resident perspective. *Tourism Management, 58*, 66–77.

Bows, A., Anderson K., & Peeters, P. (2009). Air transport, climate change and tourism. *Tourism and Hospitality Planning & Development, 6*(1), 7–20.

Buckley, R. (2012). Sustainable tourism: Research and reality. *Annals of Tourism Research, 39*(2), 528–546.

D'Alessandro, F., Menghini, S., & Pfoestl, E. (2016). Green building for a green tourism. A new model of eco-friendly agri-tourism. *Agriculture and Agricultural Science Procedia, 8*, 201–210.

Dhaka Tribune. (2017). *What is a five-star hotel in Bangladesh?*. Retrieved from: https://www.dhakatribune.com/feature/2017/07/25/five-star-hotel-bangladesh/. Accessed 25 Feb 2020.

Dwyer, L., Forsyth, P., & Spurr, R. (2016). Tourism economics and policy analysis: Contributions and legacy of the Sustainable Tourism Cooperative Research Centre. *Journal of Hospitality and Tourism Management, 26*, 91–99.

Ellis, A., Park, E., Kim, S., & Yeoman, I. (2018). What is food tourism? *Tourism Management, 68*, 250–263.

Fatema, S., Marandi, A., Zahid, A., Hassan, M. Q., Hossain, M. A., & Schüth, C. (2018). Seawater intrusion caused by unmanaged groundwater uses in a coastal tourist area, Cox's Bazar, Bangladesh. *Environmental Earth Sciences, 77*(3), 75.

Fernandes, T., & Fernandes, F. (2017). Social Media and Tourism: The Case of E-Complaints on TripAdvisor (An Extended Abstract). In P. Rossi (Ed.), *Marketing at the Confluence between Entertainment and Analytics. Developments in Marketing Science: Proceedings of the Academy of Marketing Science* (pp. 825–829). Springer.

Forsyth, P., Dwyer, L., & Clarke, H. (1995). Problems in use of economic instruments to reduce adverse environmental impacts of tourism. *Tourism Economics, 1*(3), 265–282.

Gasparatos, A., Doll, C. N., Esteban, M., Ahmed, A., & Olang, T. A. (2017). Renewable energy and biodiversity: Implications for transitioning to a green economy. *Renewable and Sustainable Energy Reviews, 70*, 161–184.

Gössling, S. (2001) The consequences of tourism for sustainable water use on a tropical island: Zanzibar, Tanzania. *Journal of Environmental Management, 61*(2), 179–191.

Gössling, S. (2005). Tourism's contribution to global environmental change: Space, energy, disease and water. In C. M. Hall & J. Higham (Eds.), *Tourism recreation and climate change: International perspectives* (pp. 286–300). Channel View Publications.

Haque, A. M. (2003). What ails tourism in Bangladesh?. *Observer Magazine*, October 6, p. 15.

Higginbottom, K. (2004). *Wildlife tourism: Impacts, management and planning* (pp. 1–14). Common Ground Publishing.

Islam, M. S. (2015) Study on factors influencing tourism: Way forward for sustainable tourism in Bangladesh. *Journal of Tourism, Hospitality and Sports, 6*, 1–13.

Islam, M. J., & Nath, T. K. (2014). Forest-based betel leaf and betel nut farming of the Khasia indigenous people in Bangladesh: Approach to biodiversity conservation in Lawachara National Park (LNP). *Journal of Forestry Research, 25*(2), 419–427.

Islam, M. W., Rahman, M. M., Iftekhar, M. S., & Rakkibu, M. G. (2013). Can community-based tourism facilitate conservation of the Bangladesh Sundarbans? *Journal of Ecotourism, 12*(2), 119–129.

Islam, S., Hossain, M. K., & Noor, M. E. (2017). Determining drivers of destination attractiveness: The case of nature-based tourism of Bangladesh. *International Journal of Marketing Studies, 9*(3), 10–23.

Jeftic, L., Matte-Baker, A., & Schomaker, M. (2006). *The state of the marine environment: Trends and processes*. Retrieved from: https://wedocs.unep.org/bitstream/handle/20.500.11822/12469/global_soe_trends.pdf?sequence=1&isAllowed=y. Accessed 14 Feb 2020.

Jovanovia, S., & Ivana, I. L. I. Ä. (2016). Infrastructure as important determinant of tourism development in the countries of Southeast Europe. *EcoForum, 5*(1), 1–34.

Kasimoglu, M. (2012). *Visions for global tourism industry: Creating and sustaining competitive strategies*. Retrieved from: https://www.intechopen.com/books/visions-for-global-tourism-industry-creating-and-sustaining-competitive-strategies. Accessed 14 Feb 2020.

Khandakar, S. (2014). Sustainable tourism development in Bangladesh: Strategies and guidelines. *World Vision, 8*(1), 176–185.

Kim, J.-Y., Hlee, S., & Joun, Y. (2016). Green practices of the hotel industry: Analysis through the windows of smart tourism system. *International Journal of Information Management, 36*(6), 1340–1349.

Kristjánsdóttir, K. R., Ólafsdóttir, R., & Ragnarsdóttir, K. V. (2018). Reviewing integrated sustainability indicators for tourism. *Journal of Sustainable Tourism, 26*(4), 583–599.

Landlopers. (2011). *Sustainable tourism: What is it, and why should we care?* Retrieved from: https://landlopers.com/2011/12/25/sustainable-tourism-definition. Accessed 14 Feb 2020.

Larson, S., & Herr, A. (2008). Sustainable tourism development in remote regions? Questions arising from research in the North Kimberley, Australia. *Regional Environmental Change, 8*(1), 1–13.

Lenzen, M., Sun, Y.-Y., Faturay, F., Ting, Y.-P., Geschke, A., & Malik, A. (2018). The carbon footprint of global tourism. *Nature Climate Change, 8*(6), 522–528.

Leung, D., Law, R., Van Hoof, H., & Buhalis, D. (2013). Social media in tourism and hospitality: A literature review. *Journal of Travel & Tourism Marketing, 30*(1–2), 3–22.

Light, D. (2017). Progress in dark tourism and Thanatourism research: An uneasy relationship with heritage tourism. *Tourism Management, 61*, 275–301.

Lim, C., & McAleer, M. (2005). Analyzing the behavioral trends in tourist arrivals from Japan to Australia. *Journal of Travel Research, 43*(4), 414–421.

Mas-Pla, J., Ghiglieri, G., & Uras, G. (2014). *Seawater intrusion and coastal groundwater resources management. Examples from two Mediterranean regions: Catalonia and Sardinia*. Retrieved from: https://bit.ly/2Rw1syV. Accessed 14 Feb 2020.

Ministry of Environment and Forests. (2009). *Climate change and tourism industry*. Retrieved from: https://www.iucn.org/sites/dev/files/import/downloads/tourism.pdf. Accessed 1 Mar 2020.

Mondal, M., & Haque, S. (2017). SWOT analysis and strategies to develop sustainable tourism in Bangladesh. *UTMS Journal of Economics, 8*(2), 159–167.

Mondal, M., & Haque, S. (2018). Social, economic and environmental sustainability: The implications to develop sustainable tourism in Bangladesh. *East West Journal of Business and Social Studies, 6 & 7*, 44–64.

New Age. (2019). *Well-planned waste disposal can save Cox's Bazar*. Retrieved from: https://www.newagebd.net/article/86714/well-planned-waste-disposal-can-save-coxs-bazar. Accessed 1 Mar 2020.

Novas, N., Gázquez, J., MacLennan, J., García, R., Fernández-Ros, M., & Manzano-Agugliaro, F. (2017). A real-time underground environment monitoring system for sustainable tourism of caves. *Journal of Cleaner Production, 142*, 2707–2721.

Pan, S.-Y., Gao, M., Kim, H., Shah, K. J., Pei, S.-L., & Chiang, P.-C. (2018). Advances and challenges in sustainable tourism toward a green economy. *Science of the Total Environment, 635*, 452–469.

Park, D.-B., & Yoon, Y.-S. (2009). Segmentation by motivation in rural tourism: A Korean case study. *Tourism Management, 30*(1), 99–108.

Peeters, P., & Dubois, G. (2010). Tourism travel under climate change mitigation constraints. *Journal of Transport Geography, 18*(3), 447–457.

Pennington, J. W., & Thomsen, R. C. (2010). A semiotic model of destination representations applied to cultural and heritage tourism marketing. *Scandinavian Journal of Hospitality and Tourism, 10*(1), 33–53.

Rahman, W. (2012). Cultural tourism and Bangladesh: An overview. *Bangladesh Research Publications Journal, 7*(1), 6–15.
Roy, S. C., & Roy, M. (2015). Tourism in Bangladesh: Present status and future prospects. *International Journal of Management Science and Business Administration, 1*(8), 53–61.
Serra-Llobet, A., & Hermida, M. A. (2017). Opportunities for green infrastructure under Ecuador's new legal framework. *Landscape and Urban Planning, 159*, 1–4.
Setyaningsih, W., Iswati, T. Y., Nuryanti, W., Prayitno, B., & Sarwadi, A. (2015). Low-impact-development as an implementation of the eco-green-tourism concept to develop Kampung towards sustainable city. *Procedia-Social and Behavioral Sciences, 179*, 109–117.
Shackley, M. (2006). Wildlife tourism: Impacts, management, and planning. *Annals of Tourism Research, 33*(3), 868–869.
Shakil, A. M. (2016). *Sustainable tourism development in Bangladesh, a case study of Cox's Bazar, Bangladesh*. Bachelor's thesis. Kokkola: Centria University of Applied Sciences.
Shih, Y.-H., Shi, N.-X., Tseng, C.-H., Pan, S.-Y., & Chiang, P.-C. (2016). Socioeconomic costs of replacing nuclear power with fossil and renewable energy in Taiwan. *Energy, 114*, 369–381.
Smith, M. K. (2015). Tourism and cultural change. In C. M. Hall, S. Gossling, & D. Scott (Eds.), *The Routledge handbook of tourism and sustainability* (pp. 193–202). Routledge.
Sorupia, E. (2005). *Rethinking the role of transportation in tourism*. Retrieved from: https://pdfs.semanticscholar.org/784f/53cd6f79c2497076ab612b36d7dcf44636db.pdf. Accessed 7 Mar 2020.
Sultana, S. (2016). Economic contribution of tourism industry in Bangladesh. *Journal of Tourism, Hospitality and Sports, 22*, 55–54.
Swarbrooke, J. (1999). *Sustainable tourism management*. CABI.
The Conversation. (2018). *Overtourism: A growing global problem*. Retrieved from: https://theconversation.com/overtourism-a-growing-global-problem-100029. Accessed 14 Feb 2020.
Tourism Notes. (2020). *Eco tourism*. Retrieved from: https://tourismnotes.com/eco-tourism/. Accessed 1 Mar 2020.
Tourism & Transport Forum Australia (TTF). (2012). *Tourism taxes in Australia passenger movement charge explained*. Retrieved from: https://www.ttf.org.au/wp-content/uploads/2016/06/TTF-Passenger-Movement-Charge-Explainer-2013.pdf. Accessed 7 Mar 2020.
United Nations Environment Programme (UNEP) & United Nations World Tourism Organization (UNWTO). (2005). *Making tourism more sustainable: A guide for policy-makers*. Retrieved from: https://bit.ly/2RuQPfJ. Accessed 1 Dec 2020.
United Nations World Tourism Organization (UNWTO). (n.d.a). *Sustainable development*. Retrieved from: https://www.unwto.org/sustainable-development. Accessed 14 Nov 2020.
United Nations World Tourism Organization (UNWTO). (n.d.b). *Tourism in the 2030 agenda*. Retrieved from: https://www.unwto.org/tourism-in-2030-agenda. Accessed 14 Nov 2020.
United Nations World Tourism Organization (UNWTO). (n.d.c). *Global and regional tourism performance*. Retrieved from: https://www.unwto.org/global-and-regional-tourism-performance. Accessed 14 Nov 2020.
Upal, N. H. (2015). Environmental hazards and conservation approach to the biodiversity and ecosystem of the St. Martin's island in Bangladesh. In N. Raju, W. Gossel, & M. Sudhakar (Eds.), *Management of natural resources in a changing environment* (pp. 259–269). Springer.
Viljoen, W. (2012). Aid for trade and the green economy in Africa. *Bridges Africa, 1*(2), 4–6.
Williams, J., & Lawson, R. (2001). Community issues and resident opinions of tourism. *Annals of Tourism Research, 28*(2), 269–290.
World Commission on Environment and Development (WCED). (1987). *Our common future* (Vol. 17, pp. 1–91). Oxford University Press.
World Economic Forum (WEF). (2009). *Towards a low carbon travel & tourism industry*. Retrieved from: http://www.greeningtheblue.org/sites/default/files/Towards%20a%20low%20carbon%20travel%20&%20tourism%20sector.pdf. Accessed 25 Feb 2020.

World Food Travel Association (WFTA) (2020). *The economic impact of food tourism*. Retrieved from: https://worldfoodtravel.org/news-the-economic-impact-of-food-tourism/. Accessed 1 Dec 2020.

World Travel and Tourism Council (WTTC). (2018). *Travel & tourism economic impact 2018 world*. Retrieved from: https://www.wttc.org/-/media/files/reports/economic-impact-research/regions-2018/world2019.pdf. Accessed 14 Feb 2020.

World Travel and Tourism Council (WTTC). (2019). *Travel & tourism economic impact 2019 world*. Retrieved from: https://www.wttc.org/-/media/files/reports/economic-impact-research/regions-2019/world2019.pdf. Accessed 14 Feb 2020.

Worldometer. (2016). *Bangladesh CO^2 emissions*. Retrieved from: https://www.worldometers.info/co2-emissions/bangladesh-co2-emissions/. Accessed 2 Feb 2020.

Chapter 23
The Principles and Practices of Sustainable Tourism Investments and Development in Bangladesh

Samshad Nowreen and Md. Mohiuddin

Abstract The tourism industry of Bangladesh is growing very rapidly in recent years though international tourist arrival is very limited compared to the neighboring countries. However, domestic tourist flow has been increasing excessively. This book chapter reveals the status of sustainable tourism indicators of Bangladesh guided by the World Tourism Organization (UNWTO) and Department of International Development (DFID). It has been found that the local people of the topmost revenue-generating destination of Bangladesh are slightly dissatisfied with mass tourism development, yet, it varies for different factors. In terms of social and environmental concerns, local people are dissatisfied with conventional tourism practices. However, they are very satisfied with the increase of diversified livelihood and economic multiplication. On the other hand, livelihood capital is used limited to sustainability in the four major sectors of the tourism industry for the country. The main reason for it could be depicted lack of theoretical understanding of research. However, the tourism industry of the country is still in the booming stage. Therefore, the application of effective knowledge with proper investments and development practices could improve the industries involved in the tourism business in a sustainable manner. Thus, this industry could rise and bring about decent employment offerings in a sustainable manner to everyone (i.e. from the community to investors or the government). Moreover, these practices will improve the capacity of the destination's inhabitants to protect the environment and ensure community empowerment.

Keywords Sustainable tourism · Principles · Practices · Bangladesh

---

S. Nowreen (✉)
Department of Tourism and Hospitality Management, University of Dhaka, Dhaka, Bangladesh
e-mail: nowreen22@gmail.com

Md. Mohiuddin
Department of Geography and Environment, Jagannath University, Dhaka, Bangladesh

A. Hassan (ed.), *Tourism in Bangladesh: Investment and Development Perspectives*, https://doi.org/10.1007/978-981-16-1858-1_23

Introduction

Tourism is a complex industry having an economic multiplication impact on a country's economy. It is known as one of the topmost revenue-generating industries in the world and is regarded as a complex industry because of the economic multiplication and product along with its services produced from this industry makes it unique. In 2019, the tourism industry contributed worth of 10.3% of the world's total Gross domestic product (GDP) from direct, indirect and induced economic activity; one in every ten jobs was created from this industry (World Travel & Tourism Council, 2020). Moreover, 4.3% of investments are contributed from total investments. However, until the 1980s most people had been thinking and practicing conventional mass tourism only where exploitation of nature and resources at adapting platform of stages of sustainable tourism development. Ironically, several studies have identified that conventional tourism models fail to benefit the local poor people and minimize the gaps between the social classes of poor and rich (Zeng, 2018).

However, the concept of sustainable development emerged in 1987 and continued to expand further. Researchers and experts identified the negative impact of mass tourism on society and the environment, and the economy in the long run. It has been found that tourism industries contribute hugely to global climate change (Bramwell & Lane, 2008). A report of the World Tourism Organization (UNWTO) (2019) identifies that only the transportation sector of the tourism industry accounts for 5% of global carbon emissions (World Tourism Organization, 2019). Furthermore, it is obvious that other tourism business sectors are responsible for contributing more to carbon emission and global climate change. This climate change triggers any kind of environmental damage by destroying ecological cycles and biodiversity loss in the earth system. Yet, various alternative types of tourism, which are viewed as sustainable tourism, have been created after the extremely high competition of sun-lust tourism (Yfantidou et al., 2018). There are some confusions to define sustainable tourism for the wide range of its practices and scopes. Still, more research is needed to be done on the developing economies where the tourism industry is burgeoning. In the case of Bangladesh, research and development of the concept of sustainable tourism are at the initial stage with very minimal understanding and practice.

The tourism industry in the country is also not well developed compared with its neighboring countries. In addition, along with doomed tourism development and persisted degradation of the destination through various natural disasters and poverty-driven populations, it threw local communities in a vulnerable position and created a poor image to the international tourists (Islam & Carlsen, 2013). Not only natural calamities, increasing trend of population growth and unemployment in the urban areas of the country also fosters an unattractive environment to the tourists. Various sectors involved in the tourism industry are yet to define sustainability goals in the tourism industry. Moreover, the theoretical framework for defining sustainable tourism principles in the context of Bangladesh needs to be identified.

Therefore, some guiding principles and practice mechanisms for sustainable tourism development are illustrated in this chapter.

Methods and Materials

A mixed method had been used to collect and analyze the data for writing this book chapter. To understand the concept of sustainable tourism development, a thorough literature review was conducted by considering the four keywords (i.e. sustainable tourism, Bangladesh, Practice and Principles) to search peer-reviewed journal articles, books, or other reliable sources. Most related literature had been considered for review. It guided a way for the primary survey to the major tourism industry stakeholders such as public, private, communities, and tourists. A total of 60 questionnaires had been surveyed, and samples were collected using a convenient sampling method. The questionnaires were targeted to get answers of livelihood assets utilized in four major sectors in the tourism industry as (1) Accommodation, (2) Recreation or entertainment, (3) Food services, and (4) Transportation. For conducting the survey using scale ranging 0 to 4 meant "poor practices of capital", and 4 meant "best practices of capital", respectively. Five capitals of asset utilized in the tourism industry as 1. human (i.e. skill, education, expertise, good health, etc.), 2. physical (i.e. infrastructure, accessibility, production input, etc.), 3. natural (i.e. air, water, soil, energy, etc.), 4. financial (i.e. investment, incentive, levies, etc.), 5. social (i.e. social network, authority, agreement, association, etc.) capitals of the specific destination and the country as a whole. These data were further analyzed using the sustainable livelihood capital assessment technique of the United Nations Development Program (UNDP) (2017). This model was inherited from the Department of International Development (DFID) (1999).

Furthermore, a questionnaire survey was conducted to the local residents of Bangladesh Cox's Bazar Sadar Upazila (closer to the beach area), with a convenient sampling method. A total of 200 questionnaires had been surveyed after finalizing the meeting with them. Demographic characteristics included age (measured as a continuous variable), gender, education, race/ethnicity, and income; all measured as categorical variables. Finally, community attachment was measured by two variables: whether the respondent lived in the area for a longer time or respondents are living in the area because of tourism. Hence, locals living for tourism business and for the shorter period were exempt from the questionnaire survey. Again, impacts have been measured using five-point agreement scales and also with a series of close-ended questions of neutral statements asking for their own perceptions of directionality (Maddox, 1985). For example, they are asked to indicate whether environmental and traffic conditions are worsened or improved as a result of tourism. Responses were solicited on five-point scales where one equaled to tourism greatly decreases, and five greatly increases the level of satisfaction with a central neutral point, thereby allowing directionality to be established by respondents. The questionnaire also included measures designed as potential predictor variables of attitudes.

After collecting the data and information, it had been evaluated using statistical methods. Finally, results from the two types of the survey gave a view to the sustainable tourism principles and practices of Bangladesh. It also revealed the gaps and potentials sectors to be developed.

Development of Sustainable Tourism Concept

Tourism is one of the overarching sectors of development worldwide as it embraces multispectral industries within it. Unlike other industries, tourism has in situ nature, which makes it unique, and risky as well. Tourism is growing so fast globally beyond any other sector. Among the four stages of tourism development, the recent era is the fastest growing era because of globalization, ease of access to information, quicker and safer medium to travel, and economic affluence. Tourism contributes 3.2% of world GDP, with an increasing trend of tourist flow worldwide. As a whole, domestic tourists contribute to this sector ten times more than international tourists. Besides, Weaver (2006) identifies the recent trend of more international tourists traveling from developed to developing countries. Theorists of modernization propound tourism by stating that, in the developing countries, tourism creates more jobs, improves infrastructure, and above all, improves the quality of life (Khan, 1997). In contrast, development theorists criticize mass tourism because it mainly benefits the capitalist economy rather than host countries' society, environment, or livelihoods (Urry, 2004). Moreover, income from tourism industries arrives in industrialized countries from the developing nations (Liburd, 2011; Place, 1995).

To give solutions to the environmental and social problems generated from tourism industries, environmentalists, along with the international development organizations, came up with the concept of "Sustainable Tourism". To achieve sustainability in tourism industries and build adaptation capacity with negative impacts of climate change, there have been a number of interpretations of sustainable tourism. However, debate on Climate Change accelerates focus on the impact of tourism on the environment and research on sustainable tourism (Ruhanen et al., 2015). After its emergence, research on sustainable tourism is huge and stable, and sometimes synonymous with ecotourism or nature-based tourism.

Ironically, the concept of sustainable tourism developed depending on several time spans. Jafari (1989) divided these time spans into four broad groups, and those are: 1) Advocacy (1950–60), 2) Cautionary (1970–1980), 3) Adaptancy (1980–1990), 4) Knowledge-based platform (1990 and after). Mass tourism had been popular until the emergence of the concept "Alternative Tourism" in the 1980s, where negative impacts of mass tourism were evident on local livelihoods and natural resources. Mosedale (2011) critically remarks that, although diverse economic practices are very important in the capitalist economy, there has been very little research done on diverse economic practices, which raised the question of precarious local livelihoods (Gibson, 2009; Mosedale, 2011). Alternative tourism embraces alternative capital-centric or noncapital tourism (i.e. gift-giving, voluntourism, etc.). In this format, other forms of capital (i.e. social, natural, human, and physical capital) are considered very important along with financial capital. Homestays, voluntourism, and cooperative tourism approaches are various types of alternative tourism that diversifies economies by generating multiple economic activities. Alternative tourism also promotes nature in a non-consumptive approach, which constructs win-win development strategies for developing or underdeveloped rural areas.

Even though several criticisms exist, sustainable tourism gained importance in the tourism industry for some of its recent success stories in several countries, whether implemented by international agencies, government, or private corporations. Furthermore, different sustainable tourism approaches (such as community-based tourism, rural tourism, agri-tourism, ecotourism, etc.) are found to bring positive outcomes. Among these different types of tourism, ecotourism is the most prominent and widely used term in sustainable tourism development because of its nature and multi-dimensionality (Blamey, 2001). Managing national parks as recreation centers, Protected Areas (PAs) or Marine Protected Areas (MPAs) can ensure sustainable development of ecosystem services by visitors use management (Miller et al., 2019). It also identifies carrying capacity limits to acceptable change.

Sustainability in the tourism system is required to incorporate resistance with the vulnerability to the anthropogenic or natural disaster even in a mature and well-planned tourism destination for the resilience of their products and services (Weaver, 2006). When the whole world is going through a major depression in the tourism industry created by COVID-19, tourism could be sustainable by limiting it within carrying capacity and cope with the shock.

Because of public pressure for conventional tourism, massive voluntary development of sustainability is improbable, though climate change and other environmental pressure (Buckley, 2012). However, forecasting negative impact upon the earth and the future of the tourism system, the United Nation's Sustainable Development Goals (SDGs) targeted to adopt specific goals to fulfill and ensure sustainable tourism development. Among those goals Goal 8, 12, 16 are directly linked with sustainable tourism. Besides this, sustainable tourism could bring about economic growth and employment; sustainable consumption and production; peace, justice, and strong institutions for the inclusive and pro-poor community of a country that are major targets of SDGs. Nonetheless, environmental development by utilizing more responsible behavior from producers and consumers and social development by the betterment of the community is marked in the objectives by 2030 (United Nations Development Program, 2020).

The minimal model for sustainable tourism interrelationship among time, capital and environment could be found where profitability, compatibility, and sustainability remain within a dynamic system (World Travel & Tourism Council, 2020). Though the model has weakness of not inclusive to labor intensiveness, it shows correlation with capital maximization with environmental degradation in the tourism industry with time. On the other hand, among several sustainable tourism frameworks, one important is proposed by Csete and Szecsi (Njorge, 2014) as:

"Regional Tourism Sustainable Adaptation Framework (RTSAF)" that provides four aspects (i.e. environmental, social, economic, and institutional) of climate change adaptation strategies, and five steps of adaptation processes of tourism. The RTSAF framework can be used as a very important tourism adaptation tool in climate-sensitive destinations (Njorge, 2014). However, this framework cannot be considered as an explicit solution to the regional climate change vulnerability because it did not address sustainable livelihood goals. Sustainable livelihood generation is one of the main targets for local development within the SDG's 17 goals. It can create poverty reduction, maintain equity, foster opportunities for a decent job, and finally, economic growth for the locals.

Tourism in Bangladesh

It is found that sustainable tourism development is poorly defined in academia or policy papers in many developing countries, with very limited development practices (Carter et al., 2015). Moreover, in developing economies, theoretical practices are also minimal. Bangladesh being a developing country has 24.3% population (150 million) living under the poverty line (where per capita GDP is 2048 US$) and 55% labor force (Bangladesh Bureau of Statistics, 2018). In the country, the unemployed population is more than 68 million; additionally, revenue sources are sought-after (World Bank, 2017). The country is one of the most vulnerable countries to climate risks, both from existing and future and very frequent natural disasters. The relative severity of disasters decreased substantially since the 1970s. However, due to improved macro-economic management, the poor's increased resilience significantly progressed by disaster management. In the long list of potential impacts from climate change, the agriculture sector's risks stand out as the most important one (Winston et al., 2010). These bring disasters at a different time, which repulses tourists from abroad.

Ironically, Bangladesh is facing rapid urbanization where the population migrates from rural to urban areas for more employment opportunities, though there are limited jobs in the cities. At the present population of Dhaka city is 12 million, and its density is 23,234 per sq. km. with a growth rate of 3.60% (Bangladesh Bureau of Statistics, 2020). Pollution, less accessibility, insecurity, less information, and poor waste management are the major constraints to making Dhaka a tourist destination. Cultural degradation is another phenomenon due to modernization in the era of industrialization of the country. Thus tourism industry of Bangladesh is still lagging behind to attract international tourists. Along with international tourists, emerging domestic tourists directly contributes 2.2% of the total GDP, generating only 0.4% of GDP from visitor's export. Creating 1.8% tourism-related jobs is considered an underdeveloped industry (World Travel & Tourism Council, 2017). The tourist's destination of the country experiences the most unplanned conventional tourism type because, in the peak season, tourists exceed far beyond carrying capacity limits. This creates an unemployment problem during the off-season. It also creates social stress and fosters cultural degradation by commercializing culture, cultural assimilation, or overarching ethical degradation due to more profitability.

The country is comprised of major natural destinations, for example, world's longest continuous sea beach Cox's Bazar, one of the largest mangrove forests names the Sundarbans, numerous wetlands, hilly regions, evergreen floodplain, and so on. Among these destinations, Cox's Bazar is the most popular and top-rated destination known for conventional tourism, where around 35,000-40,000 tourists visit every day. However, in the peak season, these tourist number increases from 100,000 to 150,000. Every day on an average 10,000 tourists visits the mangrove's outer skirts on 200 boats. In the deep forest, 2000–5000 tourists visit depending on the intensity of depth. These data were collected from an in-depth interview of the president and secretary of Tour Operators Association in Bangladesh.

In contrast, the economic value of the Sundarbans' cultural services as measured by tourism shows that the number of tourists and annual revenue has been increasing. A growing number of tourists has visited the Sundarbans for recreation in recent years. Around 1,19,260 tourists visited in the Sundarbans from 2009 to 2010, comprising only 2160 international tourists (Uddin et al., 2013). Tourism friendly infrastructures are also necessary to enhance tourism in the Sundarbans. Other than tourism, "worship" and "educational research on the Sundarbans" might be highly valued cultural services of Sundarbans, which is not determined due to inadequate data. With time traveling for recreation, purposes are increasing very sharply along with tourist's expenditures throughout the country.

Principles of Sustainable Tourism Development in Bangladesh

In Bangladesh tourism industry is expanding where domestic tourists are the main market of attraction. In the last 20 years (i.e. from 1998 to 2018), tourist's spending has increased from 50 million to 4 billion US$ (Fig. 23.1). The change is rapid, and tourists' flow is very prompt where inbound tourists hold the major share of the market.

Bangladesh's major tourism industries are travel agents, tour operators, transportation, merchandiser, entertainment, or food products lacking specific sustainability indicators or standards. However, only tourists have a specific code of conduct to follow determined by UNWTO and implemented by Bangladesh Tourism Board (BTB). Similarly, there is some code of ethics for the chain hotels or large invested resort owners in the case of accommodation sectors. In most cases, there is a weak plan for abiding by sustainability initiatives (e.g. reliable and sustainable energy, responsible production and consumption etc.). Though Pacific Asia Travel Association (PATA) Bangladesh Chapter has an environmental code for sustainable tourism, no practice in the real world can be seen. In 2017, UNWTO announced the year of "sustainable tourism for development" when Bangladesh tourism corporation (BPC) translated the international code of ethics in tourism into Bengali and circulated it to the stakeholders of this industry. However, due to negligence of

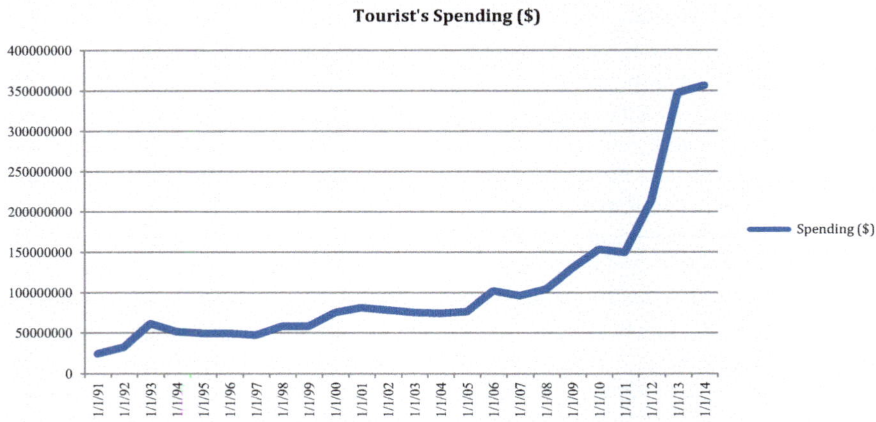

Fig. 23.1 Tourist's spending in Bangladesh (in US$). (Source: Macrotrends, 2020)

proper evaluation method of implementing code of ethics in the tourism industry, the success of ethical awareness development is unknown. There is also the absence of eco-lodges, eco-labeling, or eco-certification schemes in the country's different tourism industries.

Practices of Sustainable Tourism in Bangladesh

According to the IUCN index, Bangladesh has 51 protected areas and one marine protected area in the Bay of Bengal with three international destinations (Protected Planet, 2020). These international destinations entertain very few international tourists. Among three protected areas, two are situated in the Cox's Bazar district (i.e. Cox's Bazar Peninsula and Sonadia Island). Among the major destinations of Cox's Bazar Sadar Upazila number of tourists visit per day is five to six times more than carrying capacity (Nowreen and Mohiuddin, 2018).

Core Indicators of Sustainable Tourism determined from UNWTO (2004) shows that when any destination reaches tourists beyond its carrying capacity it brings stress to the destination by physical, ecological, social, or other stresses to the inhabitants (Bramwell, 2006). Therefore, due to mismanagement and overexploitation of the country's resources, coastal tourism experiences stress in every aspect. Moreover, the use of natural resources to produce tourism products and services is so intense. Though there are limited international tourists, an abundance of inbound tourists in one destination makes the environment intolerable to the locals. Table 23.1 shows several attributes for measuring local people's satisfaction level in the Cox's Bazar Sadar Upazila as a topmost tourist destination.

The perception about the satisfaction level of the local residents of Cox's Bazar Sadar Upazila is mixed in tourists visiting their region. A total of 37 questions had

Table 23.1 Resident's perception of Cox's Bazar as the topmost tourist destination in Bangladesh

Attributes	Perception
1. Visual attraction and appealing natural attractions	3
2. Used natural/ local resources as equipment and facilities	1.7
3. Emphasized that tourists recycle and reuse products	0.8
4. Development was integrated with the local environment/ culture	1.8
5. Nature-based recreation activities (Sunbathing, beachside walking, watersports, kayaking, bird watching etc.)	2.4
6. Regional planning	2.7
7. Accessibility	3.3
8. A feeling of belonging in the community	3.1
9. Preservation of ones way of life	3.2
10. Participation in local culture	3.5
11. Understanding of different culture	3.2
12. Community pride	4.1
13. Frequently arranged community meeting	2.6
14. Participating in community traditional crafts, skills, customs	3.2
15. Waste management	0.9
16. Traffic system	1.2
17. Crowding and congestion	1.4
18. Drug and alcohol abuse	2.1
19. Crime and vandalism	2.4
20. Urban sprawl and population growth	1.8
21. Conflicts over zoning/land use	2.1
22. Preservation of biodiversity and habitats	2.1
23. Preservation of natural areas	2.3
24. Preservation of cultural/historical sites	2.1
25. Clean air and water	1.9
26. Beautification of community	2.4
27. Nature-based recreation opportunity	2.5
28. Peace and tranquility	2.2
29. Enough decent jobs for the residents	2.7
30. Stores and restaurants owned by local residents	2.6
31. Availability of retail shops and restaurants	3.7
32. Increased value of house/or land	3.8
33. Diversified livelihood	3.7
34. Contingency planning mitigation and response	0.1
35.Quarantine precautions for air travelers (number and nature of quarantine measures; checking temperature)	0.3
36. Effectiveness of surveillance and reporting of cases	1.5
37. Public health education campaigns	2.1
Overall mean value	2.34
Variance	0.888
SD	1

Source: Questionnaire survey (2019)

been asked to them using a 5-point Likert scale ranging from 1 to 5 (i.e. 1 meant worst and 5 to best level of satisfaction). The overall value on perception is 2.34, which stands for slightly below neutral, which means local people were not satisfied with conventional tourism practices. However, standard deviation is one that directs in the cases of some attributes satisfying to the locals.

Besides, perception of financial contribution was higher as more decent job opportunities with diversified livelihoods were created because of tourism, with an increase in business, accessibility, and increasing land price. On the other hand, the overall environment has gone under degradation due to poor waste management, escalation of crowd and congestion, fewer nature-based recreation activities, poor contingency mitigation planning, or pollution control. Moreover, social degradation is burgeoned due to increased rate of crime, drug and alcohol abuse, poor traffic system, less peaceful environment, etc. Another important issue is that, like other destinations of the country, it does not have any proper and effective waste management system, neither they have waste treatment plants. In contrast, local people percept that there is potential to utilize nature-based recreation in a wise manner; tourism has increased their community pride. Moreover, local peoples' participation in tourism has increased because of understanding others' cultures. It gives a general idea of how residents of the country's top destination feel about tourism in their region and the gaps to make it sustainable.

Sustainability of Livelihood Capital in the Top Destination of Bangladesh

Exempt from urban areas, 80% of people of Bangladesh live in the rural areas where rural people in developing countries struggle to meet their basic needs. Also, lack of participation in the community's decision-making process hinders them from being empowered. However, Sustainable Livelihood (SL) is a strategy that could be utilized for poverty alleviation for the rural poor (United Nations Office of Drug and Crimes, HIV/AIDS and Crime, 2011). SL addresses environmental and social sustainability by identifying resistance capabilities with various vulnerabilities, shocks, and assets (McKeller and Smardon, 2012). In developing countries, rural poor are dependent on a subsistence economy, which can drive rural poor to areas where they cannot fulfill their basic needs (Shen et al., 2008). In developing countries, this strategy mainly supports the subsistence economy. It is clear from the figure that SL for tourism integrates rural development, sustainable environment, and tourism development. Therefore, Sustainable Livelihood Approach (SLA) could be a solution for sustainable rural development as well as sustainable tourism. To understand the relationship between SL and tourism, it is necessary to examine the tourism industry's specifics (Shen et al., 2008). Hence, poor livelihoods, even if they lack financial assets, can utilize their cultural and natural capital to involve in the tourism business (Ashley et al., 2001). Furthermore, tourism can complement other types of nature-based livelihood activities.

From the survey of tourism industry stakeholders, the following Fig. 23.2 is formed to assess the condition of the asset of four important tourism industrys of Bangladesh.

Average capitals from the above four sectors identify present assets are somewhat sustainable (with an average value of 8.5 where sustainable: 8 < = capital <12). However, if the expansion of mass tourism happens in this way, the asset might fall short and become limited sustainability to unsustainable (4.0 < =capital <8 or 0 < capital 4.0 respectively). First of all, the sustainability of livelihood capital in the accommodation sector is weighted 7.3, implying limited to sustainability. According to the tourism industry stakeholders, human capital lacks for the accommodation sector, as there is less training or human resources at the professional level. On the other hand, social capital is very less due to non-coordination among different associations or communities. Moreover, some stereotype approaches are working in the accommodation sector; and females are not encouraged to work there. This sector

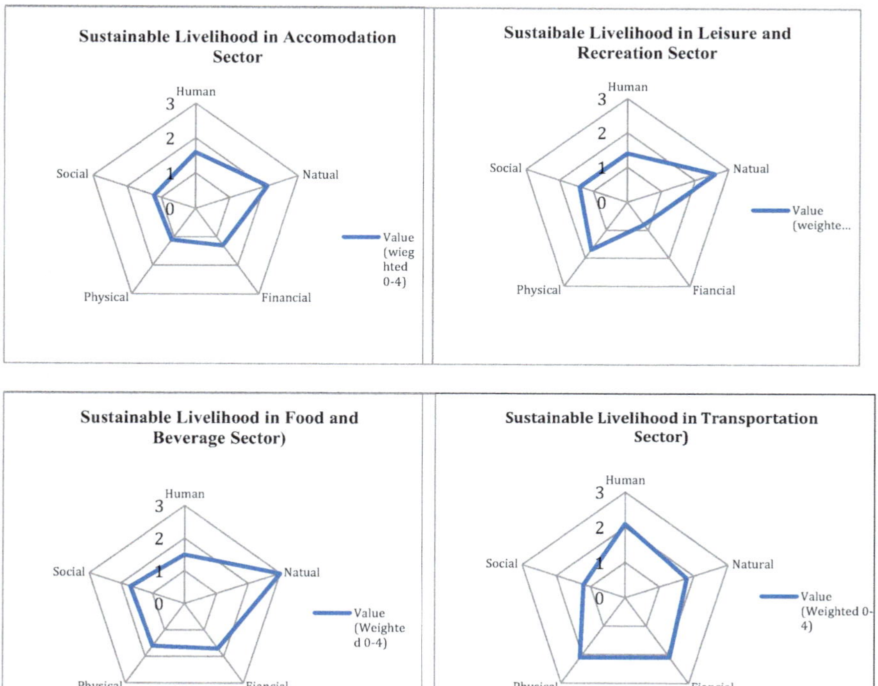

Fig. 23.2 Sustainable Livelihood Asset Model for the four major sectors of the tourism industry in Bangladesh (authors' generated) Model used from UNDP (2017). (Source: Questionnaire survey, 2019)

lacks in waste management practices and high potentials of money leakage because of investments from abroad or non-residents to the destinations.

Again for recreation and entertainment industries, livelihood capital is limited to sustainability (i.e. weighted value 7.9). Firstly, human capital is less skilled or trained for practicing diversified recreation spectrum. Research and training are also limited based on this industry. Traditional recreation practices are being faded day by day. The high seasonal influence of tourism makes it less attractive to skilled personnel to develop their careers in the recreation sector. However, natural capital is abundant with plenty of opportunities. Ironically, social capital is somewhat degrading with time by globalization and cultural assimilation, in terms of physical and financial capital is less in the tourism industry fitting to the demand. There is the least investment in the recreation and entertainment industries as well.

Similarly, natural capital is more sustainable (weighted 9.3) among any other capitals as the availability of resources about the country's food and beverage industries. More local products are used in the culinary industries. For food industries, self-sufficiency in the agricultural, livestock, and fisheries makes natural capital available. Therefore, according to the survey, the livelihood capital is sustainable until now. However, less budget and training on international food financial and human capital are less involved in these industries. More research and a positive image of culinary industries are also needed for further development of the industry. Finally, in the transportation sector, the sustainability of livelihood capital weighted 9.3. Social capital is less implied due to poor public perception of transportation industries of the tourism industry. Proper planning is required for the well-being of mankind in the transportation sector. Among five capitals, human and physical capital is used more for livelihood though it is still sustainable. But negligence of skill development or further investments for infrastructure or superstructure formulation could minimize sustainability for livelihood generation. Furthermore, social capital is invested least due to the negligence of coordination among social structures and coordination. However, natural capital utilization for creating decent employment or sustainable energy consumption is not satisfactory. Thus, sustainable livelihood capital identifies gaps in Bangladesh's tourism industry and foresees future initiatives to be taken for sustainable asset maximization in specific investment or planning process.

Direction to Sustainable Tourism Development in Bangladesh

Tourism in Bangladesh is not well developed compared to the neighboring countries. Moreover, there is a lack of sustainable tourism development principles determined by the concerned authority or well-established practices. Bangladesh's tourism policy is yet to implement sustainable policy instruments such as eco-certification, eco-lodges, waste management, environmental conservation, local empowerment, investment of micro to medium tourism industries, and so on. On the other hand, international tourists' arrival in the country is minimal, while tourism expenditure increases sharply because of domestic tourist flow. Ironically, it has been estimated that in developing countries, the emerging market is growing 3.1% times faster than their developed peers (Global Green Growth Institute, 2015). It is very crucial to make the right investment in the right place to grow. Most of the investments are happening in the urban or periphery of city areas for industrialization. Besides, the capitals and cities provide all categories of employment where the majority of the job is a low category with marginal waged labor. This makes the overall situation devastating, as people in this category live in slums or open-air with tragic sanitation or public service facilities. The unemployment rate of the workforce is 4.8% (Bangladesh Bureau of Statistics, 2020).

Furthermore, urban areas' tourism industries create seasonal employment opportunities with an increasing trend of floating people. On the other hand, there is a shortage of human resources and skilled labor force in rural areas. Besides, public services are absent in rural areas, and rural communities will suffer more, as the number of older people will increase if the economy is not properly planned. To find sustainable tourism practices in Bangladesh local people's satisfaction defines gaps of local capital invested in the tourism industry. For investment practices of coastal tourism development, natural resources should be consumed in a responsible way. Furthermore, the waste management system must be modern and effective in each and every destination. Safety and securities should also be given priority in terms of monetary instruments with effective law and order enforced. Contingency planning or responses should be embraced in the tourism development planning process as a risk management tool for disaster or vulnerability mitigation strategies. It is found that the local communities are satisfied in terms of their economic growth; however, some discrepancies still remain, as economic development needs to be more inclusive. Therefore, female and marginal groups must be incorporated in tourism industries from micro to small or medium industries of locally owned. This type of direction could be equally applied to other destinations from coastal, hilly, floodplain or watersheds of the country.

In contrast, Bangladesh's tourism industry's livelihood assets are weighted 7.8 in a range of 20, which means that it is an unsustainable practice of asset utilization. However, exurban areas can fill these gaps by sustainable asset utilization, and by providing employment opportunities to everyone, especially low to middle-income people. For example, trained human capital from all over the country could diversify their economy with their own credentials for heritage, and ecotourism could utilize human capital and participate in entrepreneurial tourism activities. Thus, the economy could be diversified and persistent. Furthermore, in a pandemic period during COVID-19 outbreak, there are very few job opportunities in urban areas, and mass people move towards rural areas that are private jobholders or daily laborer. In rural areas, due to dispersed settlement, there are huge opportunities for alternative types of economic practices.

For more sustainable financial capital development, green financing on tourism industries, disbarment of eco-friendly funds, green certification projects, increasing production and services of locally-owned tourism industries, the proliferation of value chain for agricultural, livestock or fisheries production should be given emphasized. Besides, sustainable physical capital utilization; local infrastructure, investment in green energy, and establishing forestry or environmental enterprises must be embraced in the investment or development policy and planning. Again, human capital could be sustainable in Bangladesh's destinations by stakeholder training, environmental awareness training on fishing, agriculture, strengthening local association, training public-private sector employee. Though Bangladesh has sustainable natural capital, it is not finite.

Therefore, to ensure natural capital, responsible production and consumption from micro-small to medium enterprises, eco-friendly production, coastal buffer zone creation, restoring natural coastal defense, biodiversity conservation zones, land zoning, and smart technology use should be other spectrums to consider. This can also help ensure sustainable tourism development. Finally, for sustainable social capital, females should be encouraged to participate in the tourism industry, participation in land zoning and wood management with local government, and inter-ministerial plans Shocks, trends, and seasonality influence the availability of livelihood assets in certain destinations. These initiatives of proper investment and strategic planning for development will strengthen the country for coping with natural disasters. Hence, effective policy and organizational structures in a country could minimize the access to these assets for ensuring sustainable livelihood.

Conclusion

The tourism industry of Bangladesh is expanding very rapidly. Some of the destinations in peak season are experiencing tourists beyond the expectation. Tourism in the country contributes directly, indirectly, or induced way by providing opportunities for diversified livelihood. However, there is a lack of planning for ensuring the sustainability of the tourism industry. For example, there is huge employment in the peak season, but people become jobless in the off-season. Again after any disaster, destinations become devastated by not offering any livelihood to many people. There is a lack of practice of proper waste management systems and responsible production and consumption. Moreover, there is less safety and security for the tourists. Most of the country's destinations are disaster-prone, which makes the country less attractive to international tourists. Climate change is another global concern making the country vulnerable to the natural calamity. After finding the gaps of livelihood and capitals of the destinations, more investment and development is needed in human capital development by providing training, workshops, conducting research, setting up interpretation centers, and ensuring education. The tourism industry's financial capital should be made sustainable by ensuring more investment in locally-owned tourism businesses, green investment, setting up tools for risk management, or conserving the destination's natural resources. Natural capital also needs to be sustainable by ensuring a proper waste management system, practicing eco-certification schemes. Physical capital can be sustainable through the development of local infrastructure, eco-lodges, and environment-friendly enterprises. Finally, social capital can be more sustainable by inclusive business to gender or marginal group, land zoning, or plantation. Therefore, developing sustainable livelihood by investing in all types of assets and implementing proper strategic planning will benchmark Bangladesh's sustainable tourism principles. Implementing strategic planning and effective investments to the locals will create sustainable tourism practices. Thus, conglomerations of principles and practices of sustainable livelihood and the local community's involvement will make Bangladesh's tourism industry healthy and sustainable. This will create a good image of the country internationally and decent socio-economic and environmental development.

References

Ashley, C., Roe, D., & Goodwin, H. (2001). *Pro-poor tourism strategies: Making tourism work for the poor.* Retrieved from: https://www.odi.org/sites/odi.org.uk/files/odi-assets/publications-opinion-files/3246.pdf. Accessed 17 Nov 2020.

Bangladesh Bureau of Statistics (BBS). (2018). *Report on labor force survey 2016-2017.* Retrieved from: http://203.112.218.65:8008/WebTestApplication/userfiles/Image/LatestReports/LFS_2016-17.pdf. Accessed 17 Nov 2020.

Bangladesh Bureau of Statistics (BBS). (2020). *Dhaka population 2020.* Retrieved from: http://203.112.218.65:8008/WebTestApplication/userfiles/Image/District%20Statistics/Dhaka.pdf. Accessed 18 Nov 2020.

Blamey, R. K. (2001). Principles of ecotourism. In D. Weaver (Ed.), *The encyclopedia of ecotourism* (pp. 5–22). CABI.

Bramwell, B. (2006). Monitoring for a sustainable tourism transition: The challenge of developing and using indicators. *Tourism and Hospitality Research, 7*(1), 83–84.

Bramwell, B., & Lane, B. (2008). Priorities in sustainable tourism research. *Journal of Sustainable Tourism, 16*(1), 1–14.

Buckley, R. C. (2012). Sustainable tourism: Research and reality. *Annals of Tourism Research, 39*(2), 528–546.

Carter, R. W., Thok, S., O'Rourke, V., & Pearce, T. (2015). Sustainable tourism and its use as a development strategy in Cambodia: A systematic literature review. *Journal of Sustainable Tourism, 23*(5), 797–818.

Department of International Development (DFID). (1999). *DFID Sustainable livelihood guidance sheet.* Retrieved from: https://www.ennonline.net/dfidsustainableliving. Accessed 15 Nov 2020.

Gibson, C. (2009). Geographies of tourism: Critical research on capitalism and local livelihoods. *Progress in Human Geography, 33*(4), 1–9.

Global Green Growth Institute. (2015). *Infrastructure development in the developing world.* Retrieved from: http://gggi.org/wp-content/uploads/2015/07/WP02-Green-Infra_def-and-needs.pdf. Accessed 17 Nov 2020.

Islam, F., & Carlsen, J. (2013). Tourism in rural Bangladesh: Unlocking opportunities for poverty alleviation? *Tourism Recreation Research, 37*(1), 37–45.

Jafari, J. (1989). An English-language literature review. In J. Bystrzanowski (Ed.), *Tourism as a factor of change: A socio-cultural study* (pp. 17–60). Center for Research and Documentation in Social Sciences.

Khan, M. M. (1997). Tourism development and dependency theory: Mass tourism vs. ecotourism. *Annals of Tourism Research, 24*(4), 988–991.

Liburd, J. J. (2011). Introduction to sustainable tourism development. In J. J. Liburd & D. Edwards (Eds.), *Understanding the sustainable development of tourism* (pp. 1–19). Goodfellow Publishers Limited.

Macrotrends. (2020). *Bangladesh tourism statistics 1995-2020*. Retrieved from: https://www.macrotrends.net/countries/BGD/bangladesh/tourism-statistics. Accessed 20 Nov 2020.

Maddox, R. N. (1985). Measuring satisfaction with tourism. *Journal of Travel Research, 23*, 2–5.

Mckeller, M. M. M., & Smardon, R. (2012). The potential of small-scale agro-industry as a sustainable livelihood strategy in a Caribbean Archipelago Province of Colombia. *Journal of Sustainable Development, 5*(3), 16–33.

Miller, Z. D., Rice, W. L., Taff, B. D., & Newman, P. (2019). Concepts for understanding the visitor experience in sustainable tourism. In S. F. McCool & K. Bosak (Eds.), *A research agenda for sustainable tourism* (pp. 53–69). Edward Elgar.

Mosedale, J. T. (2011). Diverse economies and alternative economic practices in tourism. In N. Morgan, I. Ateljevic, & A. Pritchard (Eds.), *The critical turn in tourism studies: Creating an academy of Hope* (pp. 194–207). Routledge.

Njorge, J. M. (2014). An enhanced framework for regional tourism sustainable adaptation climate change. *Tourism Management Perspectives, 12*, 23–30.

Nowreen, S., & Mohiuddin, M. (2018). Sustainable tourism development in disaster-prone coastal areas of Bangladesh. In H. Khatun, N. Ahmad, A. Q. M. Mahbub, & H. Kabir (Eds.), *Environment and sustainable development in Bangladesh: Geographical perspective* (pp. 163–180). University of Dhaka.

Place, S. (1995). Ecotourism for sustainable development: Oxymoron or plausible strategy? *GeoJounral, 35*(2), 161–173.

Protected Planet. (2020). *Bangladesh*. Retrieved from: https://www.protectedplanet.net/country/BGD. Accessed 24 Nov 2020.

Ruhanen, L., Weiler, B., Moyle, B., & Moyle, C. (2015). Trends and patterns in sustainable tourism research: A 25-year bibliometric analysis. *Journal of Sustainable Tourism, 23*(4), 517–535.

Shen, F., Hughey, K. F. D., & Simmons, D. G. (2008). Connecting the sustainable livelihood approach and tourism: A review of literature. *Journal of Tourism and Hospitality Management, 15*, 19–31.

Uddin, M. S., de Ruyter van Steveninck, E., Stuip, M., & Shah, M. A. R. (2013). Economic valuation of provisioning and cultural services of a protected mangrove ecosystem: A case study on Sundarbans Reserve Forest, Bangladesh. *Ecosystem Services, 5*(C), 88–93.

United Nations Development Program (UNDP). (2017). *Application of the sustainable livelihood framework in development projects*. UNDP.

United Nations Development Program (UNDP). (2020). *Sustainable development goals*. Retrieved from: https://www.undp.org/content/undp/en/home/sustainable-development-goals.html. Accessed 11 Nov 2020.

United Nations Office of Drug and Crimes, HIV/AIDS and Crime (UNODC). (2011). *Sustainable livelihoods: A broader version. Social support and integration to prevent illicit drug use*. Retrieved from: http://www.unodc.org/documents/alternative-development/SL%20Paper/Final_ebook.pdf. Accessed 12 Nov 2020.

Urry, J. (2004). Tourism, travel, and the modern subject. In S. Williams (Ed.), *Tourism: Critical concepts in the social sciences* (pp. 3–15). Routledge.

Weaver, D. (2006). *Sustainable tourism: Theory and practice*. Elsevier.

Winston, H. Y., Mozaharul, A., Ahmadul, H., Abu, S. K., Alex, C. R., Cynthia, R., David, C. M., & James, T. (2010). *Climate change risks and food security in Bangladesh*. Earthscan.

World Bank (WB). (2017). *Bangladesh*. Retrieved from: https://data.worldbank.org/country/bangladesh. Accessed 19 Nov 2020.

World Tourism Organization (UNWTO). (2004). *Indicators of sustainable development for tourism destinations: A guidebook*. WTO.

World Tourism Organization (UNWTO). (2019). *Transport-related CO_2 emissions of the tourism sector*. Retrieved from: https://www.e-unwto.org/doi/epdf/10.18111/9789284416660. Accessed 18 Nov 2020.

World Travel & Tourism Council (WTTC). (2017). *Travel & tourism economic impact 2017 Bangladesh*. Retrieved from: https://www.wttc.org/-/media/files/reports/economic-impact-research/countries-2017/bangladesh2017.pdf. Accessed 18 Nov 2020.

World Travel & Tourism Council (WTTC). (2020). *World travel and tourism report 2019*. Retrieved from: https://wttc.org/Research/Economic-Impact#:~:text=In%202019%2C%20Travel%20%26%20Tourism's%20direct,10%20jobs%20around%20the%20world. Accessed 18 Nov 2020.

Yfantidou, G., Spyridopoulou, E., Papaioannou, A., & Balaska, P. (2018). The interest of sport tourism enterprises in green practices. In V. Katsoni & K. Velander (Eds.), *Innovative approaches to tourism and leisure. Springer proceedings in business and economics* (pp. 275–287). Springer.

Zeng, B. (2018). How can social enterprises contribute to sustainable pro-poor tourism development? *Chinese Journal of Population Resources and Environment., 16*(2), 159–170.

Part VII
Challenges and the Future

Chapter 24
Investment and Development in Tourism and Hospitality Products and Service Innovations in Bangladesh: Challenges and the Future

Tanvir Abir and Md Yusuf Hossein Khan

Abstract Bangladesh has a temperate climate, natural scenery, UNESCO World Heritage Sites (WHS), and a modern democratic culture around it. Besides, the government is taking on some world-class mega-project for the country's rapid infrastructural growth. Such diversification can be a pull factor to attract international tourists to visit Bangladesh. Tourism in Bangladesh as a competitive industry is contributing to the national economy. Tourism participates continuously in the national Gross Domestic Product (GDP) per year. Tourism spending is very important for the sustainable growth of the economy and infrastructure. The main aim of this research is to identify the opportunities and obstacles behind tourism investment. The study has established several key tourism investment outcomes. Tourism can produce tremendously professional and skilled workers. It is a big economic market, and tourism can be able to earn huge foreign currencies that the country demands now. Lack of constructive attitude, weak promotional identity, poor facilities, and lack of world-class tourism firms are some of the country's tourism industry problems. To eradicate bad branding identity, massive investment is required worldwide to mark Bangladeshi WHSs and local culture.

Keywords Development · Tourism · Challenges · Future · Bangladesh

---

T. Abir
College of Business Administration, IUBAT-International University of Business Agriculture and Technology, Dhaka, Bangladesh

Md Y. H. Khan (✉)
College of Tourism and Hospitality Management, IUBAT-International University of Business Agriculture and Technology, Dhaka, Bangladesh

Faculty of Economics, University of Algarve, Faro, Portugal
e-mail: ysf.khn86@gmail.com

A. Hassan (ed.), *Tourism in Bangladesh: Investment and Development Perspectives*, https://doi.org/10.1007/978-981-16-1858-1_24

Introduction

Today tourism is the fastest rising industry in the entire world. The main economic area will be in potential global tourism. Tourism is one type of business industry in the world. Service industry production should accelerate our economic growth (Sandip, 2014). Tourism is one of Bangladesh's important industries (Elena et al., 2012). Tourism is a growing and Bangladesh wants to invest in improving the tourism industry. Bangladesh will concentrate on infrastructure growth, housing, food, and entertainment. Tourism investment and growth will pull in massive quantities of foreign currency. Tourism can generate tremendous opportunities for both our country's professional and unskilled citizens. Tourism is viewed to improve the overall economic situation, share social and cultural values and preserve biodiversity that can affect a country's economy and society.

Bangladesh is a natural-beauty area. Bangladesh has a glorious past and a rare and mixed cultural dimension. Bangladesh is an Asian nation with strong tourism potential (Hassan et al., 2020; Hassan & Kokkranikal, 2018). This region is renowned for scenic beauty from the ancient period. International visitors lauded this nation for its majestic natural scenery, rich cultural heritage, and peoples' hospitality (Ali, 2004). Few reasons which can drive visitors to come to Bangladesh in the future. Bangladesh is a natural-beauty area. Bangladesh's climate is temperate, and Bangladesh is the least carbon-emitting country. The country's political climate is pleasant, and it is a democratic republic. There are several world heritage sites in Bangladesh, such as the Bagerhat Temple, the Sundarbans, and the Paharpur Bihar Museum. It has diversified tourism attractions within a limited region. Mountains, sea beaches, slides, woods and mangrove trees, gorge, tea garden, coasts, hilly coasts, numerous plants, and flora faunas are the apotheosis of this region's natural beauty.

The Government of Bangladesh is now working on some mega-project that will draw international visitors. Major ventures are Padma Bridge, broad power plants, metro, ports, and airports. Bangladesh wants investment and growth in tourism and hospitality goods and services to reach the global tourism industry. Product or service developments apply to improvements that the consumer personally encountered and considered new; often in the case of never before seen (Hjalager, 2009). Innovations in goods or facilities are perceptible to visitors to such a degree that they might even become a consideration of the decision to purchase.

Ferdoush and Faisal (2014) showed that tourism's significance is viewed from many angles like economic, social, cultural, political, etc. After investing in tourism, it can bring economic, social, and environmental benefits, particularly in rural areas and developing countries, although it has some challenges. Salam (2014) highlighted the importance of tourism in Bangladesh along with its benefits in socio-economic development. Expertise in tourism and research paper of scholar finds some unique product and services for tourism development. Being a developing country with many popular tourist sites in Bangladesh, such as Cox's Bazar sea beach, there is a huge potential to draw visitors and grow tourism in Bangladesh. To grow its tourism industry, Bangladesh will increase expenditure behind the tourism industry and its merchandise. Good service and spectacular position will encourage the visitor to visit a specific spot. There is a variety of Beautiful location in Bangladesh.

There are several business sectors in which Bangladesh requires more investments. Among them, the tourism industry is notable to support quicker and improved methods of food production, energy and labor savings, waste management, better hygiene, faster service, and greater efficiency (Rogers, 2007). For starters, they involve automation utilizing robotics, completely different methods of preparing, and methods of cooking and flavoring powered by sensors. Tourism initiates the creation of numerous related physical and transportation facilities (Ministry of Environment and Forests, 2012). Tourist accommodation is a significant reality. Flexible accommodation requires for tourism industry growth. Hotels must offer low-priced accommodation without violating simple and necessary hotel requirements such as efficiency, cleanliness, and comfortable beds (Hall & Williams, 2008). Some hotel industry research relates to hotel amenities' distinctive qualities as creative, such as food, culture, technology, or wellness facilities (Jacob et al., 2003; Pikkemaat, 2008). Bangladesh Parjatan Corporation (BPC) has suitable guidance for tourism growth in the region. Bangladesh takes the tourism dream 2020, and this dream seeks to lead to the country's overall growth in terms of employment generation, poverty reduction, national GDP enhancement, tourism production, and conservation.

Present Investment Scenario of the Tourism Industry

Whatever governmental and tourism policy has a positive impact on the tourism industry and tourists' arrivals (Karambakuwa et al., 2011). On average, a yearly number of 552,500 tourists visited Bangladesh during the 2009–2017 period (Hossain & Wadood, 2019). In FY2019-20, the government allocated BDTk 34 billion to Civil Aviation (Biman) and the Ministry of Tourism, which is more than double the last fiscal year's allocation that indicates that the Bangladeshi government is giving importance to the tourism industry (The New Age, 2020). This evaluation was reflected in the 2019 Travel and Tourism Competitiveness Report where Bangladesh saw the largest percentage increase, from 125th among 140 countries, reaching 120th position (World Economic Forum, 2019). The budgetary allocation to the Ministry of Civil Aviation and Tourism objectives to establish a world-class aviation system and transform the tourism industry into a developed and modern economic industry (The New Age, 2020).

According to World Travel & Tourism Council (2018), Bangladesh has received 15,29,000 visitors and has earned US$ 413,00 million over the years 2006–2010 (World Travel & Tourism Council, 2018). The country has been trying to draw more visitors to its destinations from the beginning of this industry and gain more foreign currency. Statistics on this industry show that in Bangladesh, the income from tourism has increased as in 2019, BDTk 2600 crores of foreign exchange earnings were earned from tourism exports (The New Age, 2020). The beginning of 2020 was quite good for tourism as compared to 2019 as the number of tourists in January 2020 was higher than the last year but since February, there are almost no tourists anywhere due to the effects of the COVID-19 outbreak (Mason, 2020).

Present Development Scenario of the Tourism Industry

Several World Travel & Tourism Council (WTTC) reports find that travel and tourism had started adding to Bangladesh's economy (Afroz & Mahmud, 2017). The tourism industry has a small share of revenue that is contribution to economic growth to the GDP of Bangladesh's economy. Tourism industry also creating employment in Bangladesh (Ali & Parvin, 2010; Sultana, 2016).

Contribution in Economic Development

Bangladesh earned US$ 1153 million from its tourism industry during the 2009–2017 period, which is a small share of its GDP (Hossain & Wadood, 2019). Bangladesh earned an annual total of US$ 344 million in 2017, US$ 214.3 million in 2016, US$ 150.3 million in 2015, way up from US$ 79 million in 2006, from tourism revenue (Choice for Economic and Investment Research, 2019). Bangladesh's tourism industry's current size is BDTk 5000 crore in 2017, which was BDTk 500 crore in 2007 (The Daily Star, 2018). Bangladesh earned US$ 1153 million from the tourism industry during the 2009–2017 period; while the revenue from the travel & tourism industry was BDTk 850.7 billion in 2017 and on average, the yearly number of 552,500 tourists visited Bangladesh during the 2009–2017 period (World Tourism Organization, 2000).

Contribution to Employment Generation

In 2013–14, tourism directly contributed around 2.1% to Bangladesh's GDP and supported 1.3 million jobs (Planning Commission, 2015a). Tourism and hospitality created 1,138,500 jobs across the country in 2015 (2.0% of the total workforce), and this is expected to rise to 1,155,500 (2.0% of total employment) by 1.5% in 2016 (The Bangladesh Monitor, 2017). This encompasses the job opportunities of hotels, travel agents, airline companies as well as other services for transportation services (excluding services for commuters). For instance, perhaps it involves the arrangement of the leisure and entertainment industries that are supported directly by tourists. According to the WTTC (2018), Bangladesh's travel & tourism industry generated approximately a yearly 2.23 million jobs during the 2011–2017 period. Tourism generates both formal and informal jobs. The most common job places are hotel and motel, cafe & restaurant, shopping malls at the tourist site, transport, tourist guide, tourism agency, local handicrafts, etc. By 2026, Travel and Tourism is expected to bolster 2,894,000 jobs (4.1% of total employment) over the period, an increase of 1.9% annually (World Tourism Organization, 2000; World Travel & Tourism Council, 2018).

Future Investment Potentials in the Tourism and Leisure Projects in Bangladesh

Tourism is considered to be a major field of business and commerce. Tourism plays a significant part in environmental and social sustainability, in addition to economic benefits. The study showed that the tourism industry's growth is affected by cultural, environmental, and social factors (Khan et al., 2019). In 2010, Bangladesh's overseas tourist arrivals hit a historic 303,000 with a growth pace of 13.48% relative to 2009 and nevertheless, in foreign tourism receipts, BDTk 7947.00 million was registered in the same year, leading to a marginal decrease of 3.46% compared to 2009 (Ishtiaque, 2013). The Government of Bangladesh plans to build 13.2 million jobs in the 7th Five Year Plan and increase GDP growth by 2020 to 8%. Policy focus on job prospects by the development of young people into human capital under this 7th Five Year Program (Planning Commission, 2015b). The tourism industry can play a significant role in achieving the goal by 2020. This industry will at the same time employ a greater number of people. This industry included both professional and unskilled workers.

Government of Bangladesh is targeting to build ten million jobs across 100 economic zones in Bangladesh. Here, the Bangladesh Parjatan Corporation will contribute at least 0.5 million youth by offering training and seeking employment. According to the Bangladesh Bureau of Statistics (2019), Bangladesh's estimated working-age population is 106.1 million. The real labor population is 62.1 million individuals; 59.5 million workers work in both the public and private sectors; 2.6 million people are currently unemployed. Such disadvantaged young people may be involved in tourism in various fields, with adequate preparation.

Tourism is a major guiding factor for economic development. Tourism will gain huge amounts of foreign currency. Tourism will offer work growth. Tourism generated 1,984,000 jobs directly in 2014, which is projected to rise to 2,492,000 jobs directly by 2025 (World Travel & Tourism Council, 2018).

World Tourism Organization (UNWTO) (2001) estimates that there will be 1561.1 million global visitor arrivals in the year 2020. The same number for South Asia is expected to reach 18.8 million by 2020. Around the same period, the organization estimated that the number of tourist arrivals in Bangladesh in the following years would be 0.415 million. This also estimates that, by 2020, tourism earnings will grow to US$ 2 trillion a year. Tourism is a branch of the utilities. Factors such as easy accommodation, no travel limits, the provision of knowledge on specific tourism destinations, and modern marketing methods led to the increase in the world's total number of visitors.

Following World Travel & Tourism Council (2018), the overall travel and tourism share is 4.3% of Bangladesh's GDP, 6.4% in 2028. Again, Travel and tourism are projected to include 1,178,500 work openings directly in 2017 (1.8% of total employment); and that is anticipated to increase by 3.0% in 2018 to 1,214,000 (1.9% of total employment), which includes jobs by hotels, travel agencies, airlines, and other passenger transport services excluding commuting services. This is reported that travel and tourism will specifically account for 1,648,000 workers by 2028, with a 3.1% growth in the next 10 years.

Recently, the World Economic Forum (WEF) (2019) released its 2019 Travel and Tourism Competitiveness Study, which offers an overview of regional developments and the travel and tourism industry's status worldwide. Bangladesh rated 120 among the 140 countries considered for the study. We are at the foot of the table if ranked against South Asia, where India (34), Sri Lanka (77), and Nepal (102) are based. Nearly half of the overall tourists arriving in Bangladesh are Indians. The remaining visitors are mainly from countries in Asia. Although 5% of visitors come from the United States, these estimates include the US people of Bangladeshi descent. Just 7% of visitors come from all over the country. The Americas (i.e. North and South America) invested US$ 334 billion in 2018, and Europe invested US$ 570 billion in tourism. The estimated global investment behind tourism corresponds to US$ 1.7 trillion. This ensures that the Americas and Europe combined makeup 53% of Bangladesh's missed foreign tourism market (Fig. 24.1).

Bangladesh can be supported by the products of internationally renowned tourist destinations. But capital expenditure in Bangladeshi tourism product branding is extremely limited. Completely different investment for brand awareness alone will be a big initiative for tourist destination development in Bangladesh. Tourism is a service-based business. Investment in world-class services in terms of affordable and standard accommodation and food needs to be made for foreign tourists. Investing in the development of flexible policy on tourism and taxation is the government of Bangladesh. Large number share of the investment needed to develop the infrastructure. For tourism, additional facilities include roads, water systems, gardens, parks, and museums. Tourism infrastructure can be considered as the physical environment designed and erected to accommodate the visitors. Some authors have established a strong relationship between tourism development and infrastructure in theory (Adebayo & Iweka, 2014).

Investment in developing online platforms regarding tourism branding and online payment is very important for the tourism industry. Online reviews have a significant influence on the buying behavior of tourists as according to Filieri and McLeay (2013). Sales of online advertising show important growth rates. Global online travel sales in 2016 represented US$ 565 billion and are projected to increase to US$ 756 billion in 2019 (Statista, 2017).

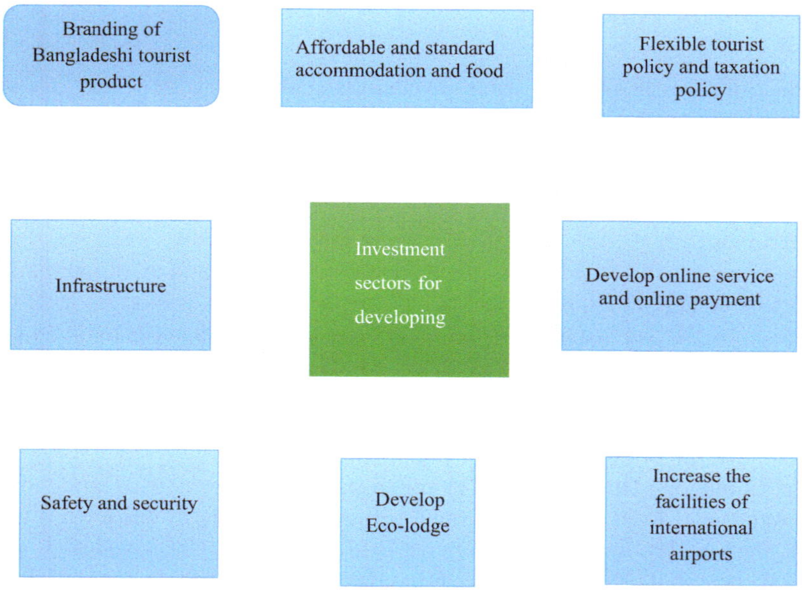

Fig. 24.1 Sub-industry of investment for tourism in Bangladesh. (Source: Alauddin et al., 2014; Das & Chakraborty, 2012; Gaggioli et al., 2013; Hassan & Burns, 2014; Majumder & Hossain, 2011; Roy & Roy, 2015).

The tourism industry is affected by security and security. Safety and wellbeing are at the core of this industry. Bangladesh needs enormous investment to guarantee tourist safety and security. Health is one of the key factors deciding the attractiveness of a country's tourism industry. To keep up with Bangladesh's rapidly growing commercial aviation over the last several years, Bangladesh needs to upgrade its Airport facilities and its relevant infrastructure without further preparation.

Future Trends of Development Policy and Strategy of Tourism in Bangladesh

Competition is increasing across the globe in the tourism industry. Developing innovative and catchy tourism products is vital for survival. Bangladesh has a range of necessitated emergency construction tourism items. Tourism items are Cox's Bazar sea beach, cultural resources, archeological sites (Mahasthangarh, Bagerhat Shait Gambud Mosque, Paharpur, and so on), ecotourism items (Sundarbans, Teknaf, Chittagong Hill District, Tanguar Hoar, etc.), tourism destination commodities and relevant.

Development Policy

Bangladesh's archaeological sites (Mahasthangarh, Bagerhat Shait Gambud Mosque, Paharpur, etc.) can be the perfect spots for tourists from all over the world. UNESCO has listed several sites as World Heritage Sites. Bangladesh needs to boost connectivity to the site by the road here, invest in naming the archaeological sites and build additional tourism facilities. Else, Bangladesh would lose out on its visitor hopes.

Sundarbans is the world's largest mangrove area. Last year a lot of international visitors came here to enjoy the Sundarbans' natural beauty. Ecolodges need to be built to enjoy the real beauties of the Sundarbans. Bangladesh needs to invest in the construction of ecolodge with restaurant standards. Investments are required to improve connectivity, technology application to support and develop vulnerable regions from catastrophes. Binding local culture to the Sundarbans natural beauty would be the time-oriented activity for tourism growth.

The Chittagong Hill Tracts (CHT) are one of the country's finest attractions. A large number of local visitors and foreigners are noticeable in this area in recent times. Sajek Valley, Rangamati's Hanging Bridge has already accepted itself as our country's best spot. Any business needs to grow here. However, there are some suggestions as small and bumpy roads need to be changed, the place needs to be marketed and indigenous culture encouraged, visitor health and protection assured. Throughout Bangladesh haors and rivers are made of natural elegance. Tanguar and Hakaluki Haor need a branding and tourism promotion. The concept of ecotourism can be implemented here for tourism growth.

Cox's Bazar sea beach is the world's largest of its kind. Every year a significant number of international visitors, including locals, come here to see the beach's natural beauty. Any business will grow here for the development of the tourism industry as soon as possible. Improve transportation or convenient access; establish land and water transport facilities for sightseeing and cruising; build water sports facilities; create cultural activities traditional ethnic culture. And promoting Kuakata beach as an attractive tourism destination and providing the tourist facilities required; developing and promoting a portion of Kuakata (Gangamati) beach into an Exclusive Tourist Area; emphasizing the promotion of tourism-related goods.

Challenges for Investment in Tourism in Bangladesh

Tourist destinations are incredibly heterogeneous and monolithic; such facilities were challenging to transport (Hossain & Nazmin, 2005). COVID-19 current moment is a significant seething problem or potential danger for the entire worldwide tourist industry. Now it seems the COVID-19 impacts a significant proportion of people in Bangladesh. COVID-19, like most of the other countries of the world, would also be a huge challenge to Bangladesh's tourism industry after that. Most of the other challenges at present are like below:

Budgetary Challenge

Tourism is not well considered as a separate development department until now. Tourism is started marching with commercial aviation, and this department receives the rest of the budget. The unequal distribution of many concerning tourism is a major challenge for tourism development. There are insufficient convenience sampling and a lack of supply of competent personnel for the future of the country's tourism industry.

Mega Projects

BDTk 34,266 crores have been allocated for seven mega-projects in the proposed budget for the fiscal year 2020–21 (Ministry of Finance, 2020). The Government of Bangladesh's main concern is to accomplish the mega project within the given. Other areas of the economy like tourism development remain invisible to the authority for this mage projects.

Poverty

Unlike other nations, Bangladesh has been throwing an economic crisis for COVID-19 recently. Investment in infrastructure in tourism is not a major concern to the authority in this respect. Bangladesh's poverty rate came down at 21.8%. In contrast, as per the lower poverty line or people living in extreme poverty, the poverty rate came down to 11.3% in 2018, according to the latest data of the Bangladesh Bureau of Statistics (Bangladesh Bureau of Statistics, 2019). Importance needs to be placed on the RMG and disbursement sectors to achieve the GDP growth benchmark. Foreign visitors will not come to Bangladesh to keep social distance and protect an individual from COVID-19 transmission. For this reason, tourism investment remains unconcerned issues for the authorization.

Unemployment

Right now, a large proportion of educated youth in our country are unemployed. 2.6 million living below the poverty line countries in the world, according to the latest data from BBS labor force survey. The unemployment rate has shifted slightly, from 4.3% in 2013 to 4.2% in 2019, and approximately 46% of all unemployed youth are university graduates (The Financial Express, 2020). Despite higher economic growth in recent years, the country has failed to create adequate jobs, especially for the rising young population. The labor force is declining even as capital-intensive businesses pick over the positions of labor-intensive industries. The government could not invest in the tourism industry for this internal unemployment issue. It is also a tourism investment challenge in Bangladesh.

Furthermore, Bangladesh has a variety of tourism problems concerning tourism. The problems are in: first, its network is not well established in Bangladesh; second, few constructive solutions to tourism; third, the transport infrastructure is not adequately built, such as highways, airways, rivers, and waterways; fourth, Bangladesh's tourism image is too weak on the international stage; fifth, the annual temperature in Bangladesh shrinks with the consequences of climate change; sixth, there is no strategy which is simple and well defined; seventh, a deficit in field control; eighth, the lack of tourism support ventures; ninth, a large number of visitors from overseas with little yield; tenth, heavy tax systems which are prohibitive; eleventh, few adequate rail services; and finally, lack of major international firms.

Challenges for Development

Poor Bangladesh Image

Bangladesh's name is in the category of poorest countries in the world. The tropical monsoon season, experienced in Bangladesh with heavy summer rains, is immediately followed by floods and landslides, often with devastating consequences. The key problems for tourism development in Bangladesh are those kinds of negative directors and producers.

Missing the Branding

There is a strong connection between the integration and the development of tourism. Bangladesh requires massive branding to improve the tourism and travel industry. Tourism plays an essential role in establishing a strong brand image. Branding investment is very relatively uncommon in Bangladesh regarding tourism.

Technical Know-How

In Bangladesh tourism, is often accompanied by the absence or weakness of advertising, promotion, and mass national attention due, in far too many case scenarios, to poor wireless communication and activation energy. Several other recognizable opportunities for developing the tourism industry in our country are the lack of well-trained, skilled, and efficient workforce.

Weak Infrastructures

Well-structured Tourism infrastructure is not enough for our country. Innovation and entrepreneurship in specific tourism industries, such as hotel infrastructure, hospitality services, tourist information systems, transportation, and communication systems, will increase. Bangladesh thus finds it difficult to provide the human rights standards of resources and equipment that tourists require.

Investment in Travel

While investment in services was very well-established capital investment in industrialized economies, it is still lagging in developing countries such as Bangladesh. Capital expenditure in service-oriented projects, especially in tourism, is often perceived as a high-risk task in most developing countries.

Bangladesh's significant barriers to sustainable tourism are: first, insufficient state expenditure distribution of the funds; inadequate infrastructure; shortage of new and adequate leisure and tourism facilities; negative national image abroad; human Resources shortages in the tourism industry; waiver issues; and foreign direct investment shortfalls.

Suggestions for a Tourism Investment Scenario

Implementation of the National Tourism Policy (NTP) can be the best solution & suggestion for tourism development and investment. Relevant authorities need to act immediately to solve some of the issues mentioned in the 2010 NTP. These suggestions are: first, the expenditure required to draft master plans for state, regional, and zonal growth of the tourism industry. Second, investment is required to establish an Exclusive Tourist Zone (ETZ) to attract foreign tourists. Third, classify tourist products according to international demand; develop and promote products according to market demand. Fourth, ensure the participation of the private sector in the development of tourism. At the same time, the government assumes the role of facilitator and develops tourist attractions and the industry as a whole, with joint government and private sector efforts. Fifth, prepare the adequate ground for local and foreign investment; provide additional facilities for waiving tax and lending money where necessary. Sixth, investment in Bangladesh's vigorous tourism campaign will attract foreign tourists. Seventh, engage local authorities in the development, expansion, and management of tourism. Eighth, conserve local culture and history, turn them into tourism goods and initiate advertisement and marketing activities. Ninth, developing and maintaining tourism products through the development of ecotourism while maintaining the ecological balance. Tenth, establish and expand quality tourism and hospitality training institutes, and develop skilled human resources. Eleventh, ensure overall tourist safety. Twelfth, building a better communication channel for the Bangladesh International Tourism Organizations with governmental and private tourism associations. Thirteenth, to create a tourism-friendly environment in Bangladesh and market its tourism potential at home and abroad. Fourteenth, research on the tourism industry, survey, and analysis of international markets supports the small tourism-based industries and provides them with the right direction. Fifteenth, to create general knowledge of security, growth, and exploration of tourism. Sixteenth, to develop and provide the right directions for human resources for the tourism industry by creating training facilities. Finally, modernizing and streamlining airports, air routes, and air services.

Conclusion

Tourism products and services in Bangladesh have a tremendous ability to draw most visitors from across the world. Tourism provides an essential and enticing instrument for Bangladesh's economic growth. Tourism also ranks among the world's leading economic sectors. Investment in tourism will be important for tourism growth in Bangladesh. Tourism has made a significant contribution to Bangladesh in terms of jobs, foreign exchange, and infrastructure growth and the branding of Bangladeshi tourism goods. Weak branded identity, weak facilities, a challenging and dynamic tourism strategy are some of tourism's challenges. The tourism industry requires to develop strategies and policies long before investment needs.

References

Adebayo, K. A., & Iweka, A. C. O. (2014). Optimizing the sustainability of tourism infrastructure in Nigeria through design for deconstruction framework. *American Journal of Tourism Management, 3*(1A), 13–19.

Afroz, N., & Mahmud, S. (2017). Analyzing the problem and prospects of ecotourism: A review on Bangladesh. *IOSR Journal of Business and Management, 19*(2), 59–65.

Alauddin, M., Shah, M. G. F., & Ullah, H. (2014). Tourism in Bangladesh: A prospects analysis. *Information and Knowledge Management, 4*(5), 67–73.

Ali, M. (2004). *Bangladesher Sandhanay*. Student Ways. [In Bengali].

Ali, M. M., & Parvin, R. (2010). *Strategic management of tourism sector in Bangladesh to raise gross domestic product: An analysis*. Retrieved from: https://orp.aiub.edu/FileZone/abewp/orpadmin-2010-048589862668081/AIUB-BUS-ECON-2010-04.pdf. Accessed 7 July 2020.

Bangladesh Bureau of Statistics (BBS). (2019). *Bangladesh statistical yearbook 2018* (p. 35). Government of the People's Republic of Bangladesh.

Choice for Economic and Investment Research (CEIC). (2019). *Bangladesh tourism revenue*. Retrieved from: https://www.ceicdata.com/en/indicator/bangladesh/tourism-revenue. Accessed 7 July 2020.

Das, R. K., & Chakraborty, J. (2012). An evaluative study on tourism in Bangladesh. *Developing Country Studies, 2*(1), 17–27.
Elena, M., Lee, M. H., Suhartono, H., Hossein, I., Rahman, N. H. A., & Bazilah, N. A. (2012). Fuzzy series and Sarima model for forecasting tourist arrivals to Bali. *Jurnal Teknologi, 57*(1), 69–81.
Ferdous, J., & Faisal, H. M. (2014). Tourism potentiality and development of Bangladesh: Applicability of pragmatic governmental management policy. *Journal of Management and Science, 4*(1), 71–78.
Filieri, R., & McLeay, F. (2013). E-WOM and accommodation: An analysis of the factors that influence travelers' adoption of information from online reviews. *Journal of Travel Research, 53*(1), 44–57.
Gaggioli, A., Pioggia, G., Tartarisco, G., Baldus, G., Corda, D., Cipresso, P., & Riva, G. (2013). A mobile data collection platform for mental health research. *Personal and Ubiquitous Computing, 17*(2), 241251.
Hall, C. M., & Williams, A. M. (2008). *Tourism and innovation*. Routledge.
Hassan, A., & Burns, P. (2014). Tourism policies of Bangladesh-A contextual analysis. *Tourism Planning & Development, 11*(4), 463–466.
Hassan, A., & Kokkranikal, J. (2018). Tourism policy planning in Bangladesh: Background and some steps forward. *e-Review of Tourism Research (eRTR), 15*(1), 79–87.
Hassan, A., Kennell, J., & Chaperon, S. (2020). Rhetoric and reality in Bangladesh: Elite stakeholder perceptions of the implementation of tourism policy. *Tourism Recreation Research, 45*(3), 307–322.
Hjalager, A. -M. (2009). Cultural tourism innovation systems – The Roskilde festival. *Scandinavian Journal of Hospitality and Tourism, 9*(2/3), 266–287.
Hossain, B., & Wadood, S. N. (2019). Potential unexplored? Tourism and economic growth of Bangladesh. *Journal of Tourism*. https://doi.org/10.26650/jot.2020.6.1.0031
Hossain, M. A., & Nazmin, S. (2005). Service strategies for the tourism industry in Bangladesh: An empirical study. *Development Review, 17*.
Ishtiaque, A. N. A. (2013). Tourism vision 2020: A case of Bangladesh tourism with special emphasis on international tourist arrivals and tourism receipts. *Journal of Business Studies, XXXIV*(2), 13–36.
Jacob, M., Tintoré, J., Aguiló, E., Bravo, A., & Mulet, J. (2003). Innovation in the tourism sector: Results from a pilot study in the Balearic Islands. *Tourism Economics, 9*(3), 279–295.
Karambakuwa, R. T., Shonhiwa, T., Murombo, L., Mauchi, F. N., Gopo, R. N., Denhere, W., Tafirei, F., Chingarande, A., & Mudavanhu, V. (2011). The impact of Zimbabwe tourism authority initiatives on tourist arrivals in Zimbabwe (2008–2009). *Journal of Sustainable Development in Africa, 13*(6), 68–77.
Khan, M. K., Teng, J. Z., Khan, M. I., & Khan, M. O. (2019). Impact of globalization, economic factors and energy consumption on CO2 emissions in Pakistan. *Science of the Total Environment, 688*, 424–436.
Majumder, M., & Hossain, T. (2011). An appraisal of tourism industry development in Bangladesh. *European Journal of Business and Management, 3*(3), 287–298.
Mason, P. (2020). *Tourism impacts, planning and management*. Routledge.
Ministry of Environment and Forests. (2012). *Bangladesh: Rio + 20: National report on sustainable development*. Retrieved from: https://sustainabledevelopment.un.org/content/documents/981bangladesh.pdf. Accessed 7 July 2020.
Ministry of Finance. (2020). *Economic transition and pathway to progress: Budget speech FY2020-FY2021*. Retrieved from: http://customs.gov.bd/files/Budget%20Speech%20 2020-21_English_Final%20Print.pdf. Accessed 7 July 2020.
Pikkemaat, B. (2008). Innovation in small and medium-sized tourism enterprises in Tyrol, Austria. *The International Journal of Entrepreneurship and Innovation, 9*(3), 187–197.

Planning Commission. (2015a). *6th five year plan FY2011-FY2015: Accelerating growth and reducing poverty.* Retrieved from: https://www.globalpartnership.org/sites/default/files/sixth-five-year-plan-2012-_part_1.pdf. Accessed 7 July 2020, pp. 201–202.

Planning Commission. (2015b). *7th five year plan FY2016- FY2020: Accelerating growth, empowering citizens*. Retrieved from: https://www.unicef.org/bangladesh/sites/unicef.org.bangladesh/files/2018-0/7th_FYP_18_02_2016.pdf. Accessed 7 July 2020.

Rogers, S. (2007). Innovation in food service technology and its strategic role. *International Journal of Hospitality Management, 26*(4), 899–912.

Roy, S. C., & Roy, M. (2015). Tourism in Bangladesh: Present status and future prospects. *International Journal of Management Science and Business Administration, 1*(8), 53–61.

Salam, R. (2014). Tourism & Socio-Economic Development: Focus on Bangladesh, a destination country. *International Journal of Business and Management (Toronto), 1*(2), 88–98.

Sandip, S. (2014). Competitive marketing strategies for tourism industry in the light of "Vision 2021" of Bangladesh. *European Journal of Business and Management, 6*(4), 210–220.

Statista. (2017). *Digital travel sales worldwide from 2014 to 2020 (in billion U.S. dollars).* Retrieved from: https://www.statista.com/statistics/499694/forecast-of-online-travel-salesworldwide/. Accessed 7 July 2020.

Sultana, S. (2016). Economic contribution of tourism industry in Bangladesh. *Journal of Tourism, Hospitality and Sports, 22*, 55–54.

The Bangladesh Monitor. (2017). *Economic contribution of travel & tourism to national economy on the raise despite neglect.* Retrieved from: https://www.bangladeshmonitor.com/front_page/14/Economic-contribution-of-travel-&-tourism-to-national-economy-on-the-rise-despite-neglect. Accessed 7 July 2020.

The Daily Star. (2013). *Tourism booming with economy*. Retrieved from: https://www.thedailystar.net/business/news/tourism-boomingeconomy-1673479. Accessed 7 July 2020.

The Financial Express. (2020). *Unemployment still a key challenge.* Retrieved from: https://thefinancialexpress.com.bd/views/views/unemployment-still-a-key-challenge-1585151594. Accessed 7 July 2020.

The New Age. (2020). *Will the budget help to recover the tourism industry?*. Retrieved from: https://www.newagebd.net/article/109325/will-the-budget-help-to-recover-the-tourism-industry. Accessed 7 July 2020.

World Economic Forum (WEF). (2019). *The travel & tourism competitiveness report (2019): Travel and tourism at a tipping point.* Retrieved from: http://www3.weforum.org/docs/WEF_TTCR_2019.pdf. Accessed 7 July 2020.

World Tourism Organization (UNWTO). (2000). *Tourism highlights 2000.* Retrieved from: https://www.e-unwto.org/doi/pdf/10.18111/9789284403745. Accessed 7 July 2020.

World Tourism Organization (UNWTO). (2001). *Tourism 2020 vision: South Asia* (p. 11). UNWTO.

World Travel & Tourism Council (WTTC). (2018). *World travel & tourism council report 2009–2018.* Retrieved from: https://www.wttc.org/economic-impact/country-analysis/country-data/UNWTO-United. Accessed 7 July 2020.

Chapter 25
Investment Barriers and Opportunities in the Tourism and Hospitality Industry of Bangladesh

Rashed Al Karim and Azizul Hassan

Abstract This study mainly focuses on the investment barriers and opportunities in the Bangladeshi tourism and hospitality industry. The study then portrays the overall view of the Bangladeshi tourism industry along with different available tourism types. The current study identifies the tourism and hospitality industry as one of the national economy' key players. The study explores several investment barriers and opportunities in the tourism and hospitality industry of Bangladesh as well as provides recommendations. The study is mostly focused on the theoretical analysis of various secondary sources such as scholarly research articles, government reports, reports of diplomats, national and international tourism-related organizations, and newspapers. Eventually, this study is expected to guide relevant authorities of the Government of Bangladesh (GoB) in expanding the tourism and hospitality industry through domestic and foreign direct investments.

Keywords Tourism and hospitality industry · Foreign investment · Investment barriers · Investment opportunities · Public-private partnership

R. Al Karim (✉)
School of Business Administration, East Delta University, Chattogram, Bangladesh
e-mail: alkarim.rashed@gmail.com

A. Hassan
Tourism Consultants Network, The Tourism Society, London, UK

A. Hassan (ed.), *Tourism in Bangladesh: Investment and Development Perspectives*, https://doi.org/10.1007/978-981-16-1858-1_25

Introduction

On the word of United Nations World Tourism Organizations (UNWTO) (2019), through generating billions of dollars in exports and providing millions of jobs, the tourism and hospitality industry has developed itself as a major contributor to the global economy through its resilient economic activities. Therefore, as the leading industry, the impact of tourism and hospitality industry on the global economy is growing considerably (de Leaniz & del Bosque, 2015). In Bangladesh, tourism is the most prospective, vibrant, and leading industry, which comprised 3.0% of the total GDP in 2019 and simultaneously created a total of 1858.9 employments which was 2.9% of total employment in Bangladesh in 2019 (World Travel & Tourism Council, 2020). In the fiscal year 2019–20, the Government of Bangladesh (GoB) has allotted BDTk 34 billion in Civil Aviation (Biman) and Tourism Ministry, which was almost twice compared to the last fiscal year indicating the importance given by the present GoB for the tourism industry. This was expressed in the 2019 in a Competitiveness Report on Travel and Tourism, in which Bangladesh jumped from 125th to 120th position out of 140 countries. Regardless of the improvement, this is still a dismal ranking, with Pakistan the only Asian nation trailing Bangladesh. Furthermore, the vast majority (97%) of tourism revenue is generated domestically.

According to World Travel & Tourism Council (2020), the total domestic tourist spending was 96% in 2019, while the international tourist spending was only 4%, which indicates that Bangladesh is not a popular destination among foreign tourists. Moreover, consistent with the United Nations Department of Economic and Social Affairs (UN DESA) (2018), Bangladesh is expected to graduate from the group of least developed countries (LDC) by 2024. Although this may show the Bangladesh economy's evolution, the graduation may cause the garment industry to decline irreversibly, losing USD 2.7 billion in export earnings every year after graduation (DATABD.CO, 2020). Accordingly, before its biggest source is depleted, Bangladesh needs to start searching for new sources of foreign earnings. The tourism industry may, therefore, be a major source of foreign currency earnings if properly nurtured as Bangladesh's tourism industry is growing at a rate of 3% of total GDP in 2019, with a contribution of 10.4% to global GDP (World Travel & Tourism Council, 2020). Considering the above-mentioned issues, the main objective of this chapter is to highlight the investment barriers of the Bangladesh tourism and hospitality industry and concentrate on the investment potential in this industry.

Overview of Bangladesh Tourism Industry

With 147,570 square kilometers of areas as well as 160 million people, Bangladesh is one of the most densely populated countries. The country has an emerging economy in Southern Asia whose West, North, and North-east part is bounded by India, Myanmar bounds the South-east part, and the Southern part is bounded by the Bay of Bengal. The climate of the country is sub-tropical with six seasons. Due to its distinctive geographical location, the country is blessed with amusing natural beauty such as the world's largest mangrove forest, the world's longest natural beach, hill tracts areas, tea estates, islands, and many archaeological resources and as a result of the tourism industry of Bangladesh is always considered one of the most promising industry in the world (Das & Chakraborty, 2012).

Regardless of this, the industry is not doing well. Just recently, the GoB seemed to realize the potential that the tourism industry may hold if given the required support. In the last decade or so, the industry has evolved in a haphazard manner. However, for the country to join the global competitive arena, a much more concerted and strategic involvement is needed. Tourism, on the other hand, was not recognized an industry for the first 28 years of its history until being recognized as such in 1999. Although growth has not been spectacular since then, expansion has shown the industry's enormous potential. In the last decade, Bangladesh's economy has grown at a rate of 5-6.5% where the RMG industry acting as the primary source of foreign currency earnings. The risk of being too reliant on a single industry for foreign currency inflows is very high, and it can have harmful consequences if that industry is struck by a crisis or recession. The GoB is working hard to diversify its export portfolio. The tourism industry, like the export industry, may be a promising option that provides a variety of other economic benefits, such as job creation, infrastructural developments, and multiplier effects on the growth of related industries and eradicating poverty.

Recently, the GoB recognized the importance of the contribution of the tourism industry to its economic, social, and environmental circumstances. Consequently, the GoB has decided to combine both the tourism industry and the private corporation into one institute to enhanced coordination, better growth, promotion, and marketing of the tourism industry in Bangladesh (Rahman, 2012). The "Bangladesh Parjatan Sangstha" was established shortly after the liberation war of 1971, with the aim of improving the tourism industry. In 1973, it was renamed Bangladesh Parjatan Corporation (BPC). Consequently, as part of the GoB's First Five Year Plan, they drew up a five-year plan for the industry's growth (1973–78). The GoB was supposed to provide the essential facilities and improve the country's natural beauty in a way that would appeal to visitors under that programme in addition to boost the tourism industry. Unfortunately, due to a variety of roadblocks, including ministerial legacy issues, neither of these proposals have been carried out properly by management (The Daily Star, 2009).

Though the Bangladesh Parjatan Corporation (BPC) remains the country's leading public-sector organization, the private sector has also made significant investments in the sector (Hassan et al., 2020; Hassan & Kokkranikal, 2018; Hassan & Burns, 2014). By developing several hotels or tour operators, these private sector entrants have penetrated the tourism industry. Presently almost thirty-two tour operators are functioning in the country, some of whom cater to domestic tours, whereas the others compromise both local and outbound tours. Moreover, those thirty-two private tour operators created the association named "Tours Operator Association of Bangladesh (TOAB)", to perform their regular activities more competently as well as to maintain communications with BPC as well as the GoB regarding their rational rights, and also to endorse the country's holiday business together. In addition, "Bangladesh Tourism Board" established in 2010, and also affiliated with United Nations World Tourism Organization (UNWTO), is another central association to support the enlargement of the tourism industry to meet the robust demand for the private industry and the tourism professionals.

Types of Tourism in Bangladesh

Tourism in Bangladesh can be divided into six different categories, according to Bangladesh Parjatan Corporation (2007): site tourism, business tourism, office tourism, study tourism, religious tourism, and other tourism types.

Site Tourism

Tourists come to Bangladesh to discover the country's natural beauty. According to the most recent data, these tourists account for the majority share (46%) of the tourist inflow into the country. This demonstrates the importance of preserving and protecting these areas, as well as providing the requisite infrastructure development assistance.

Business Tourism

Approximately 42% of tourists come to Bangladesh for the purpose of the business. Many foreigners visit Bangladesh to work or operate their companies, as a few well-known international corporations have established regional offices in the country. Significant number of potential buyers and their delegates are expected to visit the country on a regular basis due to the ever-expanding RMG industry.

Office Tourism

Tourists who came to Bangladesh for office tourism made up just 1.5% of the total number of visitors. This kind of tourism is like business tourism in several ways, but it mostly caters to the representatives of 101 multinational corporations having offices in Bangladesh. Many of these visitors come to take tours of their local offices and to assist their business organizations.

Education Tourism

The idea of education tourism refers to foreign students' enrolment in educational institutions in the country. Besides, international conferences and lectures are held by several universities and other graduate-level institutions, and attendees from all around the world attend. These international delegates all fell into the category of educational tourism, accounting for 2.4% of all visitors who visited the country.

Religious Tourism

Bangladesh is a Muslim-majority country that hosts the Bishwa Ijtema, the biggest Muslim gathering since the Holy Pilgrimage to Mecca. This annual event alone draws huge number of visitors from the Muslim world, accounting for 3.4% of Bangladesh's overall tourist inflow. The number of other ethnic communities in the world is much lower. Their gatherings, on the other hand, do draw visitors, but on a smaller scale.

Other Tourism Types

Apart from the abovementioned category of tourism, Bicycle traveling, Boat sharing, Cultural tourism, Rural tourism, River cruising, Nautical tourism, Halal tourism, Sports tourism, Slum tourism, Wildlife tourism are some of the other forms of tourism available in Bangladesh, according to Bangladesh Parjatan Corporation (BPC).

Sketch of Bangladesh Hospitality Industry

Previously due to the reluctance of the private industry, the hospitality industry, one of the vastly investment concentrated industries, in Bangladesh was mainly dependent on the investment of the public industry. Nowadays, the private industry is considered the most vibrant contributor in the country's hospitality industry due to the huge amount of investment and involvement in boosting this industry. Aside from luxury hotels in different cities, the private sector has made major contributions to the country's enrichment in another industry of hospitality, notably, holiday resorts (The Financial Express, 2017). For several years, the capital city of Dhaka relied heavily on Pan Pacific Sonargaon and Dhaka Sheraton (now Ruposhi Bangla) for high-end lodgings. At present, more than a few five-star hotels such as the Westin Dhaka, Radisson Blue Water Garden Hotel, Dhaka Regency Hotel & Resort,

Hotel Sarina, and recently opened the world-famous hotel brand Le Meridien (in November 2015) (source: www.prothom-alo.com) have enriched the hospitality industry of the capital city Dhaka and around 1800 5-star quality rooms are obtainable from all those hotel and resort. Additionally, there are approximately seven other hotels which are offering around 500 3 to 4-star quality rooms, such as Hotel Orchard Plaza, Hotel Washington, Best Western La Vinci, Lake Shore Hotel & Apartments, Platinum Suites, Royal Park Residence, Ascott the Residence. In addition to these 3 to 4-star quality rooms, there are approximately 200 one to two-star quality rooms available in various guesthouses (The Financial Express, 2017).

Chattogram, Bangladesh's second-largest city and a port on the Bay of Bengal, has a long tradition of maritime trade. Prior to independence, Chattogram had just one 5-star hotel, the Radisson Blue, which opened in January 2015. The Peninsula, Hotel Agrabad, Avenue, and Harbour View are among the city's many 4-star hotels while the St. Martin's and Asian SR as are the 3-star hotels. Cox's Bazar, 160 kilometers south of Chattogram and home to the world's longest natural sea beach, is the most popular tourist destination in Bangladesh. Before independence, Cox's Bazar was a popular tourist destination. Cox's Bazar has a variety of 3 to 5-star hotels and guest rooms to accommodate both domestic and international visitors. St. Martin's Island, located in the Bay of Bengal about 25 kilometers south of Cox's Bazar, has seen a boost in tourism and has several guesthouses to support the industry. In addition, other tourist cities such as Sylhet, Khulna, and Bogra have good western-style hotels. As a national tourism body, Bangladesh Parjatan Corporation (BPC) also runs 18 medium-sized hotels in the major cities and tourist destinations of the country. However, as opposed to private-industry services, these are increasingly seen as unattractive to travelers due to poor service efficiency (Hussain & Leishman, 2013).

Investment Barriers in the Bangladeshi Tourism and Hospitality Industry

From the establishment time to the present, Bangladesh Parjatan Corporation (BPC) has been looking after the tourism and hospitality industry in Bangladesh under the Ministry of Civil Aviation and Tourism, and this industry has faced multi-faceted issues. In consequence of that multidimensional concern, Bangladesh's tourism and hospitality industry has failed to introduce itself as a tourist destination. As a result, the tourism and hospitality industry is not embryonic, as there is no specific and compact measure in this land of natural beauty to increase foreign exchange earnings and generate jobs through this industry. Nonetheless, the following are the key obstacles to foreign investment in the growth of Bangladesh's tourism and hospitality industry, according to the observations of several previous research studies:

Unplanned and Improper Growth of the Tourism and Hospitality Industry

Bangladesh tourism and hospitality industry is facing several difficulties in communication, promotion, safety, improper implementation of laws, decreasing international tourists due to its unplanned and improper growth. Some previous literature also found similar problems in this industry. For example, Shamsuddoha and Nedelea (2008) have mentioned in their study that GoB never highlighted this industry like the others. However, this industry has tremendous potentials to earn domestic and international customers as well as has the opportunity to attract foreign investors. Moreover, in Bangladesh, there is no international standard tourism education to develop skilled manpower for this industry. There is a serious lack of management in this industry at national, regional, and local levels (Rahman et al., 2010). This industry suffers from a severe shortage of management at the state, provincial, and local levels (Rahman et al., 2010). If Bangladesh is to capitalize on the potentials of the tourism and hospitality industries for national economic growth and social change, it must implement a strong strategy. Furthermore, unplanned construction near tourist attractions has degraded the natural beauty of the region. Many unplanned commercial establishments, such as hotels, motels, and shopping malls, have ruined the atmosphere of this area, for example, in Cox's bazaar sea beach and Saint Martin Islands (Tuhin & Majumder, 2011; Rahman et al., 2010), which may distract and demotivate foreign investors.

Scarcity of Financial Incentives

Due to the dearth of financial incentives, local and foreign investors are not fascinated to invest in Bangladesh's tourism and hospitality industry. The GoB's priority industries for foreign investment are the country's infrastructural development, energy industry, Information and communications technology industry, etc. The GoB wants to invite foreign investors with some financial offers such as tax exemption, tax holiday, custom duty waiver, stamp duty waiver, low-interest loan facilities licensing facilities, etc. Previous investigations also found financial rewards or incentives as a strong factor in attracting foreign investments. For instance, Park, Chidlow, and Choi (2014) stated in their research that economic benefits through tax savings, reduced regulatory afflictions, and duty-free export, low-interest loan facilities motivate investors.

Dearth of Joint-Venture Partnerships

Joint-ventured partnerships still not popular in Bangladesh, particularly in the tourism and hospitality industry. In a developing country like Bangladesh, due to budget constraints, the GoB must give more importance to the social welfare and critical services industry than to the tourism and hospitality industry. For this reason, public tourism agencies are typically unable to offer competitive services equivalent to the private industry. In this situation, public-private partnerships are more appropriate in developed countries for the growth of this field. The GoB has also set up a public-private partnership office in Bangladesh to promote public-private partnership investment. They already have a guide for approving public-private partnership projects. As an organization under this ministry, Bangladesh Parjatan Corporation (BPC) works closely with the PPP Office to generate ideas about future projects. In Bangladesh, the joint-ventured collaboration or partnership is already practiced to serve rural people, creating job opportunities, reducing poverty from the country, for example, Grameen Danone Food Joint Venture (established in 2006); Grameen Schneider Electric Joint Venture (established in 2012); Grameen Veolia Joint Venture (established in 2008).

Lack of Promotional Activities

Bangladesh Parjatan Corporation (BPC) does not promote Bangladesh's tourism and hospitality industry frequently in newspapers, TV advert, or magazines, which one of the potential barriers to foreign investment. To build and upgrade our websites to increase international tourist traffic, we need to take pragmatic steps. It is also essential to share tourism and hospitality industry related information on common and popular websites in order to update and advertise in different ways. For instance, country like Thailand, Malaysia, and Indonesia has websites in twelve languages, while country like Cambodia has tourism websites in eight languages. In Bangladesh, we only have one language, English, on our websites (Alam, 2018). Besides, only during Cricket World Cup 2011, the Bangladesh Parjatan Corporation undertook some promotional activities to promote country's tourism industry, where Bangladesh was one of the co-hosted countries. The activities included the creation, for the first time, of a TV commercial promoting Bangladesh and the launching of a Tourism Board website. Nevertheless, the Bangladesh Parjatan Corporation (BPC) has not taken any substantial steps to promote tourism in Bangladesh after the World Cup.

Lack of Sufficient Security Measures

The most critical challenge for the growth of the tourism and hospitality industries in any tourism destination, like Bangladesh, is security and safety. For international visitors, security concerns are a problem in Bangladesh. For traveling to Dhaka and the Southeast part of Bangladesh, the US State Department has given a warning on the grounds of crime, terrorism, and kidnapping (DATABD.CO, 2020). Bangladesh, however, has increased its ranking to 105th from 123rd, according to the World Economic Forum's Safety and Security Index of 2019 and indicating that Bangladesh has been working to eradicate safety and security issues.

Lack of Cooperation from Concerned Authorities

The lack of cooperation from the concerned authorities is one of the key barriers to the investment environment in Bangladesh, not only for the tourism and hospitality industry but also for all the other industries as well. Here the term 'concerned authority' refers to the various national and local governmental departments or agencies such as Bangladesh Parjatan Corporation, Department of Environment (DoE), Bangladesh Power Development Board (BPDB), Water Supply and Sewerage Authority (WASA), law enforcement authority and the local government authority, etc. Several previous studies accused all those departments or agencies of not providing necessary services regarding electricity, water supply, natural gas supply, discharge of solid waste and sewage, to the hotel industry, sometimes because of bribe or extortion or other facilities, and sometimes their lack of coordination and workforces. Belal (2015) said that it is difficult for government officials in Bangladesh to ensure conformity without the appropriate support from their political leaders. The lack of cooperation from the government officials is a serious social concern for the tourism and hotel industry, and sometimes this lack of cooperation interrupts the hotel's business operation.

Political Disturbances

The political disturbance is one of the uncontrollable obstacles to the investment climate in the tourism and hospitality industry. According to Islam and Pattak (2017), a smooth and stable political state will play an important role in a country like Bangladesh's overall growth where democratic government and their policies regarding socio-economy, human resource development, unemployment, foreign direct investment, and environment can help gain economic sustainability. But, from the last few years, political fiascos like strikes, scuffles, labor unrest, reckless activities by some uncontrolled political leaders, etc., have become some of the hindrances to the socio-economic development of the country (Hossan et al., 2012). And because of these concerns, all the business industries of the country, including the hospitality industry, are facing numerous problems. Particularly, in the hospitality industry, the number of foreign tourists is decreasing day by day, and the foreign investment is also almost zero in this industry.

Departmental Power Exploitation

"Departmental Power Exploitation" refers to the corruptions, lengthy bureaucratic systems, bribe, extortion, etc. practiced by the various governmental departments such as Bangladesh Tourism Board, Bangladesh Parjatan Corporation, Department of Environment (DoE), Bangladesh Power Development Board (BPDB), Water Supply and Sewerage Authority (WASA), and government regulatory authority. Power exploitation by the employees of above mentioned governmental agencies or departments is pulling back the growth of the hospitality industry. All these misdeeds are standard practice in Bangladesh since its independence in 1971, and nowadays, the ordinary people of Bangladesh viewed corruption is as "a way of life" (Zakiuddin & Haque, 2002). According to a survey conducted by the Bangladesh Unnayan Parishad, 95% of participants claimed that the police is the most corrupt agency, followed by customs, excise and taxes, administration, and the judiciary. In addition to this, around 62% of respondents mentioned that government officials are primarily responsible for the corruption in Bangladesh (Zakiuddin & Haque, 2002). Besides, doing business in Bangladesh includes a very high hidden cost, and the bureaucrats are principally responsible for this increased hidden cost of doing business in Bangladesh due to their corrupt practices (Saha, 2014). In the same way, the scholars who did their study in the Bangladesh context, most of them highlighted the corruption and bureaucratic process practiced by several governmental agencies. For example, Haque and Azmat (2015), Sobhani et al. (2009), Belal (2008) mentioned that widespread corruption, bureaucratic difficulties are some of the key reasons for which the country is still lagging in comparison to other developing countries. Equally, Islam (2012) has described widespread political corruption, as well as a lack of political and social stability, as major obstacles to the tourism and hospitality industry's growth.

Investment Opportunities in Bangladesh Tourism and Hospitality Industry

Bangladesh's economy has grown steadily over the last decade, with an average gross domestic product (GDP) growth rate of 7.86% in 2018. According to the Centre for Economics and Business Research (CEBR), by 2033, it is expected to grow by an average of 7%. Bangladesh's inflows of foreign direct investment (FDI) have already established the country as a flourishing investment hub. Bangladesh received USD2.58 billion in foreign direct investment at the end of June 2018. Significant FDI investments in the transportation, storage, utility, and communications industries are seen as a result of the GoB's mega-projects. There are many positive things about investing in Bangladesh today. Bangladesh offers preferential trading opportunities and export programs to countries that invest here. Bangladesh has taken several remarkable steps, such as the development of tourist accommodation and entertainment facilities in the region, to draw more and more foreign investment in the tourism industry and to achieve consistent success in this market. In 2010, the GoB recognized tourism as an industry and established a new National Tourism Policy to promote domestic and foreign tourism. Bangladesh offers attractive and promising opportunities to build multi-faceted tourism facilities around tourist destinations for foreign investment. Private investment in the tourism industry is gradually increasing because of the GoB's pro-foreign and private business-friendly policies. Foreign investment in Bangladesh's tourism industry is welcoming at all times by the GoB. There are several government initiatives to invest in this industry that are followed by:

In order to encourage international investment, the GoB passed the "Tourism Protected Area and Exclusive Tourist Zone (ETZ) Law" in 2010. Nearly 35 thousand acres of land in various parts of the country have also been declared as protected tourist areas under this statute.

Numerous different geopolitical advantages for foreign investment exist in Bangladesh, as greater connectivity increases a future destination's competitiveness. Bangladesh is connected to all major world destinations, and many luxury and low-cost airlines fly into the country (The Daily Sun, 2017).

The GoB welcomes investment strategies aimed at facilitating investors' entry into the secondary and tertiary industries. Several incentives have been introduced for investment in certain industries, including the tourism and hospitality industry, as part of the GoB's liberal policy regime. Examples of such benefits comprise (i) tax exemptions, (ii) import duty waiver, (iii) double taxation regime, (iv) ease of profit, (v) investing in the stock market, (vi) capital repatriation, and (vii) preferential benefits.

Local banks may provide full working capital loans to foreign investors or companies. Loan amount and interest rates are decided by negotiation. Other incentives offered by the GoB include offers citizenship with a minimum investment of USD 500,000 or a conversion of USD 1,000,000 to any recognized financial institution (non-repatriable) and permanent residency with a minimum investment of USD 75,000 (non-repatriable) (Rashid, 2019).

The GoB has recently reinvigorated the Public-Private Partnership (PPP) project as a means of overcoming infrastructure investment constraints in Bangladesh, and it is seen as a key mechanism for the country's tourism growth. It appears to be an excellent way of engaging the private industry in fostering tourism and thereby growing investment. Following the introduction of the PPP Guideline in 2010, several public-private partnership initiatives are undertaken in the tourism industry as:

- Plan to construct a Tourism and Entertainment Village at Parjatan Holiday Complex, Cox's Bazar
- Plan to establish an Exclusive Tourist Zone (ETZ) on 1164.8522 acres of land in Sabrung, Teknaf Upazila, Cox's Bazar
- Construction of watchtower in Kuakata
- Establishment of a 5-star Hotel and other facilities at the existing motel compound of BPC at Sylhet
- Construction of an international level tourism complex at the BPC's Motel Upal compound in Cox's Bazar

In addition to these, there are many segments of the tourism and hospitality industry for investment such as:

- Establishing 5-7 star hotels in Dhaka, Chattogram, Khulna, Mymensingh, and other Bangladeshi tourism hotspots
- Inauguration of a cable car connecting Cox's Bazar and Teknaf
- Development of international quality tourist amenities in the Chattogram Hill Tracts area
- Construction of recreational tourism facilities in different places of Chattogram division
- Investment in airlines to connect all the tourist destinations to the capital of Bangladesh and setting up and operating mega cruise ship facilities from Cox's Bazar to Phuket, Bali, and Singapore (The Daily Sun, 2017).

Bangladesh is earnestly seeking foreign investment for all of these projects. Moreover, the above-mentioned points indicate that Bangladesh's tourism industry is a very prospective market for foreign investment, and surely in a short time, the investment would return benefits. Therefore, prospective investors from around the world are encouraged to visit Bangladesh and discover new business opportunities in the country's nascent tourism industry. Bangladesh has an investment-friendly climate. Bangladesh, like other South Asian economies, has been able to build a favourable investment climate in the tourism industry.

Recommendations for Concerned Authorities to Attract Foreign Investments

Several steps have already been taken by the GoB to improve road communication and air links, especially for tourism-based areas, to provide a much better service for tourists. Based on the prior discussion following are the recommendations for the GoB along with tourism and hospitality industry related concern authority:

- To formulate, amend, and execute the laws and policies related to civil aviation and tourism.
- Airports should be democratized, as well as air routes and air services should be synchronized.

- To do proper branding of the tourism and hospitality industry, because this industry will be one of the major sources, after RMG industry, in earning foreign currency.
- To take initiatives for the modernization of existing tourism infrastructure and development of related infrastructures.
- To ensure a safe and secure aviation system
- To take initiatives for the expansion of inbound tourism
- To take initiatives for tourism-related human resource development
- Development and marketing of tourism products,
- To research for the expansion of the tourism and hospitality industry.
- Finally, though investment incentives are important in attracting investors, they can never be a replacement for a favourable investment environment. To attract investors, the GoB can deliver the following services, in conjunction with the United Nations Conference on Trade and Development (UNCTAD) (2009):
- To allow duty-free imports of raw materials and machineries
- To offer training incentives to locally employed employees
- To permit faster depreciation on environmentally sustainable properties
- To support in the long-term leasing of commercial property
- To provide credit assurances to small businesses organization
- To include subsidies or tax incentives to investors in high-growth zones.

Conclusion

As the tourism industry is an investment-hungry marketplace, investors from all over the world will soon arrive and will be benefited accordingly. The Ministry of Civil Aviation and Tourism will provide a one-stop service for prospective investors. Bangladesh's tourism industry has seen significant domestic investment in the form of hotels, motels, resorts, theme parks, amusement parks, and cruise ships, among other things. Some PPP mega tourism projects are also set to open in Bangladesh soon, which will undoubtedly attract foreign direct investment in the country's tourism and hospitality industries. Furthermore, like in other countries, growing job opportunities in the tourism and hospitality industry is now playing a major role in the country's economic growth. Akin to other developing countries such as Spain, Switzerland, Malaysia, Thailand, UAE, and Singapore, the tourism industry of Bangladesh will be considered as one of the major sources of foreign currency earning. However, diverse incompatible circumstances such as political unrest, lack of proper information, shortage of international airports, poor service quality are imposing a great threat to the potential of the tourism industry. Accordingly, interrelated authorities along with the GoB are requested to look into these concerns for the expansion of the tourism and hospitality industry, in addition to enhancing foreign direct investment by attracting potential investors across the globe.

References

Alam, J. (2018). Problems and prospects of tourism industry in Bangladesh: A case of Cox's bazar tourist spots. *International Journal of Science and Business, 2*(4), 568–579.

Bangladesh Parjatan Corporation (BPC). (2007). *Bangladesh tourism vision 2020*. BPC.

Belal, A. R. (2008). *Corporate social responsibility reporting in developing countries: The case of Bangladesh*. Ashgate Publishing.

Belal, A. (2015). Social and environmental accountability in developing countries. In D. Jamali, C. Karam, & M. Blowfield (Eds.), *Development-oriented CSR* (pp. 153–166). Greenleaf Publishing.

Das, R. K., & Chakraborty, J. (2012). An evaluative study on tourism in Bangladesh. *Developing Country Studies, 2*(1), 17–27.

DATABD.CO. (2020). *Tourism: A possible new driver for the economy of Bangladesh*. Retrieved from: https://bit.ly/3oURx1O. Accessed 3 Nov 2020.

de Leaniz, P. M. G., & del Bosque, I. R. (2015). Sustainability: A competitive advantage in the tourism industry. In K. H. Collins (Ed.), *Handbook on tourism development and management* (pp. 39–58). Nova Science Publishers.

Haque, M. Z., & Azmat, F. (2015). Corporate social responsibility, economic globalization and developing countries: A case study of the readymade garments industry in Bangladesh. *Sustainability Accounting, Management and Policy Journal, 6*(2), 166–189.

Hassan, A., & Burns, P. (2014). Tourism policies of Bangladesh – A contextual analysis. *Tourism Planning & Development, 11*(4), 463–466.

Hassan, A., & Kokkranikal, J. (2018). Tourism policy planning in Bangladesh: Background and some steps forward. *e-Review of Tourism Research (eRTR), 15*(1), 79–87.

Hassan, A., Kennell, J., & Chaperon, S. (2020). Rhetoric and reality in Bangladesh: Elite stakeholder perceptions of the implementation of tourism policy. *Tourism Recreation Research, 45*(3), 307–322.

Hossan, C. G., Sarker, M. A. R., & Afroze, R. (2012). Recent unrest in the RMG sector of Bangladesh: Is this an outcome of poor labour practices? *International Journal of Business and Management, 7*(3), 206–218.

Hussain, S. S., & Leishman, D. (2013). *Bangladesh: Hotel Restaurant and Institutional (HRI) 2013*. Retrieved from: http://files.eacce.org.ma/pj/1372826254.pdf. Accessed 3 Nov 2020.

Islam, S. M. (2012). *Tourism marketing in developing countries: A study of Bangladesh*. Doctoral Dissertation. University of Strathclyde.

Islam, M., & Pattak, D. C. (2017). Impact of macro environmental factors on garments industry that drives export in Bangladesh. *Studies in Business and Economics, 12*(2), 100–111.

Park, B. I., Chidlow, A., & Choi, J. (2014). Corporate social responsibility: Stakeholders influence on MNEs' activities. *International Business Review, 23*(5), 966–980.

Rahman, M. S. (2012). Exploring tourists perception: The case of Bangladesh. *Tourismos: An International Multidisciplinary Journal of Tourism, 7*(1), 81–98.

Rahman, M. L., Hossain, S. N., Miti, S. S., & Kalam, A. A. (2010). An overview of present status and future prospects of the tourism sector in Bangladesh. *Journal of Bangladesh Institute of Planners, 3*, 65–75.

Rashid, M. (2019). *Destination Bangladesh*. PwC Bangladesh.

Saha, S. K. (2014). Corruption and good governance: The case of Bangladesh. *SUST Journal of Social Sciences, 22*(2), 45–52.

Shamsuddoha, M., & Nedelea, M. O. (2008). Exploring tourist spots for promoting investments in tourism industry of Bangladesh. *Revista de turism-studii si cercetari in turism, 6*, 5–9.

Sobhani, F. A., Amran, A., & Zainuddin, Y. (2009). Revisiting the practices of corporate social and environmental disclosure in Bangladesh. *Corporate Social Responsibility and Environmental Management, 16*(3), 167–183.

The Daily Star. (2009). *Tourism potential*. Retrieved from: https://www.thedailystar.net/news-detail-81436. Accessed 3 Nov 2020.

The Daily Sun. (2017). *Tourism attractions of Bangladesh and investment opportunities*. Retrieved from: https://www.daily-sun.com/post/221928/Tourism-attractions-of-Bangladesh-and-investment-opportunities. Accessed 3 Nov 2020.

The Financial Express. (2017). *Emerging tourism industry in Bangladesh*. Retrieved from: https://today.thefinancialexpress.com.bd/print/emerging-tourism-industry-in-bangladesh-1511940387. Accessed 3 Nov 2020.

Tuhin, M. K. W., & Majumder, M. T. H. (2011). An appraisal of tourism industry development in Bangladesh. *European Journal of Business and Management, 13*(3), 287–299.

United Nations Conference on Trade and Development (UNCTAD). (2009). *Promoting foreign Investment in Tourism*. United Nations.

United Nations Department of Economic and Social Affairs (UN DESA). (2018). *Revision of world urbanization prospects*. UN DESA.

World Tourism Organization (UNWTO). (2019). *World tourism barometer*. Retrieved from: https://www.unwto.org/barometer. Accessed 3 Nov 2020.

World Travel & Tourism Council (WTTC). (2020). *Bangladesh 2020 annual research: Key highlights*. WTTC.

Zakiuddin, A., & Haque, W. (2002). *Corruption in Bangladesh: An analytical and sociological study*. Transparency International Bangladesh.

Conclusion

Azizul Hassan

This book aims to present the past, present, and future of the Bangladeshi tourism industry's investment and development contexts when the tourism industry here represents relevant ancillary facilities (i.e. transportation, policy, products and services, hospitality, sustainability, etc.). This book's contents have seven parts (i.e. theoretical and contextual synopsis, the present context, institutional involvement, product and services, facility offers, sustainability practices, and challenges and the future) in 24 chapters excluding the introduction. These chapters comprehensively and critically describe Bangladesh's tourism industry's possibilities on the ground of investment and development.

Findings of this book's affirm that Bangladesh's tourist attractions are diverse, including forest and wildlife, archaeological sites, beaches, historical mosques and monuments, resorts, picnic spots, etc. The country can provide opportunities for many different tourism activities to cater to almost all sorts of demands of varying tourist types. The history of tourism in Bangladesh has offered peace, comfort, and prosperity. The country has stayed as the land for tourists who like to wander, rest or roam around for mind refreshment and feast their eyes with the resources-rich for tourism activities that nature fondly bestows.

Bangladesh is the home of the largest mangrove forest (i.e. the Sundarbans), the longest unbroken sandy sea beach (i.e. Cox's Bazar), and the world. The country stays as the South and Southeast Asian countries' connecting point. Thus, Bangladesh offers abundant resources for domestic and foreign tourists with its historical sites, scenic beauty, unique cuisine, ethnic diversity, rich heritage, profound religious sentiments, and more. The essentiality of tourism in Bangladesh is recognized internationally in a few recent stylized data and facts highlighted in

A. Hassan (✉)
Tourism Consultants Network, The Tourism Society, London, UK
e-mail: azizulhassan00@gmail.com

A. Hassan (ed.), *Tourism in Bangladesh: Investment and Development Perspectives*, https://doi.org/10.1007/978-981-16-1858-1

many international organizations (the United Nations World Tourism Council, the World Travel & Tourism Council, the United Nations, etc.).

Bangladesh's primary tourism infrastructures are instead the country is an element of excessive concern for both the foreign and local tourists. Tourism infrastructures are yet to be adequate with quality and development. A significant lack of well-equipped recreational facilities exists in Bangladesh to attract international tourists. The country also has limitations in quality hotels and resorts, well-trained tourist personnel to guide tourists, attractive entertainment facilities (i.e. museums, shopping malls, theme parks, movie theatres, etc.). A significant number of the Bangladeshi population cannot read, speak or write English to guide tourists to the desired destination or unable to take the initiative to help them at the time of their needs. Furthermore, the lack of community support comes as a critical development barrier to this industry. Local communities can hardly realize the tourism industry's values and potentials and the tourists.

Accommodation facilities for tourists in the Bangladesh tourism industry need more investment and development. Bangladesh requires to increase affordable tourist accommodation. Supported by the advent of services like Airbnb, tourists' creation of proper tourist accommodation should not require strenuous efforts. Owners of Airbnb listings need to have sufficient accountability for law enforcement. In Bangladesh, transportation systems and facilities need refurbishment, ranging from the Biman Bangladesh Airlines to the most local transports. The national airliner Biman

Bangladesh can be encouraged to follow international airliners' success cases and evolve accordingly. The conventional local means of transportation like rickshaws can attract tourists that search for cultural authenticity. Drivers of such vehicles need to become able to converse in basic English. Security can be a serious concern. The law and order system and situation are critical to the development of this industry very often. Bangladesh has been gradually developing the safety and security index in the most recent years that is expected to be achieved with ease.

At the national level, the lack of plan and policy (master plan) by the GoB to attract investment to develop innovative technology applications, modern recreation facilities, marketing, promotional activities, updated weather forecasting techniques, etc. are vital ones in such regard. Thus, this is very apparent that effective efforts are required to be implemented by the GoB for overcoming the barriers to develop the tourism industry of Bangladesh for diversifying its economy and attaining sustainable economic development. Chapter-wise findings of this book are presented below:

The book's first chapter serves as an introduction. With a summary of all chapters' abstracts, this chapter emphasizes on Bangladesh's investment and growth prospects.

Chapter 2 highlights Bangladesh's tourism and hospitality industry, which eventually reveals that the country has been offering FDI benefits in this field. Tourism is known to be the means of supporting business enterprises in many countries around the world. Investments in the tourism and hospitality industry is related to the acquisition, opening or relocation of established enterprises or investments in construction or renovation projects. For both general economic and industrial growth, such as the tourism and hospitality industry, positive decision-making for foreign public and private sector investments is critical. Awareness in these areas is still very insignificant in the sense of a developing nation, such as Bangladesh. Nonetheless, there are problems, but this chapter provides several recommendations.

The results of Chap. 3 identify some drawbacks: lack of educational knowledge, improper guidance, and lower financial investment from private and public institutions that have kept this industry unattended for several decades. The chapter also highlights the COVID-19 pandemic and devises a model for resolving the challenges of sustainable tourism and hospitality management. The chapter investigates the tourism-led-growth followed by the evolving trends of the investment and development on adventurous, new, and community-based tourist places such as Smart tourism and Augmented Reality eco-tourism.

In Chap. 4, results show that considerable investment and development in Bangladesh's tourism industry are required that can facilitate the integration of the economy with tourism for better financial and technical outcomes.

The findings of Chap. 5 identify both external and internal factors relevant to investments that can improve the existing situation. The research makes more valid suggestions and arguments and expects to create knowledge for potential investors and offer some suggestions for the policymakers and academics having interests in tourism.

Chapter 6 presents that Bangladesh has many tourism product and service offers, mostly nature-based. The development of tourism services and products in the country is gradually developing with the support of investment flows and technology application. There is a lack of required investments for modernization. This research offers some suggestions for tourism service and product investment in Bangladesh. The study expects that tourism products and services with policy support can appear useful tools for Bangladeshi economic growth.

Chapter 7 finds that major public tourism agencies and organizations (i.e. BPC, BTB, Biman Bangladesh, MoCAT) are at present working on tourism-focused projects in the coastal zone of Bangladesh to attract both tourists and investors. This research suggests that these agencies need an evolving long-term sustainable incentive system to attract domestic and foreign investors willing to invest in major tourism development projects in Bangladesh. This chapter addresses the ways public agencies play important roles in the investment and development of the Bangladeshi tourism industry.

Findings of Chap. 8 reveal that banks and financial institutions are required to design traveler-friendly products and services. However, the Bangladeshi banking system has minimal tourism-related products such as credit card facilities. The chapter finds that most lending facilities are loans either for SME or for consumers. Thus, the chapter suggests proper banking rules and regulations for putting in place for integrating tourism-friendly products and services. Eventually, the partnership between the bank and the tourism industry can contribute to promoting the Bangladeshi tourism industry.

Chapter 9 findings identify some particular products and services that can play a significant role in generating revenue from the tourism industry. This research makes suggestions to promote the investment and development of tourism products and services and support the policymakers.

Chapter 10 notices that in the most recent years, Bangladesh's NBT destinations are becoming more popular to domestic visitors. This chapter also finds that the contributions of NBT in the tourism industry of Bangladesh by generating new job opportunities and contributing GDP are significant. On the ground of the current status, NBT has more development potentials. NBT thus facilitate sustainable development in the country and can warrant solving many different challenges for harvesting more. However, this demands more political commitments supported by coordinated and integrated efforts from the government's relevant ministries. More promotion of environmental education and awareness, active and responsible participation of private actors, efficient marketing strategies for attracting more international tourists, management of these destinations, and practicing tourism governance to improve NBT development and its investment are useful.

Chapter 11 identifies some specific features of Bangladeshi cuisine as food variety, enriched culinary traditions, the unique blend of spices, etc. A country's present and past can be represented by its gastronomy as well. This can offer insights into ancient gastronomic traditions, farming practices, and customs and manners. This research labels Bangladesh as an underdeveloped food tourism industry that is majorly influenced by ethno-social diversity and its indigenous communities. This chapter analyses potential and offers recommendations for food tourism to be established and developed as the Bangladeshi tourism industry's central pillar.

In Chap. 12, identifies a total of four factors that affect medical tourism in Bangladesh based on existing literature and theories. (e.g. healthcare quality, institutional and cultural factors, macroeconomic, and individual) and several subfactors for medical tourism. The chapter finds that for promoting medical tourism in Bangladesh, intensive planning, proper coordination among various units (private and public), public awareness, and human resources quality (e.g. doctors, nurses, medical technologists, etc.) are some requirements. Also, suitable private and public investment in infrastructure is essential and prerequisites to promote medical tourism in Bangladesh. This chapter offers some policy implications and guidance for initiating medical tourism for a better and sound healthcare system in Bangladesh, both for the foreign domestic and patients.

Chapter 13 finds that this chapter determines some particular features as the key facilitators through SWOT analysis. GoB initiatives, Strategic Marketing Plan Model application, financial investment, new promotional activities, ecotourism, adaptation of advanced technologies, the collaboration of the healthcare sector and medical institutions are essential to unlock the full potentials of this multi-billion industries. The chapter finally highlights the strategies required to shape Bangladesh as the world's leading medical tourism destination.

Chapter 14 argues that tourism has become a large industry in the last few decades. The authors mention some selected countries (i.e. Thailand, India, Indonesia, Sri Lanka, and Kenya) that can reconstruct the tourism industry through proper management and capacity utilization. The authors then argue that Bangladesh can become famous in tourism because of her natural resources. The country's agriculture can be used for the tourism industry's development. The authors then suggest some policies that they believe can contribute significantly to Bangladesh's economic development. Results of the research present that there are potentialities for agri-tourism in Bangladesh and the country, capable of earning huge currencies from agri-tourism with the support of proper measures taken.

Chapter 15 shows that agri-tourism can support livelihood generation. This can also leave positive impacts on the investment environment in rural areas and create job opportunities. The chapter emphasizes policy supports for ensuring sustainable agri-tourism in Bangladesh.

The secondary data source based Chap. 16 comprehensively discusses the issues, prospects, and challenges of garments tourism in Bangladesh. In general, the chapter found that for reviving the garments tourism, the GoB should consider taking aggressive measures for promoting this industry's branding followed by the establishment and incentivize of the entrepreneurs, making safe travel for foreigners and international tourists.

The results of Chap. 17 furnish that even ecotourism in Bangladesh becomes popular day by day due to the tourists' growing interest. Still, it remains a newer form of tourism. Bangladesh has great potential to develop ecotourism when it can be viewed as the hub of ecotourism destination. This research can support policymakers and marketing practitioners in making new policies and strategies on ecotourism products to foster ecotourism in Bangladesh.

Findings of Chap. 18 assess both positive and negative economic, social, cultural, and environmental impacts. The chapter affirms that no initiative has been taken to find out the optimal solution by using SO, WO, ST, and WT strategies. This research suggests that the SO strategy is the best available option for Bangladesh's tourism industry.

Chapter 19 chapter affirms that this industry is a large source of revenue that needs direly needs better management for revenue collection. This research explores revenue management practices and techniques for the Bangladeshi tourism industry with new projects, infrastructures and marketing activities undertaken by the GoB, and private organizations. The chapter also focuses on the effects of COVID-19 with other barriers on international tourists' mobility that results from low revenue collection.

Chapter 20 finds out the lack of appropriate tourism education as a significant reason for the lack of tourism professionals. Human resources in tourism need to be created through a proper tourism education system. The authors mention that most educational institutions in the country are not well equipped and developed to generate tourism professionals through the existing education system because of inadequate investment. The chapter results show that the shortage of different stakeholders' cooperation and insufficient investments are the reasons for the improper development plan for tourism education. This research suggests that tourism education needs a rapid transformation from the primary to tertiary level, including updated curriculum, program offers, pedagogy, andragogy, research, training, and learning environment. This chapter concludes by stating that the mentioned measures are necessary to convert human resources to human capital and create more significant impacts on its overall development.

Chapter 21 findings show that more investments in Green tourism from new investors, tourist agencies, financial institutions, and GoB can provide customized Green tourism services, contribute to economic growth, protect the environment, and culture of tourist areas Bangladesh.

Chapter 22 shows that sustainable tourism in the country supports increasing tourist awareness and helps to implement policies for the tourism industry.

Chapter 23 finds that tourism can produce a professionally skilled workforce, and this industry can attract for earning foreign currencies. This study suggests the eradication of lousy branding identity and huge investments. Lack of constructive attitude, weak promotional identity, inadequate facilities, and lack of world-class tourism firms are some of our country's tourism industry problems.

Chapter 24 identifies some specific challenges (i.e. lack of constructive attitude, weak promotional identity, poor facilities, lack of world-class tourism firms, etc.) and presents relevant data, information, and facts to overview the future of investment and development of the tourism industry of Bangladesh.

Chapter 25 spots that the tourism industry is one of the key contributors to the national economy. The chapter also explores different investment barriers and opportunities in the Bangladeshi tourism and hospitality industry and offers some recommendations. This research can guide relevant authorities of the GoB to expand the tourism and hospitality industry through domestic and foreign direct investments.

As an industry, tourism emerged from a comparatively small-scale activity to the leading enterprise in the world. The tourism industry, in general, is a sort of collaborative activities and services that cover transportation, food, accommodation, entertainment, shopping, hospitality services, etc. All such activities can enhance the economic development process by developing entrepreneurial skills and tourism infrastructures, creating jobs, improving the balance of payments, earning foreign exchanges, and more revenues. Tourism is a significant industry for economic development, having substantial roles in Bangladesh's national economy. This industry can stay resilient to face a crisis when the demand for tourism products and services is sustained.

To develop the tourism industry through investment and development, Bangladesh needs effective research studies and policy implementations supported by the fiscal incentives. Following tourism's increasing popularity across the world, Bangladesh's tourism industry can support the economy moving to the later stage that can be the agenda for future research studies.

The manufacturer's authorised representative in the EU is Springer Nature Customer Service Centre GmbH, Europaplatz 3, 69115 Heidelberg, Germany. If you have any concerns regarding our products, please contact ProductSafety@springernature.com

Printed and bound by CPI Group (UK) Ltd, Croydon, CR0 4YY
30/09/2026
02225248-0014